DIGITAL PLAY

Digital Play
The Interaction of Technology, Culture, and Marketing

STEPHEN KLINE,
NICK DYER-WITHEFORD,
AND GREIG DE PEUTER

McGill-Queen's University Press
Montreal & Kingston · London · Ithaca

© McGill-Queen's University Press 2003
ISBN 0-7735-2543-2 (cloth)
ISBN 0-7735-2591-2 (paper)

Legal deposit second quarter 2003
Bibliothèque nationale du Québec

Printed in Canada on acid-free paper.

This book has been published with the help of a grant from the
Canadian Federation for the Humanities and Social Sciences, through
the Aid to Scholarly Publications Programme, using funds provided
by the Social Sciences and Humanities Research Council of Canada.
Funding has also been received through SSHRC grant no. R2298A04,
The Canadian Video and Computer Game Industry.

McGill-Queen's University Press acknowledges the support of the
Canada Council for the Arts for our publishing program. We also
acknowledge the financial support of the Government of Canada
through the Book Publishing Industry Development Program (BPIDP)
for our publishing activities.

**National Library of Canada Cataloguing
in Publication Data**

Kline, Stephen
    Digital play: the interaction of technology, culture, and marketing /
Stephen Kline, Nick Dyer-Witheford, and Greig de Peuter.
    Includes bibliographical references and index.
    ISBN 0-7735-2543-2 (bnd)
    ISBN 0-7735-2591-2 (pbk)
    1. Electronic games industry. 2. Electronic games – Social aspects.
    1. Dyer-Witheford, Nick, 1951–   11. de Peuter, Greig, 1974–
    III. Title.
    HD9993.E452K58 2003   338.4'77948   C2002-905794-9

Typeset in Sabon 10/12
by Caractéra inc., Quebec City

Contents

Illustrations   vii

Acknowledgments   ix

1 Paradox Lost: Faith and Possibility in the "Information Age"   3

PART ONE   THEORETICAL TRAJECTORIES   27

2 Media Analysis in the High-Intensity Marketplace:
The Three Circuits of Interactivity   30

3 An Ideal Commodity? The Interactive Game in Post-Fordist/
Postmodern/Promotional Capitalism   60

PART TWO   HISTORIES: THE MAKING OF
A NEW MEDIUM   79

4 Origins of an Industry: Cold Warriors, Hackers, and Suits,
1960–1984   84

5 Electronic Frontiers: Branding the "Nintendo Generation,"
1985–1990   109

6 Mortal Kombats: Console Wars and Computer Revolutions,
1990–1995   128

7 Age of Empires: Sony and Microsoft, 1995–2001   151

8 The New Cyber-City: The Interactive Game Industry in the
New Millennium   169

PART THREE　　CRITICAL PERSPECTIVES　193

9　Workers and Warez: Labour and Piracy in the Global Game
　　Market　197

10　Pocket Monsters: Marketing in the Perpetual Upgrade
　　Marketplace　218

11　Designing Militarized Masculinity: Violence, Gender, and the Bias
　　of Game Experience　246

12　Sim Capital　269

　　Coda　Paradox Regained　294

　　Notes　299

　　Bibliography　331

　　Index　357

Illustrations

TABLES

1 Genre by Unit Sales, 1998   254

DIAGRAMS

1 The Circuit of Capital   51

2 The Three Circuits (a)   51

3 The Three Circuits (b)   52

4 The Three Circuits of Interactivity   53

5 Contradictions in the Three Circuits of Interactivity   54

6 The Three Circuits of Interactivity in the Mediatized Global Marketplace   58

7 The Interactive Game Industry   172

Acknowledgments

When we first embarked on this project, relatively little scholarly attention was devoted to computer and video games and the emergence of an interactive entertainment industry. Yet today this book is part of a rapidly expanding commentary on the computer and video games industry, including, among others, Michael Hayes, Stuart Dinsey, and Nick Parker's *Games War*, J.C. Herz's *Joystick Nation*, Steven Poole's *Trigger Happy*, Douglas Rushkoff's *Playing the Future*, and David Sheff's *Game Over*. We have used their work extensively, but in reading their accounts, perceived the need for a more critical analysis of this new medium.

Stephen Kline wishes to thank his children, Daniel Shennen and Meghan Sarah, for alerting him to the importance of this new medium, and graduate students Brent de Waal and Jackie Botterill who helped him transfer his parental concerns about Leisure Suit Larry into a research program supported by the Social Sciences and Humanities Research Council of Canada. Diana Ambrozas and Brent Stafford made important early contributions to the development of this research project and David Murphy has provided encouragement and technical support throughout. He also wishes to thank the many colleagues who have listened with patience and commented generously on various aspects of this work in progress.

Nick Dyer-Witheford thanks many friends in the Media, Information and Technoculture program of the Faculty of Information and Media Studies at the University of Western Ontario for their ongoing support and inspiration: Bernd Frohmann, Gloria Leckie, and Catherine Ross for welcoming him to the program that provided an exciting matrix for work on this book; Carole Farber, Tim Blackmore, and Jacquie Burkell for collegial exchanges about webs, war, and women online; and Zena Sharman, David Kinlough, Anthony Martin, Dave Kudirka,

Edward Fraser, and many other students, undergraduate and graduate, for discussions of digital play that made teaching truly interactive. He also thanks his wife Anne, and his daughters, Adrienne and Miranda, for supporting the toil behind this book and also for not getting too interested in computer and video games – one of their many shining examples of autonomy from the compulsions of virtual capitalism.

Greig de Peuter thanks students and faculty in the School of Communication at Simon Fraser University for facilitating an environment for critical thought. For providing intellectual support and personal encouragement in equal intensity, Greig extends special thanks to Albert Banerjee, Mark Coté, David Murphy, and Katharine Perak; your words were vital in seeing this project through to completion. He also thanks his partner, Sheila, for her role in moving this project from the virtual to the actual. Finally, he thanks his parents for buying him an Atari 2600.

At McGill-Queen's University Press, we wish to thank Aurèle Parisien, for his enthusiastic support for the project from the outset. Claire Gigantes, our copyeditor, has provided the thorough reading that a multiauthored manuscript requires, and Joan McGilvray has helped usher the book through the final stages of publication. We would especially like to express thanks to our two anonymous reviewers for their insightful remarks and helpful suggestions.

DIGITAL PLAY

I

Paradox Lost:
Faith and Possibility
in the "Information Age"

Technology, in short, has come of age, not merely as a technical capability, but as a social phenomenon. We have the power to create new possibilities, and the will to do so. By creating new possibilities, we give ourselves more choices. With more choices, we have more opportunities. With more opportunities, we can have more freedom, and with more freedom we can be more human. That, I think, is what is new about our age. We are recognizing that our technical prowess literally bursts with the promise of new freedom, enhanced human dignity, and unfettered aspiration.[1]

Is it wicked chance or synchronicity that made *Titanic* one of the blockbuster movies of the twentieth century? That great ship was, after all, not just an impregnable vessel but a much-heralded communication medium symbolizing all that was progressive in an era of industrialization in which the conquest of space and time had become essential to the expansionary economies of trade and the global circulation of knowledge and people. So much so that the accelerating speed of travel became the obsession of its builders, who bragged about their superior engines as much as about the scale and opulence of their ship. Throwing caution to the wind, the White Star Line's owners and operators came to believe that they had triumphed over nature and eliminated all risk and unforeseen danger from transatlantic travel. Their arrogance expressed itself in a blind faith in the power of technology and a cheery optimism that they had transcended history at the dawn of a new age. The Hollywood love-boat story of ill-starred romance on that fatal voyage was promoted to bring teenagers into the cinema to defray the enormous costs of a lavish production. But the melodramatic tale

of star-crossed lovers would not have distracted all the viewers from
the spectacular but class-divided catastrophe brought about by the
technological arrogance of those who designed and sailed the Titanic.
The same technological arrogance permeates the euphoric descriptions
of the contemporary "information age."

DANCING ON THE HEAD OF A PIN

Technology can no longer be viewed as only one of many threads that form
the texture of our civilization; with a rush, in less than half a century, it has
become the prime source of material change and so determines the pattern of
the total social fabric.[2]

In the so-called information age, each new series of computer chips,
each smart appliance, and each domestic communication technology is
successively celebrated by the captains of wired capitalism as ways
of shrinking our world while expanding our freedom. Nicholas
Negroponte, cofounder and chairman of the Massachusetts Institute of
Technology (MIT) Media Lab, optimistically promises that computers
are transforming the whole field of technology: "The access, the mobility, and the ability to effect change are what will make the future so
different from the present."[3] Pointing to the many social changes
already resulting from cybernetic invention, Negroponte believes that
"like a force of nature, the digital age cannot be denied or stopped."[4]
With the patented certainty of those who can read social change directly
from a silicon chip, he sees our troubled world being transformed by
cybernetic communications: "Early in the next millennium your right
and left cuff links or earrings may communicate with each other by
low-orbiting satellites and have more computer power than your
present PC. Mass media will be redefined by systems for transmitting
and receiving personalized information and entertainment. Schools will
change to become more like museums and playgrounds for children to
assemble ideas and socialize with other children all over the world. The
digital planet will look and feel like the head of a pin."[5] In the writings
of futurists like Negroponte, the computer is an icon of technologically
driven social change – an emblem of the more participatory, democratic, creative, and interactive world that is allegedly being delivered
to us just by the power of the silicon chip.

For all the breathless excitement of such statements, their vision is
in fact quite old. It is not just that faith in technological progress has
been a main theme of Western, and particularly North American,
culture from the era of railways to the age of nuclear power: the most
recent version of this technocreed, celebrating the new world created

by digitization – the "computer revolution," the "information society," the "wired world," the "global village" – has been around for a surprisingly long time.

In the 1960s Daniel Bell argued that the dynamism of American society marked a major historical transition from an "industrial" to a "post-industrial" society. In the post-industrial economy the production, processing, and communication of information in services and cultural commodities replaced the production and distribution of natural resources and industrial goods as the key sector of the economy. These changes were associated with a new "intellectual technology" of computers and telecommunications, with the increased importance of managing technoscientific innovation, and with the growing role of "knowledge workers" who could communicate, accumulate information, and ultimately process the flow of data in the competitive market economy.[6] Bell saw the new postindustrial society that he prophesied as liberating in some ways and problematic in others. His ideas were to be championed by others in a far more simplistic way.

One of these popularizers was Alvin Toffler, whose notion of a "Third Wave" of technological progress won wide attention in the 1980s – the very time when the personal computer (not to mention the video game console) was appearing in many homes.[7] History, claimed Toffler, taught that technological invention was a powerful force for changing the whole of society. The growth of agricultural techniques constituted the first wave and manufacturing technologies the second. Crucially, however, it was communication technologies – the "infosphere" – that would precipitate the third and most radical wave of social change.[8] Toffler argued that computer applications were solving the crisis experienced in advanced industrial societies. Industrial-era technologies, such as the mechanized assembly line and mass media, encouraged rigid hierarchies, harsh class divisions, and generally depersonalized mass societies. Toffler was especially suspicious of the complicity of the mass media in perpetuating oppression and homogenization: "centrally produced imagery, injected into the 'mass mind' by the mass media, helped produce the standardization of behaviour required by the industrial production system."[9] But the third wave would challenge all of that. The primary message of the new order of computerized society was flexibility and what Toffler called "demassification." Computers and telecommunications gave us multipurpose tools for processing information – and bestowed a new openness and adaptability on the humans who use them. The third-wave technology promoted adaptation in users and consumers rather than the bending of man to the rhythms and routines of the machine. The ramifications of flexible media would be felt in more democratic interactions between workers and employers, between governments and industry, and between producers and consumers. According to Toffler: "This Third Wave of historical change represents not a straight-line extension of industrial society but a radical shift of direction, often a negation, of what went before. It adds up to nothing less than a complete transformation at least as revolutionary in our day as industrial civilization was 300 years ago."[10]

Invoking a modernist conception of progress, the idea that digital technology is bringing in an entirely new and better social order is familiar, not to say stale. In its basic form it is rooted in a version of what is called "technological determinism" – the idea that new machines drive social, political, and cultural change. The plough gave us agricultural society, the steam engines the industrial society, the computer the information society, and so on. This is a pretty simple idea – too simple, say many critics, who point out that it neglects the way political, economic, and cultural factors in turn shape the capacity for and direction of technological change. In *Under Technology's Thumb* (1990), William Leiss says that the central problem with technological determinism is that it inscribes our collective destiny within an ironclad logic of human possibility.[11] Leiss takes the techno-hypsters to task for their simple-minded theories about the relationship between technological innovation and social change: "Strictly speaking, there are no imperatives in technology. The chief mistake ... is to isolate one aspect (technology) of a dense network of social interactions, to consider it in abstraction from all the rest, and then to relate it back to that network as an allegedly independent actor."[12] Within these deterministic discourses, Leiss says, "social theory dissolves into commonplaces."[13]

The cultural theorist Kevin Robins hits a similar note when he observes that the belief in the coming information age "is driven by a feverish belief in transcendence" and demands a profound leap of faith, "a faith that, this time round, a new technology will finally and truly deliver us from the limitations and frustrations of this imperfect world."[14] He goes on to say: "There is a common vision of a future that will be different from the present, of a space or a reality that is more desirable than the mundane one that presently surrounds and contains us. It is a tunnel vision. It has turned a blind eye on the world we live in."[15]

But to understand the full nature of this myopia, we cannot rest with just a critique of technological determinism. To really take the hype out of "technological hyperbole" we have to look at a more recent twist to the information age argument, one that links faith in digital technologies to faith in the free market.[16] In exploring this link we will discover some important *continuities* between previous historical epochs and our present one – despite its technological novelties.

WIRED REVOLUTIONARIES
AND CORPORATE MILLENARIANS

The era of Java Enterprise Computing has arrived. No longer must we be tied to a single master. Today we consider the following to be inalienable and available to all. The right to harness technology to stay, not just one, but also several steps ahead of the game. The right to a new computing dynamic with the vision to take you into the future. And not only do you have the right to information technology that works the way you want it to, you have the right to change it at will. It is your due, now is the time to realize significant return on your technological investment. It is not simply about systems, it's about the emancipation of information. Java Enterprise Computing is here and it will set you free.
Stretched across this two-page ad is the text: "LIBERTY!"

As we look more closely at those leading the headlong charge to realize the "freedom," dignity, and efficiencies buried deep in the circuitry of digital media – whether Sun and Oracle, Microsoft and AOL, or Nintendo and Sony – we notice that their optimistic promotional discourse elides the interests of corporations and consumers in their vision of the information highway. "It's about the emancipation of information," bubbles the Java ad. Yet we are not so much experiencing a "revolution" in domestic media as a corporate battle over the provision of services to consumers. Java, we must remember, is not exactly about splitting the atom or discovering a cure for cancer. It's a programming language for connected multimedia computing – the networked exchanges of digital sound, video, text, and graphics that enable the distribution of multimedia products, interactive marketing services, Web-TV, and interactive gaming. So although the Java ad claims that digitized media are freeing us from a "single master," it seems that the "emancipation of information" turns out to be another opportunity for profit by corporations looking to "realize significant return on your technological investment." To the vanguard of this revolution, then, the Internet is first and foremost a corporate battleground wherein the future patterns of communication will be set and won or lost at great profit. "No blood will be shed in this revolution," but certainly a lot of dollars will be diverted as more and more of our culture is commodified.

This is the essence of the so-called "New Economy" based on information technologies, whose proponents, alongside speculation about the latest developments in fibre optics and virtual reality, often make excited reference to the works of the eighteenth-century economist Adam Smith. The basic ideas of Smith and his followers are familiar. In the market the self-interested activities of innumerable individual buyers and sellers result, as by the operation of an "invisible hand," in the best possible allocation of social resources.[17] The competition amongst enterprises ensures that goods sell at a just price and that the market is responsive to the needs and wishes of purchasers, thus maintaining "consumer sovereignty." This competitive activity also gives entrepreneurs a unique spur to increase efficiency – for example, by technological innovation. This in turn makes capitalism an engine for constant economic growth. Provided the operations of the "invisible hand" can be kept free from distorting influences, particularly the meddling of governmental regulators, market society can combine freedom and prosperity, serving our collective interest. "*Laissez-faire*" – let it be, let the market work its magic – became the slogan of Smith's economic followers.

The actual development of industrial capitalism over the course of the nineteenth and twentieth centuries – with its cycles of boom and bust, concentration of corporate power, and class conflict – has led many to question this depiction of market forces. Over the last quarter-century, however, Smith's economic theory has undergone a revival. One key element in this resurrection is the marriage of *laissez-faire* economics with the information revolution.[18] The wedding involves several propositions. The first is simply that capitalism is a social system uniquely capable of technological change. Entrepreneurial competition makes the free market the only system that could generate the scientific advances that climax in the information revolution. When socialist regimes, particularly the former Soviet Union, collapsed in 1989 it was widely suggested that such state-run economies were too rigid and authoritarian to produce the innovations necessary for postindustrialism, or to allow its citizens access to the communications channels central to the creation of an information economy. From this point of view, robots, artificial intelligence, and computer networks are the special children of the free market.

The argument goes further. If the information revolution cannot exist without the market, the market cannot reach its full perfection without the information revolution. The original model of the invisible hand proposed by Smith and his followers relied on "perfect information." That is to say, buyers and sellers were assumed to have full knowledge about each other's actions and capacities. Many of the actual problems of markets could, in this view, be ascribed to blockages in information flow. Blatantly unjust transactions or vast accumulations of monopolistic power, for example, arose because buyers and sellers simply did not have adequate knowledge about market opportunities and alternatives. But communication through high-tech networks removes these obstacles. Computers, fax, cellphones, and media of all sorts create a constant, multichannelled flow of information between buyers and sellers, companies and suppliers, entrepreneurs and consumers – and in doing so give an unprecedented nimbleness and precision to the dancing fingers of the invisible hand.

This is the essence of "friction-free capitalism," a fashionable concept among the e-commerce optimists and one that figures prominently in Bill Gates's reflections in his biography, *The Road Ahead* (1995). Ignoring the ironies that such words invite in the mouth of the information age's most aggressive monopolist, Gates celebrates the movement of business into cyberspace for fulfilling Smith's dream of a world of "perfect knowledge" or "perfect information," a prerequisite for "perfect competition." Gates promises us "a new world of low-friction, low-overhead capitalism, in which market information will be plentiful and transaction costs low."[19] Freed by digitization from its rigidities and imperfections, the market passes into a veritable paradise of exchange, becoming "a shopper's heaven."[20]

The "new economics" thus combines information revolution doctrine and free-market logic in a brilliantly complementary union. Technologies must be promoted because doing so fosters free markets; markets must be freed to spur technological innovation. The two sides of the argument circle each other in perpetual orbit. But the basic formula is that free markets plus information technology equal health, wealth, and social progress. It is a formula that has, over the last five or so years, been endlessly repeated. For example, Frances Cairncross's panegyric to the global village, *The Death of Distance* (1997), foresees our digital prosperity emerging from the newly linked global multimedia telecommunication networks that constitute "the single most important force shaping society in the first half of the next century."[21] Like so many others, Cairncross cites the growth of e-commerce as the model of things to come. The economic implications of connectivity must first work their way through the corporate world before they bring the promised peace and harmony. "The changes sweeping through electronic communications will transform the world's economies, politics, and societies – but they will first transform companies."[22] That is why Cairncross opens her book not with a careful analysis of how the new costing of global communication will affect democracy, cultural diversity, or family life but by itemizing thirty ways that telecommunications pricing can change business strategy. In short, the information revolution is a management revolution. That is what matters most about the information highway, claims Cairncross: "it makes the market system – since the collapse of communism the dominant method of allocating resources around the world – work better."[23]

It is in the promotional language of *Wired*, the digerati's favourite glossy magazine, that we find the most euphoric celebration of "new economy" ideology. According to the magazine's "Encyclopedia of the New Economy":

When we talk about the new economy, we're talking about a world in which people work with their brains instead of their hands. A world in which communications technology creates global competition – not just for running shoes and laptop computers, but also for bank loans and other services that can't be packed into a crate and shipped. A world in which innovation is more important than mass production. A world in which investment buys new concepts or the means to create them, rather than new machines. A world in which rapid change is constant. A world at least as different from what came before it as the industrial age was from its agricultural predecessor. A world so different its emergence can only be described as a revolution.[24]

"Free markets" are "central to this process," the writers explain: "but to say that the new economy is about the unprecedented power of global markets to innovate, to create new wealth, and to distribute it more fairly is to miss the most interesting part of the story." More importantly, "markets are themselves changing profoundly" because "working with information is very different from working with the steel and glass from which our grandparents built their wealth." An information economy is "more open" and "more competitive" than that of the industrial age. The rise of Microsoft, for example, is, according to the authors, "testimony to the power of ideas in the new economy." The results of these changed conditions are "new rules of competition, new sorts of organization, new challenges for management."

The change in economic paradigm is "redefining how we need to think about both good times and bad." We do not know, the authors admit, "how to measure, manage, compete in the new economy." Indeed, "we don't know how to oversee it, or whether it ultimately needs oversight at all." Another thing we don't know is "where – or how – the revolution will end." Just about all that we do know, apparently, about the new economy is that "we are building it together, all of us, by the sum of our collective choices." It remains clear to *Wired*, however, that the ineffable logic of *laissez-faire* and fast modems is propelling the march of human progress across its next frontier; "Read on pioneer," instructs the author of the "Encyclopedia."[25]

What the old and the new versions of technological hyperbole have in common is a failure to account for historical context and social forces. For *Wired*'s promise of peace and prosperity ultimately depends not just on the wonders of computers but on the ways technological possibility is conceived, appropriated, designed, and, most importantly, sold. But what distinguishes current new media boosterism from earlier information age rhetoric is that our faith is being grounded in raw technological determinism as well as in a new social theory whereby communication media and free markets are a *determinative unity*. Communications technology and free markets make up the dancing dialectic of a new corporate millennialism that will allegedly galvanize our consumer economy, revive our flagging democratic culture, and repair our wounded environment.

THE RISE OF INTERACTIVE PLAY

Searching for a case study on which to test the digerati's claims, we might remind ourselves that in the closing years of the twentieth century the film industry was not the only megamedia complex telling the story of the *Titanic*'s fatal voyage. Another, more futuristic entertainment business was already exploring the saga. *Titanic: Adventure in Time*, a computer game by the multimedia developer CyberFlix, promised its purchasers not merely the spectacle of nautical disaster but virtual immersion in it, as a British secret agent seeking to retrieve priceless stolen documents on the doomed ship: "On a star-filled night in 1912, you stand witness as a graceful giant slides reluctantly into oblivion. Massive decks and huge propellers rise at grotesque angles as deep metallic groans split the still night, momentarily drowning out the cries of the dying. Water churns, and still-glowing portholes vanish into icy blackness as you recall what led you here. The world will be changed forever, and you are the sole possessor of the knowledge of how things might have been."[26] Launched in 1996, the game actually preceded the movie. *Titanic: Adventure in Time* found a spot among the top-selling ten computer games for 1997 and 1998; worldwide rights were bought by a software sales and consultation firm called, appropriately enough, Barracuda; copies of the game could still be found on store shelves in 2002.

This might seem a mere footnote to a movie success story. But the interactive game industry, comprising video and computer games, rivals film in terms of its global revenues and impact on popular culture.[27] Writing of the way video games swept into North American culture in the early 1990s, Michael Hayes and Stuart Dinsey suggest that "the effect on the consumer ... [was] comparable to that generated by the Golden Age of Hollywood in the 1930s, 40s and 50s."[28] In fact, the US interactive-play business now matches Hollywood in economic power. According to the high-tech business journal *Red Herring*, the interactive game industry's revenues for 1999 topped $8.9 billion, compared to US movie box office receipts of $7.3 billion. The journal notes, however, that this figure is somewhat deceptive, since Hollywood generates far larger revenues thanks to various "synergetic" linkages such as pay-per-view TV, video and DVD rentals and sales, etc. Once these are taken into account, the global film industry took in some $47.9 billion; even if home and arcade gaming were added together, worldwide gaming revenue would only be $30 billion. On the other hand, the game industry is growing much faster than the film business; a typical prediction is that in the US alone it will climb from eight billion dollars in 2000 to twenty-nine billion dollars in 2005, these numbers roughly doubling worldwide.[29]

Four decades have seen the digital game transformed from the whimsical invention of bored Pentagon researchers, computer science graduate students, and nuclear research engineers into the fastest-expanding sector of the entertainment industry. Lara Croft, the shapely heroine of *Tomb Raider*, is today amongst the hottest of current media celebrities, experienced through the more than forty million Sony PlayStations sold worldwide; playgrounds across the continent are swept with competing *Pokémon* and *Digimon* epidemics; business analysts scan the virtual communities coalescing around games such as *Quake*, *Ultima*, and *Everquest* as trailblazing e-commerce business models. No longer produced in garages by youthful geeks, a video game can now take a team of up to fifty specialized artists, writers, designers, animators, and programmers – working on expensive game engines with a commitment of financial resources of up to ten million dollars spread over two or three years. Hit titles like *Doom*, *Mortal Kombat*, and *Tomb Raider*, though less expensive to produce than blockbuster movies, reap profits on an even larger scale, potentially generating revenues in the hundreds of millions.

Interactive games are now viewed as the leading edge of a significant entertainment industry spanning very different technological platforms: home video game consoles, personal computers, Internet play, portable and wireless devices, arcade and virtual reality theme parks. Digital play practices have gradually colonized our homes, pockets, and cyberspace, becoming a daily habit for millions of people. According to a study supported by the Interactive Digital Software Association (IDSA), the industry's major promotional organization, over sixty percent of Americans, or about 145 million people, "play interactive games on a regular basis."[30] Traditionally, youthful males under eighteen have been most attracted to this exciting form of entertainment, a pattern that our own research suggests still prevails, many boys playing on average one hour a day, and some heavy players languishing up to three hours a day at their consoles.[31] IDSA, however, claims recent dramatic breakouts from this traditional young male market niche, both in terms of age, the reported average age of an interactive game player now being twenty-eight, and in terms of gender, females allegedly making up forty-three percent of players.[32] Whatever the precise composition of the gaming audience, it is large, with the more than two hundred million games sold in 2000 equivalent to two for every household in America.[33] Some years ago Allucquère Rosanne Stone suggested that "it is entirely possible that computer-based games will turn out to be the major unacknowledged source of socialization *and* education in industrialized societies before the 1990s have run their course."[34] As we enter the third millennium, this prophecy seems well on the way to being realized.

The video and computer game industry also exemplifies the globalizing, transnational logic of twenty-first-century capital. Although founded in North America, many of its major corporate contenders are Japanese companies – Nintendo, Sega, Sony. The market for interactive games is today almost equally divided between North America, Europe, and Japan. Although the bulk of industry revenues comes from these bastions of advanced capital, games are now disseminated all around the world, booming digital play cultures appearing in countries such as South Korea and Malaysia, and gaming networks beginning to link contestants across continents.

Interactive play thus appears as a quintessential product of digital capitalism's "new economy." Indeed, the games industry has captured the zeal of the information revolutionaries. T.G. Lewis, in his book *The Friction Free Economy* (1997), cites the "Japanese video game empires of Sega, Nintendo and Sony" as exemplifying the allegedly "non-Keynesian, non-Newtonian" logic of the new electronic marketplace.[35] Contemplating the market potential of these digital devices, Negroponte muses: "It would seem that if you are an information and entertainment provider who does not plan to be in the multimedia business, you will soon be out of business."[36] Even those who view digital capital from a critical perspective, such as the political economist Nicholas Garnham, acknowledge that game giants such as Nintendo, Sega, and Sony "are in fact the first companies ... to have created a successful and global multimedia product market."[37]

THE END OF MASS MEDIA?

Alvin Toffler, that pioneer of information age optimism, declared nearly twenty-five years ago that video games were far more than a "'hot item'" in the stores: they embodied the liberating possibilities of third-wave-era media of communication.[38] "Not only do video games further de-massify the audience," Toffler remarked, "and cut into the numbers who are watching the [television] programs broadcast at any given moment, but through such seemingly innocent devices millions of people are learning to play with the television set, to talk back to it, and to interact with it ... They are manipulating the set rather than merely letting the set manipulate them."[39] Picking up where Toffler left off, in the work of silicon utopians, video and computer games are hyperbolically celebrated for their "interactivity."

This is a term loosely applied to any media in which the audience technologically intervenes to structure its own experience. The claims are by now familiar: digital games are interactive media *par excellence* because their entertainment value arises from the cybernetic loop between the player and the game, as the human attempts, by the movement of the joystick, to outperform the program against and within which he or she competes. This feedback cycle is often represented as a dramatic emancipatory improvement over traditional one-way mass media such as television and its so-called "passive" audiences. Against mass culture's hegemonic embrace through its broadcast technologies, digital media devices and content will liberate us because their audiences structure their own experience in a triple sense: through technological empowerment, consumer sovereignty, and cultural creativity. The digerati, at their most celebratory, use interactivity to declare the mass media model, and the mass culture and system of corporate power that go with it, overthrown. The subject at the centre of it all, the interactive gamer, becomes the apogee of consumer sovereignty: the "prosumer," in the rhetoric of Toffler.[40]

Young people's growing fascination with interactive play is itself one of the clearest signs that the digital era is well underway. As Negroponte wistfully declares: "It is almost genetic in its nature, in that each generation will become more digital than the preceding one."[41] Although many "adults" might believe "these mesmerizing toys turn kids into twitchy addicts and have even fewer redeeming features than the boob tube," video games are a sign, according to the wired revolutionaries, that digital media producers are being responsive to the next generation.[42] No longer submitted to the whims of television moguls, children's virtual drives down the information highway anticipate a future where popular culture will be accommodated to their own deepest desires. They would lead us to believe that the demassified media lead inexorably to the democratization of cultural production, as media corporations are plugged into youthful consumer wants, and as gamers define their own paths through the narratives of interactive games. But that is not all. Negroponte says that "electronic games teach kids strategies and demand planning skills that they will use later in life."[43]

For many of the silicon utopians, there is a certain glee in viewing the generational divide that is beginning to appear between the young, who understand and adapt to interactive media more quickly, and the older generation, who foolishly resist the inevitable wiring of the world. It is therefore with enthusiasm that Japanese management guru Kenichi Ohmae, prophet of a "borderless world," speculates on the generational implications arising from this medium's rapid diffusion to youth in Japan.[44] "Nintendo kids," Ohmae asserts, "are making new connections with the tens of millions of their peers throughout the world who have learned to play the same sorts of games and have learned the same lessons."[45] The "web of culture," he says, "used to be spun from the stories a child heard at a grandparent's knee. Today it derives from ... children's experience with interactive multimedia. Nintendo kids, whose neighborhood is global, will increasingly use technology to participate in the global economy ... For Nintendo kids, such transactions will be part of everyday life."[46]

Commenting on the popularity of video gaming in Japan, Ohmae notes "a cultural divide growing between these young people and their elders."[47] But he is enthusiastic about this emerging culture of video gaming, which he believes lessens the social isolation of this generation. He goes on to speculate:

That experience has given them the opportunity, not readily available elsewhere ... to play different roles at different times, of asking the what-if questions they could never ask before ... Perhaps most important, Nintendo kids have learned, through their games, to revisit the basic rules of their world and even to reprogram them if necessary. The message, which is completely alien to traditional Japanese culture, is that one can take active control of one's situation and change one's fate. No one need submit passively to authority.[48]

Like Ohmae, many silicon apostles speak of the empowerment of global youth culture as "connectivity" and "interactivity" begin to reverse the passivity, alienation, and isolation created by the mass broadcast technologies of the past. New media challenge authority and promote entrepreneurial attitudes, they claim, making the generational divisions the stepping-stone into the future of globally wired capitalism.

Douglas Rushkoff's 1997 book *Playing the Future: What We Can Learn from Digital Kids* typifies the optimistic reading of the effects of the multimedia revolution on youth culture. While Rushkoff stresses the capacity of youthful audiences to use mass media content to their own ends, it is a view that tends to vilify TV as a linear technology that subordinates its audiences to a broadcast organization. Audiences participate in the narratives by identifying with characters within a preprogrammed flow of meaning. In contrast, the video game player, the argument goes, can manage the flow of meaning from the screen, choosing characters and navigating game spaces at will. Interactivity for this digital generation is paramount, Rushkoff argues, for now, "the kids ... rather than simply receiving media, are actively changing the image on the screen. Their television picture is not piped down into the home from some higher authority – it is an image that can be changed" by them.[49] This new media breaks the one-way flow of meaning associated with broadcasting. Online gaming especially transforms play into "a group exercise in world creation where reality is no longer ordained from above, but generated by its participants."[50] Rushkoff goes on to say: "Fully evolved video game play, then, is total immersion in a world from within a participant's point of view, where the world itself reflects the values and actions of the player and his community members. Hierarchy is replaced with a weightless working out of largely unconscious preoccupations."[51]

Rushkoff advises parents not to worry about the future playgrounds that the digital empires of Nintendo and Microsoft are constructing: "While their parents may condemn Nintendo as mindless and masturbatory, kids who have mastered video gaming early on stand a better chance of exploiting the real but mediated interactivity that will make itself available to them by the time they hit techno-puberty in their teens."[52] He imagines a digitally empowered generation emerging from the global multimedia matrix where interactivity and connectivity become the forces of generational liberation – and the training ground for jobs in the digital sector. Teenagers from around the world, Rushkoff claims, now assemble in virtual communities, using networked multimedia to make their own culture, playing online games, and socializing in chat-rooms. For Rushkoff, today's "screenager sees how the entire mediaspace is a cooperative dream, made up of the combined projections of everyone who takes part."[53] Interactive media are therefore celebrated for creating a new caste of media audience: an active subject, parting with the tyranny of mass media for the freedom of joysticks.

For Rushkoff, the digital generation is at the frontier of human liberation: "The thing we are about to become is already with us. Just look – really look – at your children for tangible proof, beyond the shadow of a doubt, that everything is going to be all right."[54] Interactive media are empowering because they put the consumer in command of media content; video games put kids in control of the flow of narrative and meaning for the first time in the history of mediated socialization. And so in the rhetoric of digital futurists, video games promise to transform the very basis of our centralized culture of the mediated mass market into a decentralized, connective, and populist republic of technology.

PERCEIVING PARADOX

Over the last couple of decades a growing number of voices have dissented from such technological euphoria. They point out that far from levelling and democratizing, the coming of the information age has been marked by growing disparities in income, global unrest, and economic instability, along with increasing corporate control and waning accountability in the cultural industries.[55] In a reaction against the inflated hyperbole of information revolutionaries, neo-Luddite perspectives have become prominent.[56] Writers of this camp revive the image of early-nineteenth-century insurrectionists who smashed the machines of early industrial capital. Today's neo-Luddites present these early radicals not as ignorant obscurantists but as intelligent and justified opponents of the dehumanizing technologies that concentrated power in the hands of commercial owners. They warn that we should exercise a similar scepticism towards rhetoric about the necessity, promises, and inevitability of societywide digitalization. In the wake of the recent economic meltdown of Internet industries these critiques have attracted renewed attention. Indeed, even Rushkoff has belatedly recanted his optimistic prognosis for the digital era.[57]

We too reject the hyperbolic optimism that believes democracy is inherent in all information technology. But we are not content with a revival of Luddism, if it amounts only to nostalgia for a predigital era. We believe that digital technologies and global markets, as well as struggles in and against both, will indeed shape the future. But understanding the process requires an understanding of paradox and contradiction, not blind faith and a deterministic bent. We set out to avoid both technophilia and technophobia and attempt a more historical, more complex, hence more balanced account of the information revolution. We do so by turning a critical eye towards the video game as just one digital invention that is already in the hands of millions of young people.

Our critique begins with a rejection of the digital euphoria of technological determinism. Arguments such as Rushkoff's conveniently ignore the process of the design and construction of gaming experiences as the transmission of meaning from "producers" to "consumers" in the context of the power relations of a market society, which are not escaped by this entertainment industry. Blindness to the complex corporate institutions, technical constraints, design processes, and marketing calculations that generate the game experience, and to the "negotiations" that take place between producers and consumers of digital games in the context of a for-profit cultural industry, leads the enthusiasts to conclude that the players construct the possibilities of their own cultural narratives and fantasies. In such one-eyed visions of interactive gaming the player is seen as defining the very rules of the game.

We also are not so quick to pit the mass media against the digital, or the supposed "passivity" of television audiences against the alleged "activity" of digital ones. Interactive gaming did not fall from the sky ready-made but rather emerged on the basis of the very mass-mediated markets and culture it supposedly surpasses. Video gaming is in many ways an offspring of television – technically, in so far as game consoles depend on the television screen for their visual display; culturally, as an extension of the privatized in-home action-adventure entertainment forms TV provided; and promotionally, as television advertising was a central element in selling the concept of gaming to children and youth. Furthermore, both television and video gaming are channels of commercialized culture, carrying a flow of commodified entertainment to youthful media audiences. Put simply, the new media are built on the foundations of the old.

So in our view the claims made by the digerati are only partially true. There is a real difference, of course, between interactive gaming and the flow of television programming: choice and responsiveness have been programmed into digital play. Critics of high-technology culture err if they fail to acknowledge the dynamism of youthful entertainment audiences or the ways media producers recognized and responded to them. The gamer chooses their characters and their teams and explores in those virtual spaces. In navigating the game's branching paths and deciding on the course of the narrative, video game players do indeed engage the virtual world as "active" audiences. Playing games is a complex psychological engagement that blends creative exploration with narrative in a form of mediated communication that infuses young people's engagements with participatory intensity. It is a dynamic cognitive activity and cultural practice that elicits a variety of audience responses: selection, interpretation, choice, strategy, dialogue, and exploration characterize the player's relationship to the symbolic contents they manipulate on the screen. Clearly, there is an important cultural shift taking place from spectators to players.[58]

But the interactive enthusiasts need to take a closer look at the degree and kind of "active" participation of young audiences in the construction of their "own" digital culture. Choosing a corridor, character, or weapon – a rail gun or a chainsaw in a *Quake* death match – can be very absorbing. But it is hardly a matter of radical openness or deep decision about the content of play. Though gamers navigate through virtual environments, their actions consist of selections (rather than choices) made between alternatives that have been anticipated by the game designers. Gaming choice usually remains a matter of tactical decisions executed within predefined scenarios whose strategic parameters are preordained by the designers.

This preprogramming is implanted at a number of levels: technologically, in the capacities and valences of the machines players access; culturally, in the nature of the scenarios and storylines chosen for development; and commercially, in the price point of the software and hardware and in the marketing strategies that shape the trajectory of the industry as a whole. Indeed, to talk about "choice" in interactive games we must also address the market processes that have an impact upon what games are made available in the first place. When young gamers sit down to play *Pokémon* on their Game Boy, or *Everquest* on their PC, entering an imaginary world that has been programmed to respond to their fascinations and desire for entertainment, they are at the point of convergence for a whole array of technical, cultural, and promotional dynamics of which they are probably, at best, only very partially aware.

Indeed, one of the main objectives of the game industry is to make sure that the player does not reflect on these forces. The *sine qua non* of game designers is described by some as the "disappearance of technology." They have learned that the enthusiasm of the gamer dissipates when characters or weapons act inappropriately, when players experience the boundary of the game space, when they are forced to interact with avatars in cumbersome ways, or when they are too quickly killed by an enemy. That is why the disappearance of the interface with computers is among the chief goals of gaming. As one game producer told us:

One of the goals of a good game design is that the user becomes completely immersed into the experience so that they are not thinking that they are interacting with a computer, they are not thinking that they are fiddling with a joystick. The technology is so seamless, the design is so seamless, that they get into the character, and they completely lose sight of their surroundings and everything. In order to convince the person that they are immersed in an experience, the technology has to be so good that it makes itself invisible.[59]

Immersed in the game, the player becomes an imaginary subject who is fighting virtual monsters in the catacombs of an infernal planet or plotting the overthrow of the simulated leaders of the Egyptian dynastic order. Precisely in that moment of suspended disbelief, the system of interactive play becomes most fetishized. The construction of that willing delusion by which the players imagine they are controlling their own fantasy defines the magic of gaming. Digital designers devote a lot of energy to understanding its mysteries.

If the mediated nature of the game experience becomes apparent, if its various technical and cultural design components impinge on the player's consciousness, it is a sign that something is wrong – that the playability of the system is deficient. And deficient playability is not only a problem of technological performance or scenario narration. It is also a problem of market value: "If the control is awkward, the ability to suspend disbelief is lost because you are faced with a poor man-machine interface. So you can never immerse yourself in the experience … So every time the frame rate drops or the character feels sluggish or behaves in a way you don't want it to behave, you are no longer in the game. You are now a frustrated consumer."[60] That is why the idea of programming user transparency for complete immersion in the gaming experience has became the Holy Grail among video game designers.

There is nothing wrong, of course, with designing an absorbing virtual play environment: we too like to get lost in a game. Neither are we advocating clumsy video games. But given that game designers devote such attention to erasing the interface from players' awareness, eliminating every trace of the produced nature of the game experience, and promoting gaming as the zone of the superfantastical where we go to be entranced, it is hardly surprising that the average video gamer remains innocent of the reasons it takes two years to research and program an updated version of *Star Wars*, and how a single game can require millions of dollars in direct investment. Few of us know much about the creative work that goes into interface design or the marketing strategies that guide the latest *Tomb Raider* game and movie. The gamer is unlikely to think much about the engineering wizardry or the history of this cultural form while he or she sits in front of the screen. At that moment, gamers are extensions of their virtual technology, unlikely to be aware of how that play was constructed for them in the mediated entertainment marketplace, or of how complex cultural biases came to be inscribed in the game.

They are also unlikely to think much about how energetically game developers have sold the play experience to them – despite the fact that marketing a game may account for up to one-third of the costs of production, or that the promotional campaigns devoted to the launch of new consoles – Microsoft xbox, Sony PlayStation 2, and

Nintendo GameCube – will probably amount to $1.5 billion world-wide.[61] As with television, game makers had to learn not only to design and sell a new medium but also to construct the very audience for that medium. Here our account focuses attention on the growing cadre of digital "cultural intermediaries" (e.g., designers, marketers) who manage the flow of digital play culture to youthful consumers. The point is not only that interactive games are now a crucial node in a web of synergistic advertising, branding, and licensing practices that spread throughout contemporary popular culture. It is also that these promotional practices work their way back into game content – so that considerations of market segmentation, branding, franchising, licensing, and media spin-offs are now present at the very inception of game characters, scenarios, and plotlines. This is a crucial aspect of what we term "digital design practices," which needs to be understood as a strategic marketing process through which the abstracted "x on the wall" at a planning session in a game maker's boardroom is transformed into a joystick in the hand of the gamer, into a mediated experience for the audience.

We cannot overlook the fact that even highly participatory multi-player online games such as *Quake* and *Everquest* are designed in ways that bear similarities with the way mass media content, such as television programs, are directed. As critical communication scholars, we find the "Don't worry, be happy" attitude of members of the digerati like Rushkoff and Negroponte disingenuous. There is a slippery slope from their conceptions of a digitally empowered player to a doctrine of the sovereign consumer that blindly accepts whatever the market dispenses as right and good. This combination of technological determinism and market idolatry is the ideology we see coming together around futurist celebrations of interactive gaming.

The paradox that is lost in such visions of digital progress is that genuinely new technocultural innovations, from cellular phones to interactive games, are being shaped, contained, controlled, and channelled within the long-standing logic of a commercial marketplace dedicated to the profit-maximizing sale of cultural and technological commodities. While interactive games are in many ways genuinely "new" media, their possibilities are being realized and limited by a media market whose fundamental imperative remains the same as that which shaped the "old" media: profit. While this encounter between digital media and capitalist markets may in part (as the new-economy gurus claim) be reshaping markets, it is also constraining and channelling the directions taken by new media. Moreover, the demassified digital media do not necessarily mark a hard break with the symptoms of a mass-mediated culture, leading automatically to cultural diversification. We have to see

the disturbances and frictions created by the intersection of new potentialities with old logics. Only by understanding the play of paradoxes –
the discontinuities and continuities in economic, cultural, and technological spheres – that is structuring digital capital can we estimate the
probable trajectory and possible alternatives for digital play culture. But
this is the contradiction that the digerati cannot come to terms with. By
shutting their eyes to the constraints that circumscribe interactive media,
they blind themselves to actual, rather than merely notional, possibilities
for change. It is against this reduction of possibility that our analysis
takes aim. Our argument is not that multimedia systems are intrinsically
oppressive, vacuous, or malign. It is rather that their potential is being
narrowed and channelled in ways that betray their promise, even as that
potential is promoted with the rhetoric of choice, interactivity, and
empowerment.

TOWARDS A CRITICAL MEDIA ANALYSIS

As Robins reminds us: "The institutions developing and promoting the
new technologies exist solidly in this world. We should make sense of
them in terms of its social and political realities, and it is in this context
that we must assess their significance."[62] Bearing this in mind, we set
out to *unpack* the interactive game historically by situating this new
media in the context of the digital marketplace. We are using "unpacking" to describe our method of uncovering the history of the games
industry and the promotional packaging that encircles digital play. In
our view, taking an historical perspective and gradually pulling back
some of the layers in the process of commercialization will help us to
gain a better understanding of the video and computer game as well
as how it reached its current status in popular culture and high-
technology capitalism. In this way, we might get a better sense of how
a particular cultural practice and cultural industry is linked up to, or
intersects with, the general dynamics of profit accumulation. In our
case study we examine the dynamic of market expansion in the information age and seek to develop a critical media analysis that can keep
pace with the integration of digital technologies and cultural industries.
    The book is divided into three parts – "Theoretical Trajectories,"
"Histories: The Making of a New Medium," and "Critical Perspectives."
In the first part we lay the theoretical foundations for our investigation.
Chapter 2 briefly surveys three streams of thought in communication
studies: media theory, political economy of communication, and cultural studies. The debates in these three streams of thought, although
unfolding from very different perspectives, have all grappled with the
consequences of the emergence of the mass-mediated marketplace on

economic, social, and cultural life. Although each provides valuable resources for understanding video and computer games, none, we suggest, is by itself adequate to the task. Indeed, tendencies to isolate these contending perspectives can sometimes be an obstacle to grasping the complexity of new media in a high-intensity market setting.

Attempting to overcome these divisions in the study of media, we draw on Raymond Williams, one of the pioneers of communication studies, whose social histories of media in our eyes provide an exemplary critical approach in that they recognize the dialectical interplay of technologies, culture, and economics. Inspired by Williams's methodology, we propose a theoretical framework to guide our critical media analysis of the interactive game. We begin with the "circuit of capital," which involves the ongoing process of making and selling commodities in order to accumulate profit. Within this overarching circuit we identify three subcircuits. Our "three-circuits model" situates digital play as it comes into being at the convergence of technological, cultural, and marketing forces in the mediatized global marketplace. In the technology circuit, we are referring to the practices of inventors, machines, and users; in the cultural circuit, to the production and circulation of meaning in video games as media "texts"; in the marketing circuit, to the communication practices that link marketers, commodities, and consumers in the gaming marketplace. Although it is useful for the purpose of analysis to distinguish between the circuits, in practice they interact in a state of dynamic process.

A critical perspective on interactive play must address the digital futurists' claim that new media represent a total break with the past and an entry into an unprecedented utopian moment of technological empowerment in the new economy. We take up this task in chapter 3 with the help of the accounts of social and technological change put forward by theorists who suggest we are experiencing a transition from "Fordist" to "post-Fordist" capitalism – or from industrial capitalism to "information capitalism." Such theories recognize the importance of digital technologies in restructuring work, play, and all forms of social interaction. But instead of seeing these technologies as marking a total break with the past, they insist that the effects of new technologies must be seen in their intersection with the continuing, and indeed intensifying, force of a global market economy predicated on the priorities of profit, commodification, consumerism, and managerial strategy. The concepts of post-Fordism and information capitalism require that we articulate the paradoxical play of continuities and discontinuities crucial to our analysis of interactive media.

Though these ideas originate with political economists, several writers have drawn a connection between the post-Fordist economy and postmodern culture characterized by simulation, hyperreality, the increasing role of design and marketing, and the ascendancy of the image – a culture of which digital play is surely an exemplar. It is also exemplary of the intensifying "promotional" ethos of contemporary capitalism, with its hypercommercialism, high-intensity synergistic branding practices, and its privileging of cultural intermediaries such as designers and game marketers. As a way of drawing together these ideas, we suggest that interactive games can be seen as an example of what Martyn Lee terms the "ideal commodity" of a post-Fordist/ postmodern/promotional capitalism – an artifact within which converge a series of the most important production techniques, marketing strategies, and cultural practices of an era. The three-circuits model of interactive gaming and the concept of interactive games as an "ideal commodity" in post-Fordist capitalism are twin theoretical reference points that we explore throughout the book.

Historical perspective is, in our view, a vital dimension of critical media analysis. In part two, "Histories," having established our theoretical orientation, we submit the development of interactive entertainment media to a critical historical analysis and, in so doing, clear away some of the smoke – and the sense of inevitability – that has muddled debates about the trajectory of digital media culture. Following Williams's suggestion that critical analysis must discern the "intentionalities" that enable and constrain both cultural and technological possibility, we examine the video game industry's technological innovations, digital design practices, and audience-building tactics as they emerged within particular historical moments and specific institutional constellations.

In chapter 4 we tell the story of how in the 1960s and 1970s interactive game technology arose from an extraordinary conjuncture of military-industrial research and hacker experimentation; of its commercialization by the first great video game company, Atari; of the passage of the new entertainment media from arcades into living-rooms; and of the catastrophic industry crash that in the mid-1980s all but liquidated the industry from North America. Chapter 5 looks at the revival of the digital play business by the Japanese company Nintendo, and at the role played by this famous video game firm, with its tiny plumber-hero, Mario, in rationalizing the design, branding, and marketing practices of the interactive game industry and creating a "Nintendo generation" familiar with digital play. Chapter 6 shows how in the 1990s Nintendo's near-monopolistic control over interactive entertainment was challenged both by other video game console makers, such as Sega, and by games developed for the personal computer, such as *Doom* and *Myst*, creating an era of competitive turbulence that made and destroyed entrepreneurial fortunes and provoked a frenzy of technological, cultural, and promotional innovation in interactive play. In chapter 7 we see how this expanding cultural industry, increasingly associated through online play with the attractions of the Internet and the lure of e-commerce business models, has attracted corporations such as Sony and Microsoft – giants of information capitalism who see in interactive gaming a critical high ground to be captured in the struggle to dominate digital markets and build multimedia empires. Part two concludes in chapter 8 with a "state of play" overview of the industry as it enters the third millennium, releasing its latest technocultural marvels – the PlayStation 2, the xbox, the GameCube – amidst the uncertainties of the dot.com crash.

Many of the stories of the interactive game industry have been told before, often by gaming enthusiasts, corporate historians, or reporters following the bizarre twists of virtual culture.[63] Our aim is not to recapitulate these sometimes rather breathless accounts but rather to expose the persistent dynamics and ongoing vectors that have made interactive play a major force in today's digitally mediated world market. Thus, our account focuses on themes such as the role of interactive gaming as a trailblazer for the entrepreneurial, innovation, and intellectual property dynamics of information capitalism; the commodification of digital play; the role of marketing in making play cultures visible and targeting them as consumer segments; game design as a cultural practice for managing the expansion of markets through the activities of cultural intermediaries; market-based negotiation with consumers and integration of audience feedback; the problems and possibilities arising from perpetual technological innovation and the recurrent exhaustion of entertainment values in an industry based on constantly renewed hardware and software; the links between digital play and the military; and the emergence of online gaming as a subsector of the video game industry that is of critical importance to e-commerce models.

The third part, "Critical Perspectives," examines controversies, tensions, and unresolved problems within the contemporary interactive play business. The so-called friction-free capitalism of the new economy has plenty of contradictions, conflicts, and perhaps even self-destructive tendencies. The sequence in which we present issues here approximates to the three dimensions of interactive gaming that we identified in our three-circuits model – technology, culture, and marketing – although we emphasize that each case involves the interplay of all three levels.

In chapter 9, we look at the unanticipated problems arising within the technological circuits of the industry from the unpredictable actions of the human subjects who both make and use digital machines: at the point of production, in terms of new forms of globalized labour unrest; in terms of use, in the problems of piracy and hacking. In chapter 10 we turn to the marketing circuit and the central role of marketing strategists in managing the video game industry's growth. The interactive gaming industry has been a pioneer in managing the brand-building and synergistic marketing techniques that are key to commodifying today's fluid youth markets. But this very success may have a long-term effect in stifling creativity and experimentation in digital play, as the imperatives of branding, synergistic connection, and massive marketing drive towards industry consolidation and focus on a handful of sure-fire licences, franchises and serialized blockbuster hits, such as Nintendo's *Pokémon*. Chapter 11 looks at the cultural circuit and the controversies about violence and gender that have always troubled the industry. We examine the design and promotional dynamics that have skewed the industry towards a culture of what we term "militarized masculinity" – a source of testosterone-niched commercial success, but also a focus of continuing criticism from those who are disturbed by the virtual carnage and the marginalization of girls and women, and perhaps now a real barrier to industry growth.

Our final chapter draws these themes together in an examination of the interactive entertainment industry's first big hit of the third millennium, *The Sims*. Examining how this game makes visible the technological, cultural, and marketing preoccupations of digital capital, we suggest that in each zone the game illuminates some of the key dilemmas of the interactive game industry: in technology, a tension between enclosure and access; in culture, a divergence between violence and variety; and in marketing, a deep rift between commodification and play. The interactive game industry now exists in a wider political-economic context of sliding markets, global turmoil, and war, navigating these tensions only with some difficulty. But in our coda, "Paradox Regained," we suggest that such contradictions provide grounds not just for criticism but also for a qualified optimism, since they mark the openings through which new forces can emerge to transform yet again the worlds of work and play.

PART ONE

Theoretical Trajectories

To propose a theoretical perspective within which to analyze interactive gaming, we draw on the conceptual toolkit of communication studies. This is a complicated undertaking, not least because communication studies is a tradition comprising many streams of thought. From the various streams we see especially valuable resources for analysing interactive gaming in media theory, political economy of communication, and cultural studies. They are helpful in taking up our task because writers in each field have debated the impact of communication media on social, cultural, and economic life. In our view, a critical media analysis of interactive gaming requires a supple mixture of concepts and perspectives drawn from each of the three disciplines; the additional challenge is to confront the divide between these perspectives.

The next chapter therefore takes a look at some of the key branches in the discipline of communication studies. Our survey begins with the influence of the Canadian tradition of media theory. The "father" of this school, Harold Innis, impresses on us the importance of considering the unique "biases" in time, space, and experience that specific media inscribe in our social practices. Extending Innis's ideas, Marshall McLuhan teaches us that media of communication are a unique category of technological development because they both mediate the transmission of culture and "disturb" patterns in the wider culture. A second, very different line of analysis, the political economy of communication, focuses on the role of media in perpetuating capitalism, on the structures of corporate ownership and control that dominate contemporary means of communication, and on the ideological filters they impose. But a third theoretical field, cultural studies, has shown in its analyses of the relationship between mediated culture and power, identities, beliefs, and values are not abstractly and automatically transmitted through media: rather, there is a complex set of negotiations through which audiences actively "decode" the meanings that have been encoded by producers in media content, making cultural texts important sites for complex contests over signification and representation.

All of these theoretical contributions offer a valuable yet partial perspective on new media in the capitalist marketplace. Juxtaposed, each reveals the deficiency of the others. Our overview indicates the need for a more multidimensional approach to critical examination of the mediated marketplace in general and interactive gaming in particular. We find valuable pointers in the analysis of Raymond Williams, one of the founders of communication studies. We agree with Williams that the cultural impact of a new medium cannot be diagnosed until we understand the historical circumstances of its development. We also greatly admire his insistence on concrete and specific studies of particular media; his suspicion of sweeping abstractions and confident reductions; his emphasis on an analysis of institutional contexts; and his insistence on the importance of human agency and intention in the shaping of technological systems. We also share his belief that critical media analysis must take account of the dialectical interplay of technologies, culture, and economics.

Williams's example persuades us that one of the keys to under-standing the emergent mediascape is to produce a more integrated, synthesized analysis of the lockstep dance of technological innova-tion, cultural practice, and high-intensity marketing. We therefore conclude chapter 2 by proposing a systematic map of the techno-cultural-capitalist matrix that we call the "mediatized global market-place." We present diagrams of the interplay between what we call the "three circuits": the circuit of culture, the circuit of technology, and the circuit of marketing. They represent a framework for crit-ical media analysis that can guide a more careful general reading of the epochal restructuring of the market economy and, more specifically, illuminate the nature of digital play.

Although it is our intention to pick up and extend Williams's method of media analysis, we also elaborate on our three-circuits model, to keep pace with our digitized object of analysis, by taking cues from some more recent theorists, whose work we discuss in chapter 3. To update critical media analysis for the era of informa-tion capitalism, we identify three necessary elaborations. First, such analysis must recognize the emergence of digital technologies, espe-cially networked computers, and their impact on markets and cultural industries. Second, it must address the changing media environment itself, especially what we describe as the postmodern-ization of culture. Third, it must reveal how the technological and cultural possibilities of new media are converted into commodities in the marketplace, taking account of a new intensity of marketing practices and the increasing prominence in the corporate workforce of a new strata of "cultural intermediaries."

In approaching these issues we have found it fruitful to consider some of the perspectives opened by the Regulation School of polit-ical economy. Their theories of a transition from "Fordism" to "post-Fordism" provides a way of locating the arrival of digital technologies and the role of new media of communication in capi-talism's changing "regimes of accumulation." To these concepts we add the broadly similar work of Tessa Morris-Suzuki, whose account of "information capitalism" analyses the role of digitization in creating a "perpetual innovation economy." Writers such as Fredric Jameson, David Harvey, and Stuart Hall have all pointed to connections between the new post-Fordist phase of capitalism and aspects of postmodern culture, with its emphasis on the primacy of image, surface, and sign. To explore these issues we call on the work of Jean Baudrillard, whose concepts of "simulation" and "hyper-reality" have obvious relevance to the digital illusions of interactive gaming. But there is another, underestimated, aspect of Baudrillard's analysis of postmodern culture that deserves greater emphasis and is also vital to analysis of digital play: his discussion of the role of new intensities of advertising, marketing, and promo-tional strategy in creating the endless whirl of symbolic creation and destruction that characterize the consumer society.

We then bring these theoretical threads together in a discussion of Martyn Lee's intriguing suggestion that for each phase in the development of capitalism it is possible to identify an "*ideal-type commodity form*" – one that embodies its most powerful economic, social, and cultural tendencies.[1] The ideal-type commodity reflects the entire social organization of capitalism at a given historical point in its development. We suggest that the interactive game can be seen as an "ideal-type" commodity exemplifying the current phase of capitalist market relations, which maintains growth through the inte-grated management of technological innovation, cultural creativity, and mediated marketing.

2

Media Analysis
in the High-Intensity Marketplace:
The Three Circuits of Interactivity

INTRODUCTION: THEORIZING DIGITAL PLAY

How should we analyze video and computer game play? As a technological experience, created by digital programs, user-machine interfaces, and telecommunications networks? As a market transaction in which we consume digital commodities produced for profit by corporate media empires? Or as a cultural text that offers players immersion in riveting stories full of fantastic characters, bizarre environments, and gripping narrative choices?

To construct a theoretical perspective for a critical media analysis of the interactive game we must begin by pooling and evaluating existing intellectual resources. Communication studies has long been occupied by debates on "mass" and "new" media, trying to assess their role in and consequences for culture and economy. As we noted in the introduction to Part One, studies of media are frequently taken up within one of three optics: media theory, which sees new communication technologies radically restructuring our most basic coordinates of experience and our perceptions of time and space; political economy, which examines new media and technology as extensions of capitalist power; and cultural studies, which treat interactive games, as well as film and television, as media texts to be read in terms of representation, narrative, and the "subject-positions" offered to audiences. These perspectives are sometimes divergent and often isolated from one another.

Based on our overview of the field, we see the need for a more multidimensional approach to interactive games. We see the various theoretical strands productively woven together in the critical media analysis of Raymond Williams, one of the earliest practitioners of communication studies. Drawing on Williams's legacy, we propose a map of the global mediatized marketplace, a map that schematically diagrams the complex interplay between the three circuits of culture, technology, and marketing.

MEDIA THEORY: TIME/SPACE BIAS

We have already criticized the technologically determinist silicon futurism that dominates so much writing about interactive games. But coming to grips with digital entertainment clearly needs *some* theory of how technological change relates to cultural transformation. We find hints towards a viable approach in the works of Harold Innis, the Canadian "father" of media theory, and his controversial student, Marshall McLuhan.

In his seminal books *The Bias of Communication* (1951) and *Empire and Communications* (1950), Innis developed a subtle account of the relationship between communication technologies and social change. It is easy to misrepresent his thought by isolating key concepts; its distinguishing feature is an insistence on the way factors work together to produce complex effects. But three of Innis's concepts are especially important to us: the "bias" of communication technologies, their role in the rise and fall of "empires," and their relation to "oligopolies of knowledge."

Innis declares that all media have "bias."[1] He does not mean that they transmit skewed information (although this may well be the case) but something much deeper. Media, Innis says, affect our whole perception of time and space. Some media are spatially biased in that they can send messages over great distances (think of electronic broadcasts). Others are time biased in that they emphasize the preservation of cultural memory (think of chiselled stone monuments). Successive media innovations cannot therefore be understood as mere improvements in the transmission of information. Rather, they involve radical reorientations in the basic coordinates of lived experience. "History," claims Innis, "is not a seamless web but rather a web of which the warp and woof are space and time woven in a very uneven fashion and producing distorted patterns."[2]

Innis was interested in uncovering the uses to which media of communication are put and their related effects on wider social relations of power. To do this Innis first emphasized that a media system's bias emerges within the specific social circumstances into which a technology is introduced. To understand this, we have to see how media are inserted in the vast agglomerations of power that Innis refers to as "empires." The Roman road system enabled the consolidation of imperial power; Latin masses and monastic manuscripts were vital

to the medieval church; telegraph systems that created rapid transcon-
tinental news reporting were essential to industrial capital. Here, Innis
shows us that media of communication both shape and are shaped by
the cultural and economic circumstances from which they emerge.

In discussing the relations of power characterizing a particular empire,
Innis used the concept of an "oligopoly of knowledge."[3] The term designates the political and cultural control exercised by a particular social
group or class through the restrictions and directives they can impose on
the way knowledge is organized, stored, and distributed. The introduction of new media can consolidate or potentially shift the terms of such
oligopolistic control. Their effects, however, are often paradoxical. A
medium that might at first seem to reinforce the power of an established
élite could eventually subvert it by undermining the relations of power
into which it is introduced. As Innis remarked, "sudden extensions of
communication are reflected in cultural disturbances."[4]

In his best example Innis shows how the printing press was first used
to print the Bible and extend the power of the Church and its control
over moral and intellectual discourse.[5] Yet eventually print subverted
the clergy's oligopolies of written knowledge, enabling the mass production and distribution of knowledge through channels outside the
Church's control. Mechanization of print accelerated public access to
previously sacred or controlled knowledge, contributing not only to the
secularization of medieval life but also to the promotion of expanded
trade routes and technoscientific discourse as key vectors of modernist
and economic expansionism. For example, the demand for news
created by print technology itself hastened the development of the telegraph and the organization of news services, laying the foundation for
mass-circulation magazines and newspapers and increasing the flow of
national advertising. In this way, Innis, who was one of the first theorists to recognize the strong connection between media and markets,
pointed out that this commercialization of communication in turn created new oligopolies of knowledge as corporate media acquired
increased power to manipulate and direct public opinion.

Innis's sweeping insights into the role of media cannot be reduced
to simplistic models that distinguish technological, economic, and
social factors. Indeed, the whole thrust of his writing is to understand
the interplay of these forces. As Marshall McLuhan observed in his
introduction to *Empire and Communications,* Innis looked for "patterns in the very ground of history and existence" and "saw media,
old and new ... as living vortices of power creating hidden environments that act abrasively and destructively on older forms of culture."[6]
His methodological premise is that only by tracing these changes
historically and in detail could we gain insight into how specific media

transmit their bias of time and space to reinforce or disturb relations of power.

McLuhan continued Innis's investigation of media as "living vortices of power" in ways that both developed and distorted his mentor's legacy. He was a theoretical *agent provocateur* whose his ideas have recently become wildly popular – so much so that we hesitate to bring them up again, especially because they have been widely recruited to prop up the very sort of technological utopianism we so strongly disagree with. Despite this (mis)use of McLuhan, his ideas about the cultural dynamics and disturbances caused by new technologies and media of communication remain important to our thinking about video games.

One of the crucial insights we take from McLuhan is his observation that technological mediation is a condition of culture. He illustrated this idea in his various studies of media of communication. Written in 1951, *The Mechanical Bride: Folklore of Industrial Man* was McLuhan's pioneering foray as an observer of the "social myths" surrounding new technologies and the emerging consumer society.[7] Using as his entry point a number of advertisements and popular media texts, his book was unique because it addressed the impact of technology, including media technologies, within the practices of everyday life. In the promotional stories told to consumers about the emerging industrial society, McLuhan saw a budding "folklore" of modern technologies, noting that there is a deep and meaningful engagement with these domesticated devices; they are our "mechanical brides."

McLuhan shows us that each technological device communicates on the level of everyday experience, creating a cultural ripple effect as people use the potential of technologies to act differently – but also to feel differently. The family of the industrial age, for example, appeared to him not so much gathered around the radio as trapped in a transition between the values of a print-based culture of the mechanical age and the oral culture of the electric era. In that sense, the advertisements he studied suggested that industrialized societies were in the midst of a cultural transformation perhaps as profound as the one Innis discovered in connection with print: "We are today as far into the electric age as the Elizabethans had advanced into the typographical and mechanical age. And we are experiencing the same confusions and indecisions which they had felt when living simultaneously in two contrasted forms of society and experience."[8]

For McLuhan, then, media of communication open up the possibility of expanding human experience. McLuhan's most important contribution may be the notion of media as prosthetic extensions of our senses. Each new medium amplifies the perceptual modalities through which it communicates, creating new sense ratios. The book extends the eye, through the visual practice of reading; the radio, as audio communication, is an amplification of the ear. But this process is not all gain. New media also entail loss, McLuhan says, because "the price we pay for special technological tools, whether the wheel or the alphabet or radio, is that these massive *extensions* of sense constitute *closed* systems."[9] For example, the book restricts communication to visual symbols, the radio to voice, music, or sound. As cultures favour communication through any specific medium they are therefore also engaging in a form of "autoamputation," truncating and narrowing the possibilities of interchange and experience.[10]

McLuhan thus pushes the theory of media bias outward from Innis's concern with societal oligopolies of knowledge to embrace the totality of human experience. He sees our whole society, culture, and subjectivity transformed by communication technologies – our bodies, our senses, our feelings and patterns of experience. The now-trite statement "the medium is the message" means that the deepest significance of any new medium of communication is "the change of scale or pace or pattern that it introduces into human affairs."[11] For example, the light bulb is a medium that disturbs the patterns of both work and leisure by extending time.

So while Innis focused on print, McLuhan was a theorist of electronic media. Mechanized technologies such as print, McLuhan claimed, had accentuated the rationalist mind/body split, the separation of economy and culture, and the isolation of individuals in society. The new electronic environment was overwhelming these divisions, however, as radio and television created new communities, new identities, new aesthetics, and new sensibilities based on collective participation in a synergetic media environment. Here his most famous idea is perhaps the prediction of an electronic "global village." Based on his observations about the speed of transmission of broadcast signals around the globe, he announced that planetary inhabitants were being brought back to the sort of mutual proximity and community that they had enjoyed in preindustrial society, "retribalized" by participation in global television.[12]

Sometimes McLuhan rose to ecstatic heights, suggesting that electronic media "have extended our central nervous system itself into a global embrace, abolishing both space and time ... The creative process of knowing will be collectively and corporately extended to the whole of human society." However, as writers such as Arthur Kroker and Heather Menzies have emphasized, alongside this techno-optimism there was a darker, more disturbing McLuhan – one who could write of how our dependence on electronic technologies was in fact making humans nothing more than the "sex organs" of the machine, the means by which technological systems reproduced themselves.[13] The real value of McLuhan's writing lies in the tension between these different possibilities.

In the context of our study of interactive games, McLuhan unexpectedly offers us insights in a little-noted chapter about games in *Understanding Media* (1964).[14] McLuhan argues that games ought to be seen as a medium of social communication. As with his theories of electronic media, the specific rules or content of play is less important than the experience of participation. For it is this attraction, "the *pattern* of a game," that reveals "the central structural core of the experience" of communal participation.[15] For McLuhan, play and games are worthy of our attention because they provide a map of much broader social forces; indeed, they are "deeply and necessarily a means of interplay within an entire culture."[16] Much of our "'adjustment' to society" requires "a personal surrender to the collective demands. Our games help both teach us this kind of adjustment and also to provide a release from it."[17] It is through games that we seek to heal the alienation and overloaded mental circuitry of the mechanical age. As a "popular response" to the stress of work and technological society in general, "games become faithful models of a culture. They incorporate both the action and the reaction of whole populations in a single dynamic image."[18]

How useful are McLuhan's eclectic, erratic ideas to our study of the video game? His claim that new media have deep consequences in structuring subjectivity not just at the level of cultural content but of perceptual process is one that has been widely taken up by digital futurists. Many suggest the video game medium is the message, the leading vector of a process of virtualization of social experience, a domestic Trojan horse for a very far-reaching set of digital dependencies and acceptances. These transformations include familiarization with new spatial experiences of on-screen navigation and of connection to cyberspatial collectivities; to a new sense of time, based on experiences of speed, reversibility, and resumability in play; and, arising at the intersection of these time-space reorientations, a gradual habituation to virtual immersion, disembodied identity, and multimedia intensity.

McLuhan's recognition that changes in media – changes to our nervous system – are traumatic events seems particularly cogent in an era of near "virtual reality" experiences in cyberspace. One has only to watch the absorption of a video game player engaged in a sports game such as NFL *Fever* or NBA *Live* to get an intuitive sense of his ideas. Such games translate a common recreational experience from the corporeal to the virtual level; full body agility, strength, and speed, contact with the elements, and physical confrontation are shifted into issues of hand/eye coordination, screen navigation, and pixels. Put bluntly, sport video games invite kids to sit in front of a screen rather than go out and kick a ball around (or even cheer in the stands). Arguably, this is just one more point along a gradient that has seen the division of amateurs from professionals, spectators from players, stadiums from television sitting-rooms. Yet it is an important acceleration: when every mouse potato can be its own Gretzky, the rift between real bodily virtuosity and disembodied virtual reality is significantly elided.

There are, of course, different ways to tally this shift. To the degree that digitalization coexists with older corporeal skills and cultural competencies, overlapping rather than displacing, it creates an expanded sensorium. However, to the degree that it supplants rather than supplements other forms of sociality and experience, it also contains the seeds of diminishment, atrophy, or attenuation. If we accept, with McLuhan, that technocultures remake subjects on a deep level, then we should also consider his darker insights, that this process is not unequivocal gain. Interactivity, for example, may not only be empowerment and education but also loss and amputation, as digital aptitudes squeeze out or devalue other nonelectronic capabilities.

McLuhan's perceptions therefore seem crucial to understanding video gaming and virtual technologies. But as many critics have pointed out, there are major weaknesses in his work. His definition of media as prosthetics conflates technologies, media, and social forms, obscuring important differences in the social contexts and purposes in which media are developed and used. Too often his aphorisms smack of technological determinism, enabling popular writers such as Alvin Toffler and Frances Cairncross to convert his tentative probes into dogmatic, formulaic charts of the information revolution.

We have two fundamental criticisms of McLuhan. First, he pays too little attention to the relations of social power that structure media. The optimistic vision of the "global village" completely ignores the huge social divisions and power inequalities opened by electronic technologies. It skips over the divisions in wealth that separate the young North American owners of PlayStation 2s and xbox's, commanding what were once military levels of computing power from their homes, from the majority of the world's children who can never afford such gadgets. It does not register the massive resources for manipulating and monitoring consumers exercised by Nintendo or Sony, or ask about the divisions in the digital workplace where consoles and games are produced. To get some grasp of this we need to keep in mind the concerns raised by McLuhan's mentor, Innis, about "oligopolies of knowledge" and their role in the expansion and contraction of an "empire."

Second, asserting that the "medium is the message" overlooks the issue of content. Put simply, sometimes the message is the message. To understand video games, we have to look not just at how they alter our perceptions of speed and space but also at how these sensory alterations are associated with and inflected by very specific sets of meanings – about, say, gender, or violence, or consumerism. Is digital gaming liberating if it brings to young players a steady stream of *Quake*-style carnage? Perhaps. Does it matter that for decades games have been addressed overwhelmingly to male players, excluding girls and young women? These are certainly questions that go beyond a technological logic into issues of culture and economy: from cultural practices and systems of meaning to media ownership and marketing strategies. To take up these two points we therefore go beyond media theory to two other branches of media criticism: political economy and cultural studies.

POLITICAL ECONOMY:
CULTURE INDUSTRY, MEDIATED MARKETS, AND ELECTRONIC EMPIRE

Looking at recent theories about digital media, we saw how dot.com enthusiasm blends technofuturism with the revived free market economics of Adam Smith. But there is another, more critical line of analysis of the media in a market economy, one often referred to as "political economy" of communication.[19] It too has roots in the work of an ancient economist – in this case in Karl Marx's radical analysis of capitalist market society as one divided between owners and workers. Underneath the apparent free and fair exchange of the market, Marx said, lay a deep, systemic imbalance between capitalists and the labour they exploited. Marx wrote in the nineteenth century. Although he was an author and a journalist who sometimes worked for the major communication industry of his day, the newspaper business, his analysis of class conflict focused on the factory. But as media industries grew in importance thinkers who shared Marx's perception of the injustice, conflict, and crisis tendencies inherent in market society developed from his insights a scathing critique of the capitalist owned media.

These critics saw profoundly problematic ideological and cultural consequences of the interaction between the industrial capitalist system of mass production and the "one-to-many" communication systems of mass media. For example, in Europe during the 1930s and 1940s a group of Marxian intellectuals known as the Frankfurt School saw commercialized media – which they called "the culture industry" – extending the logic of mechanized capitalist production into more and more realms of everyday life.[20] Impressed with Hitler's use of radio and film for propaganda, they depicted mass media as the ultimate instruments of ideological manipulation and social control. When members of the Frankfurt School fled from fascism to North America, they continued a critique of the biases of the commercial press and the emptiness of Hollywood films. Mass advertising extended capitalist ideological control into the deepest regions of our inner life, falsifying the core of human needs. The pacifying spectacles of the entertainment sector imposed the vacuous visions of marketing managers, rather than what people really wanted. According to the Frankfurt School's most famous representative, Herbert Marcuse, the power of media over contemporary thought and feeling was so insidious and total as to render society "one dimensional."[21]

In postwar North America this line of analysis was developed by a school of political economists of media radicalized in student and anti-Vietnam War activism of the 1960s. Herbert Schiller, for example, argued that new cultural industries, especially television, exemplified the ideological control of capitalist "mind managers."[22] Media content on television was, after all, transparently simplistic, biased, and mindless – massaged by corporate barons to appeal to suggestible and passive audiences who already had so little control over their own lives that they were happy to escape into the mediated fantasies. The expanding media was a system pacifying and distracting consumers from the oppressive drudgery of their lives, the "bread and circuses" that kept contemporary workers blinded to their own exploitation.

Adapting Marx's thought to media analysis was not always easy or productive. Indeed, some parts of his writing came to seem like an impediment to critical cultural analysis. In one notorious passage, Marx had described culture as a superficial "superstructure" flimsily perched on top of, and shaped by, a "base" of industrial production that was the site of the real, material economic action.[23] This "base/superstructure" metaphor consigned newspapers, television, films, broadcasting, and, we would add, video games to a marginal role quite contrary to their growing importance in advanced capitalism. Many theorists who had been influenced by Marx (like Raymond Williams, whose work we mentioned in the first chapter) became increasingly impatient with this obsolete formulation but continued to find inspiration in other parts of Marx's work.

An example of such an alternative approach is an essay by Nicholas Garnham that we have found very useful. He proposes a model for understanding the mass media that takes as its starting point Marx's observations about the "circuit of capital."[24] As Garnham explains, Marx noted that "capital is a process which is continuous, circular and through time ... a cycle within which the moment of consumption is part of the production process."[25] Businesses must not just produce commodities but also sell them. They must find buyers who want their goods, have purchasing power, are in an accessible location, and so on. If they do so they emerge with an expanded sum of money – profit – to spend or invest and thus start the cycle again. But there is no certainty of success. If they fail, the cyclical process stalls. For an individual firm, such interruptions in the "circuit of capital" mean lost profits and perhaps bankruptcy. On the scale of an entire industry or economy – as happens in the great cyclical crises of capital – they mean widespread business failure, mass unemployment, and social crisis. As a result, the managers of market society are increasingly preoccupied by the smooth circulation of capital.

In the early phases of industrial capitalism, entrepreneurial efforts focused on production – imposing work discipline, introducing new machinery, and creating the factory system. Demands for goods such as food, shelter, and clothing could be taken for granted. The main problem in the sphere of circulation was getting the goods to market, an issue addressed by great nineteenth-century transport networks of canals, roads, railways, and steamships. But as basic needs were met, and as industry began to create an ever-greater volume and diversity of goods, finding buyers became more challenging. The market system gradually developed an increasingly complex selling apparatus that involved retailers and wholesalers, department stores, national branding, marketing, and advertising.

In this system, Garnham says, the role of the mass media can be seen under two aspects. First, media industries are themselves businesses, selling information and entertainment to consumers, with their own interest in speeding the process by which these commodities reach buyers. This has led them to a series of technological and organizational innovations aimed at saturating each moment of everyday life with opportunities for media consumption. Steps in this direction include the development of the commodity "news," with its daily (or hourly) cycle of consumption; the creation of an eye-catching tabloid press designed for rapid reading; and the emergence of radio and television – media that can be watched or listened to in the midst of other activities, at any time of day or night.[26]

Second, Garnham says, mass media are the bearers of advertising. Most media depend heavily – in the case of broadcast radio and television almost entirely – on selling the attention of audiences to advertisers.[27] In their reliance on advertising, newspapers, radio, and television are important vehicles for the overall stimulation of the needs and expectations necessary to sustain what is termed a "consumer society." In this role, "commercial" mass media accelerate the circulation of all commodities.

Garnham's observations dovetail neatly with those of researchers such as William Leiss, Stephen Kline, and Sut Jhally, who have emphasized the role of marketing in the commercialization of mass media and in creating a "high intensity market setting" that permeates multiple spheres of social life.[28] They argue that the circuit between production and consumption is mediated through a series of marketing subindustries – such as audience research, advertising design, and media buying – whose profitability depends on their success, real or imagined, in communicating with audiences that are intended to cultivate the desire and need for new commodities or brands. By serving as a "bridge" between commodity sellers and buyers, the marketing communication practices and consumers' activity in watching such advertising and engaging with other forms of marketing become a necessity for capital.[29] In this view, the market has become altogether saturated with media (and vice versa), which in turn brings us to the consolidation of a distinctly "mediatized" marketplace.

Political economy of media has produced many impressive analyses of the forces of corporate concentration and conglomeration that are today producing gigantic electronic empires. Scholars from this school have also drawn out in considerable detail the sort of ideological shaping of information flows that results from such a commercial system, arguing that it results in an emphasis on trivialized infotainment over serious social analysis, journalism skewed towards the interests of ruling élites, and exclusion or marginalization of content that criticizes corporate power. Although such discussion usually follows the broad lines indicated by the Frankfurt School, it has extended and refined their analysis. A widely known example is the model of "media filtering" through ownership, advertising, legal intimidation, reliance on conservative sources, and knee-jerk rejection of left-wing views incisively outlined by Edward Herman and Noam Chomsky in the book *Manufacturing Consent: The Political Economy of the Mass Media* (1988).[30]

As yet there has been no critical political economic analysis of the video game industry. In one respect, this just reflects a generational time lag in a discipline that was forged in the analysis of post-World War II media. Political economists have been more at home with newspapers, radio, television, film, even the music industry, than with "new media." Recently, however, analyses by Dan Schiller, Edward Herman, and Robert McChesney have pursued political economy into the digital age.[31]

They look at the role of the new corporate behemoths in structuring the so-called information highway, in transforming the common space of the early Internet into an e-commerce virtual shopping mall, and in shaping the architecture and content of broadband entertainment systems. Thus, political economist Robert McChesney refers to the "rich media/poor democracy paradox"; otherwise put, "the corporate media explosion and the corresponding implosion of public life."[32]

Games, however, are hardly mentioned in these accounts. When they are, it is too often as part of a broad-brush dismissal of the mindless entertainment content of the corporate media order. This is not adequate. As a new entertainment medium, interactive games raise a set of issues that differ from those presented both by broadcast media and by other digital media. Political economists are understandably quick to write off the hype of techno-utopians; their work is indeed a vital antidote to the sort of determinism that McLuhan often lapsed into. But we need to find a way both to maintain a critical perspective on power relations and to account for the unprecedented processes of feedback and participation presented in digital culture. Such a perspective would permit analysis of the increasingly ominous capacities for targeting, tracking, and strategic management that digitalization put in the hands of media corporations; but it would also recognize the potential for crisis in the market system as well as dissidence, transgression, and alternative practice to emerge among, in our case, video game players and workers alike.

There is also, perhaps, another reason why political economists have been slow to deal with such an important new media. Video games are an entertainment product that many people find fun – a lot of fun. The issue of pleasure has been a niggling thorn in the side of political economists of media since the time of the Frankfurt School. Commercial media clearly had the ability to consolidate market power; but did consumer culture serve the interests of capitalists only or was it also what the "masses" wanted? Is it highbrow arrogance to critique the public's enjoyment of cultural products simply because it profits business? While it is true enough (and, we think, important to point out) that the ardent *Grand Theft Auto 3* aficionado hunched over his PlayStation 2 has become a revenue source for Sony's electronic empire, or that the *Final Fantasy* fan is powerfully distracted from issues of free trade, global poverty, and ozone depletion, how much does this tell us about the attractive power, symbolic content, and pleasure of digital play? Political economy does not complete the job of unpacking the video game. We need to call in another important stream of media criticism – cultural studies.

CULTURAL STUDIES:
MEDIA TEXTS AND ACTIVE AUDIENCES

Frustrated by the limits of a political economy style of analysis, early scholars associated with the British tradition of cultural studies tried to find a more dynamic way of analysing media and popular culture. They weren't convinced that studying the mass media only in their structural aspects or seeing the media only as a brute instrument of corporate power was adequate for illuminating the complexities of mass-mediated culture, especially as it was taken up by active living human subjects. Pioneering intellectuals in this tradition, such as Raymond Williams and Stuart Hall, were heavily influenced by Marx's ideas, but they were keen to leave behind his "base/superstructure" metaphor, considering it a rather crude model for understanding something as complex as culture. Like political economy and media theory, cultural studies is multifaceted, so we'll highlight only a few aspects that we find helpful for our analysis of the video game.

Especially influential was a group of researchers in the mid-1960s to 1980s at the Centre for Contemporary Cultural Studies at the University of Birmingham in England. Despite the Frankfurt School's diagnosis of the emptiness of mass culture, for better or worse, in the new consumer societies people were spending more of their leisure time at shopping malls and watching television than going to the symphony. Challenging the very notion of "high" culture, cultural studies takes an "anti-élitist" perspective: popular culture in a capitalist society, commercial media products, working class culture, and youth subcultures were recognized as socially meaningful and academically legitimate topics of study.[33] From the cultural studies viewpoint, we need to pay attention to mainstream media and culture such as fashion, television programs, music, and video games because they are rich sources of social meaning that provide us with resources and reference points for giving significance to the world around us and for expressing and constructing our identities, our sense of who we are.

The Birmingham school of cultural studies was profoundly inspired by the variety of literary and language studies that blossomed from French "semiotics" – the study of signs. In the work of authors such as Roland Barthes, all the manifestations of popular culture, from wrestling matches to the photographs on magazine covers, were seen as "texts" generated by complex symbolic "codes," which are charged with meanings about complicated social issues and relations such as class, gender, ethnicity, and nation. Rather than being dismissed as trivial, popular culture was apprehended as a rich field of social signification. For the scholars at Birmingham, a semiotic approach offered a way out of the sterility of a Marxian analysis that unravelled the media and culture only to arrive at a predictable condemnation of capitalism.

One main concern in cultural studies is the construction of meaning and the role of mass media and popular culture in that process. Stuart Hall, who directed the Centre at Birmingham for several years, and his colleagues examined a wide variety of media "texts," such as news programs, soap operas, and Harlequin romances. They often took a textual approach to representation, examining the various connotations associated with images and narratives. From the cultural studies perspective, textual representations are never innocent; instead, they are very closely connected to systems of power in society. Once we press "start" on our *Tomb Raider, The Sims* or *Crash Bandicoot* video game we know that gaming is not simply a cybernetic relationship with a machine but also a mediated cultural text, offering to us subject-positions (e.g., the hero or the villain) and game scenarios that carry social meanings about, say, gender relations, colonialism, and consumerism. Cultural studies researchers therefore analyse how media texts give our experience of the world meaning through representations (or images) and patterns of narrative; how they offer readers, viewers, or players certain points of identification, or subject-positions, in relation to those narratives; and how they contribute to the construction (and sometimes subversion) of an everyday "common sense," a repertoire of assumptions and premises about how things are in the world at large.

The Birmingham researchers insisted that the meaning of a media text could not be simply read off the structures of media ownership. They posed a challenge to simplistic notions of media manipulation and "hypodermic-needle" theories of mass-media influence. In this regard, one of the important contributions of cultural studies was its focus not only on the analysis of media representations but also on how media audiences interpret them. For example, in his famous article "Encoding/Decoding," Hall challenged the sender-receiver model of communication, which assumed that audiences passively took in preset media meanings.[34] He suggested that while the producer of a text might "encode" certain meanings, there was no guarantee that the same meanings would emerge from the "decoding" (interpretation) performed by the audience. Although the producer might clearly encode preferred readings, decodings could deviate to a greater or lesser degree. Thus, a film might encode a dominant meaning that applauded the exploits of the heterosexual male action-hero. Audiences might accept this meaning. But they might also come up with "negotiated" readings, accepting the general framework but making qualifications (by, for example, emphasizing the homoerotic elements in the hero's relation with his sidekick), or even "oppositional" readings that completely rejected the encoded meanings (by taking sides, say, with the vixen-villainess or the evil terrorist or some other "bad" character). The audience, by no means passive, was an "active" creator of meaning and a contender in the struggle to define "common sense."[35]

The analyses that were put forward by proponents of cultural studies seemed much closer to the sources of people's fascination with and enjoyment of media and culture – even if they were deeply commodified – than did the dry analysis and often puritanical tone of political economists. When concerns with interpretation, subjectivity, and identity were linked with psychoanalytic theory, as they were in the work of film and television analysts influenced by the Freudian Jacques Lacan, they offered a way of getting at the pleasure and excitement of media experience.

How have cultural studies perspectives played out in relation to video and computer games? Here too, they have hardly been given the attention they deserve, although because of their long-standing interest in youth culture, cultural studies authors have been somewhat more alert to interactive games than political economists. Steven Poole's *Trigger Happy: The Inner Life of Video Games* (2000) draws richly and productively on semiotic analysis to decipher game narratives.[36] The legacy of cultural studies is also particularly apparent in the discussions about women and computer games, which has produced a wealth of analysis about the textual positions offered, or not offered, to women in games.[37] Certain cultural studies ideas, such as the "active audience," seem to chime with the way interactive technology gives players choices about what identity to adopt, what actions to take, and what scenarios to select. For example, there have been valuable analyses of how "strategy games" such as *Civilization* or *Sim City* can be played in ways that subvert the preferred readings of game designers.[38] We can expect more interpretations that treat interactive games as "texts to be read" for evidence of struggles over meaning and the resistive power of media audiences.[39]

In terms of analysing media, cultural studies is a significant advance over the failures of media theory and Marxism to come to grips with the specifics of media meanings. But cultural studies has shortcomings too. The active audience idea, for example, was a crucial corrective to the image of media audiences as inert couch potatoes. But like all good ideas it created its own problems. According to critics, the active audience idea gradually blurred with the much more conventional notion of consumer sovereignty.[40] That is, the assertion that people make their own meanings comes very close to the business platitude that consumers get what they want. And if everyone is really a cocreator of meaning, why should we even worry whether Time Warner/

AOL or Disney/ABC or Sony or Microsoft and a handful of other media conglomerates control most of what we watch and listen to and play?[41] Some critics worried that emphasizing the subversive agency of the audiences of media commodities conveyed the message that the capitalist market was a "[place] of diversity, play and anti-élitism."[42] If sneering at *Seinfeld*, or letting one's Sims die of starvation, can be seen as an act of defiance against consumer capitalism, the bar for what constitutes political activism has certainly been lowered.

So although cultural studies draws our attention to the social construction of meaning in video games and to the real pleasures gamers might find in digital play, its approaches have, in our opinion, three weaknesses. First, they say nothing about the specific qualities of the new media. To discuss the white-knuckle concentration of *Quake* players, bobbing and weaving as their avatars strive to survive, as a "textual reading" is surely to miss something crucial about the game experience. And if we are all already active audiences, what is this gripping "interactivity" that keeps players riveted to the blood-spattered screen or persuades them to labour lovingly for hours digitally creating new labyrinthine levels of slaughter to circulate freely on the Net? Semiotics is a vital tool, but we also need an analysis that can distinguish virtuality from textuality and suggest how the nature, extent, and limits of interactivity experienced by game players is distinct from the form of engagement offered by other media.

Second, cultural studies' accounts of video games tend to underplay the commercial structuring of the industry. Although most video game analysts who use the cultural studies lexicon recognize – who could not? – that they are dealing with a vast entertainment business, there is little discussion of the history, institutional context, structure, and dynamics of this capitalist complex and the wider process of cultural commodification. There is a risk of "erasing relations between culture and specific economic and political realities."[43] Thus, Marjorie Ferguson and Peter Golding have suggested that we analyse media not only as representations but also as cultural practices that include a nexus of textual and political-economic factors.[44]

Third, by focusing so heavily on how audiences make meaning from media texts, there is a risk of taking for granted that audiences are themselves constructed in the context of a media marketplace. The "audience" does not spring into being ready-made. As the political economist Vincent Mosco reminds us, "the very term *audience* is ... a product of the media industry itself, which uses the term to identify markets and to define a commodity."[45] For this reason we address in our account the role of cultural intermediaries such as game marketers and designers in the construction of the audience for video games.

Our search through the glittering fragments of communication studies to find perspectives that are adequate for the exploration of interactive gaming has led to somewhat mixed results. Media theory, political economy, and cultural studies all illuminate aspects of our topic. But none of these approaches is by itself sufficient. By isolating certain features of new media at the expense of others our understanding will be inevitably limited, perhaps in problematic ways. Moreover, many critics have decried the "divide" between political economy and cultural studies, even though scholars call in theory for a *rapprochement* (usually in a way that favours their side of the divide).[46] At the same time, the rich and subtle legacy in media theory of Innis – and, in a more problematic way, of McLuhan – has been largely abandoned or ignored both by political economy and cultural studies and left to be appropriated, in a propagandistic way, by the futuristic digerati. It is unfortunate that this situation exists at the moment interactive media occupy centre stage, and we believe it is time to propose a more multidimensional approach to analysing new media.

RAYMOND WILLIAMS: LOCATING MEDIA IN THEIR INSTITUTIONAL CONTEXT

In taking up the task, we draw inspiration from Raymond Williams. In our eyes, Williams's social histories of media provide an exemplary critical approach that recognizes the interplay of technologies, culture, and economics.[47] We also greatly admire his insistence on the concrete empirical historical study of media, his suspicion of sweeping abstractions and confident reductions, and his insistence that analysts recognize the importance of human agency and intention in the shaping of technological systems.

Williams was a fierce enemy of technological determinism. He called it an "untenable notion" and, singling out McLuhan as a culprit, remarked that "if the effect of the medium is the same, whoever controls or uses it, and whatever apparent content he may try to insert, then we can forget ordinary political and cultural argument and let the technology run itself."[48] For Williams, then, such theories of the media were "explicitly ideological: not only a ratification, indeed a celebration, of the medium as such, but an attempted cancellation of all other questions about it and its uses."[49] Any technological determinism ends up "[legitimating] existing media institutions" because it always represents the way things are as the necessary, inevitable outcome of features intrinsic to certain types of machinery – "of intrinsic formal properties of media rather than the effects of the power dynamics of *social relationships* and *institutional practices.*"[50] Williams therefore argued for a "radically different" perspective that would see "communication technology ... [as] at once an intention and an effect of a particular social order."[51]

He was also a critic of free market dogma. Far earlier than most, Williams recognized the importance media industries were assuming in a reconfiguration of capitalism: "The major modern communications systems are now so evidently key institutions in advanced capitalist societies that they require the same kind of attention, at least initially, that is given to the institutions of industrial production and distribution."[52] But the attention he recommended was deeply wary of corporate power over communication and culture. Williams was profoundly influenced by Marx's portrait of capitalism as a system of vast social inequality whose media and communication systems were skewed to support the profits and interests of a wealthy ruling class.

But he was as critical of his conventional Marxian colleagues as he was of free marketers and techno-utopians. In particular, he rejected a certain sort of "economic determinism" that believed culture could be simply reduced to structures of ownership and control.[53] Against this mechanistic view he proposed a more fluid understanding of the economy and culture as related but not in a way that permitted one-way arrows of causation to be drawn between them. For Williams, technological determinists who argue as if "technologies have made modern man and the modern condition" and economic determinists who see technological innovations just as "*symptoms* of change of some other kind"[54] were both wrong, because both obscure the fact that "the moment of any new technology is a moment of choice."[55] In Williams's view, the imperatives of the capitalist system set limits and exert pressures on cultural expression and technological innovation. But this is a process marked by negotiation and struggle more than a story whose ending can be completely predicted in advance by the profit motive.

The best example of Williams's approach to media is found in his *Television: Technology and Cultural Form*, written in 1974. He argued that television was not the outcome of a "single event" but depended upon earlier scientific discoveries, such as electricity and photography, and various institutional settings, such as engineering labs and the military.[56] As the idea of television took shape, various interests contended to define that technology's potential use – for point-to-point, organizational, surveillance, and broadcasting purposes. Television might have been used in many ways other than those practised in Britain and the US – to extend state domination as it was in the Soviet bloc, for example, or to promote public education. Instead, "the process of its development came to be dominated by commercial intentions."[57] But Williams stressed that the development of television as a commercial media institution was not inevitable. Even after its potential as a broadcast medium was understood, there followed a period during which its possible social forms were explored. In Britain, for example, when Williams wrote, the debate over the balance between "public service" and "commercial" control of communications was an ongoing and "unfinished struggle."[58] There is no singular or overriding use to which television is necessarily put.

One of the things we especially respect in Williams's work is his interest in media technologies in their cultural and economic aspects. This is evident in his notion of television "flow." In one sense, flow is a technological capacity of television – the ability to continually stream words and images to a receiver, without pause or interruption, fostering an experience, "watching television," that is like an electronic river of sorts. In this way, televisual flow also has cultural aspects. Unlike personal communication characterized by "specific messages to specific persons," flow is constructed around a notion of "varied messages to a general public."[59] When television "flows" into the household, it potentially ruptures previously existing social relations. Televised sport, for example, undermines the social occasions that brought people together in the public spectacles of football and soccer; it signals a dematerialization of a kind of social community and its displacement into the mediated realm of television. In this way, media are deeply cultural, reshaping practices in the realm of everyday life.

The ways such processes occur, however, are closely related to economic forces and institutional decisions. Given the growing influence of commercial funding, producers shifted their attention away from the social content of television or its impact on public debates towards extending the amount of time that audiences spent using the medium. This commercialized conception of the role of television became consolidated within broadcast management circles around the notion of "planned flow" – a particular mode of programming that was intended to "capture" the viewer as an audience for the particular network for reasons of competition and marketing, to which end it heavily favoured fictions and entertainment.[60] Thus, for Williams, analysis of television must appreciate the relationship between the structure of the commercial institution of television and the structure of experience that it brings to contemporary life. Williams's approach places in the foreground the historical constitution of media as institutional and cultural practices, and his analysis of television is thus one of the best historical accounts of the emergence of a mediatized marketplace.

Based on his observation that commercial media were rooted in a distinct economic situation, that they were linked to a particular set of intentions, and that they "flowed" into individual homes and subsequently reverberated in a range of social practices, Williams found the term "mass communication" not only inadequate but in fact a distorted way to represent television broadcasting.[61] If the emergence, use, and effects of a medium were institutional, social, and cultural, then could it really be adequately described in purely technical terms? The phrase "mass communication" implied a sense of technological determinism; in this case, a formal property of a technology (i.e., one-to-many broadcasting) is used to describe the characteristics of diverse social agents. Furthermore, Williams explained that the term "mass" was historically invoked as a "term of contempt" indistinguishable from "mob" and often deployed to marginalize oppositional social movements.[62] Williams concluded: "What is really involved in that descriptive word 'mass' is the whole contentious problem of the real social relations within which modern communications systems operate."[63] Williams preferred the phrase "social communication,"[64] which, for critical media analysts, means asking not only "who says what, how, to whom, with what effect" but also "'for what purpose?'"[65]

Williams's perspective was never bleakly pessimistic; he did not adopt a neo-Luddite position. His objective was critical but also positive. His emphasis on the different historical potentials of any given technology was intended to contribute to a social struggle for "equal access to media production [which] would allow for a more democratic culture in which people had chances to discuss issues, formulate ideas, and creatively envision their lives."[66]

It is our intention to take up where Williams left off. We model our case study of video and computer games along the lines he established in *Television: Technology and Cultural Form* where he set out to engage "a particular cultural technology, and to look at its development, its institutions, its forms and its effects, in this critical dimension."[67] We hope to do for the concept of "interactivity" in digital games what Williams did for "flow" in regard to television – making the point of convergence an analysis of the cultural, technological, and economic forces bringing a new media into being. In this way, we shall ground in our analysis the intentions and practices that underlie the more immediate experience of "playing games." Moreover, we shall ground these factors within specific institutional contexts and social settings in which media influence markets and culture. For it is in a materialist history of a new medium that we can uncover the dimensions of intentionality and conflict that ground both critical evaluation and progressive advocacy with regard to culture and technology.

MAPPING THE MEDIATIZED GLOBAL
MARKETPLACE: THE THREE-CIRCUITS MODEL

We have so far registered some of the perspectives in media theory, political economy, and cultural studies, and how aspects of them are brought together in Williams's media analysis. This review has helped us move closer to our more specific analytic objective. In our view, the key issue for understanding the emergent mediascape in which digital play is a crucial feature is to produce an integrated analysis of the lockstep dance of technological innovation, cultural diversification, and globalized consumerism. To that end, we have made a model of the intersection of these forces in what we term the mediatized global marketplace. We call it the "three circuits model." Although it may seem elaborate, we hope it clarifies what is an inherently complex process. In order to make our thinking clear, in what follows we build the model up in successive layers, starting from its most basic elements, adding in new dimensions as we proceed, then applying its theoretical structure more concretely to interactive gaming.

We start by taking a cue from the political economy of media, because we are persuaded that information capitalism represents a fundamental intensification and acceleration of processes of commodity exchange. At the very beginning (diagram 1), like Garnham, we posit a version of the "circuit of capital" – the process in which corporate production creates commodities for consumption for purchase, which in turn generates the flow of money and profits to start the cycle over again. In this circuit, media figure both as commodities (e.g., games, films, books, music, etc.) and as the vehicle for marketing other nonmedia commodities (e.g., cars, jeans, cosmetics) through advertisements, product placements, and other promotional strategies; it is important to note that media often serve as the means for marketing other media commodities (e.g., television and print advertisements for games or films). This circuit must be envisaged dynamically: it moves with ever-increasing velocity and ever-expanding scope as media enterprises use new technologies to speed up production and sales and enlarge the scope of their operations – globalization, in short.

Within this overall market cycle, we then distinguish three subcircuits (diagram 2) – those of culture, technology, and marketing. The cultural circuit involves what we can broadly call the production and consumption of cultural meanings – books, films, TV programs, music, games – the "texts" circulated by the postmodern media industries; the cultural circuit also comprises the practices or activities associated with both designing and playing games. The technology circuit involves the digital

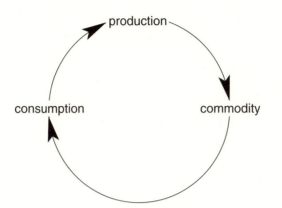
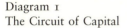

Diagram 1
The Circuit of Capital

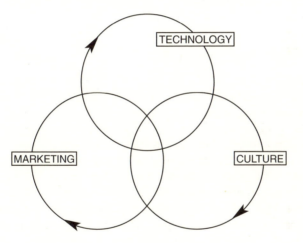
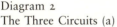

Diagram 2
The Three Circuits (a)

artefacts – computers, consoles, telecommunications, and software –
that constitute the infrastructure on which these industries also
depend. The marketing circuit involves the research, advertising, and
branding practices that are also vital to media industries in two ways
– for selling their own cultural and technological products, and for
the revenue streams from advertising other products. The intersection
of these technological, marketing, and cultural forces within the wired
market is central to our understanding of the emerging "oligopolies
of knowledge" that are propelling growth in information capitalism.

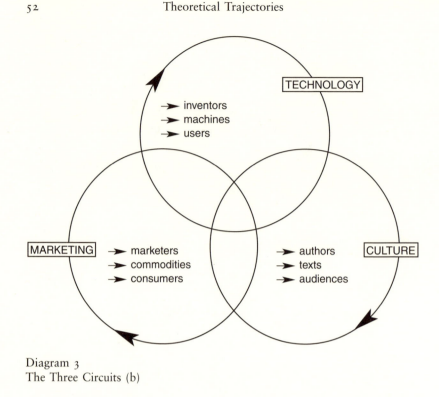

Diagram 3
The Three Circuits (b)

Each of the three subcircuits is itself a dynamic process, involving socially organized structured flows, cultural practices, and feedback loops that bind human agents and artefacts in cycles of creation, consumption, and communication (diagram 3). In the cultural moment, the circuit is the loop of meanings circulating between authors, texts, and audiences. A number of cultural studies writers have worked with the idea of the "circuit of culture."[68] Applying it to a study of the Sony Walkman, for example, Paul du Gay and his coauthors say it means to "explore how [this media technology] is represented, what social identities are associated with it, how it is produced and consumed, and what mechanisms are associated with it."[69] At the technological level, we are dealing with what Cynthia Cockburn describes as the "circuit of technology" – the complex interplay between inventors, machines, and users.[70] In the marketing circuit, we are looking at the interaction between marketers, commodities, and consumers.

Since we are now concerned specifically with computer and video games, we should make certain terminological adjustments to our three-circuits model to more closely reflect the agencies and practices of this particular entertainment form (diagram 4). The cultural circuit

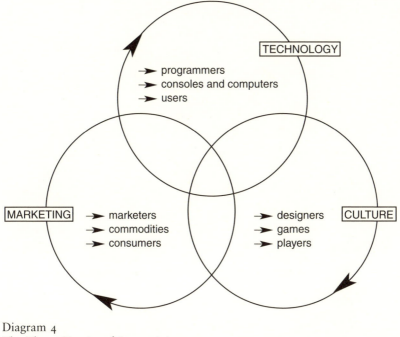

Diagram 4
The Three Circuits of Interactivity

involves not so much authors, texts, and readers as designers, games, and players; the technology circuit is constituted not just by generalized inventors, machines, and users but by programmers, computers/consoles (taken as including telecommunications connections), and users; the marketing circuit, which consists of marketers, commodities, and consumers, remains unchanged.

CIRCUITS OF INTERACTIVITY

Let us now apply our three-circuits model to video and computer game play (diagram 5). In the cultural circuit the game player is discursively positioned as a protagonist within a fictional scenario. Here we "read" the video game as a semiotic apparatus that invites players to assume an imaginary identity, or, to use a more technical term, "interpellates" them in a particular "subject-position."[71] The immediate moment of on-screen play is analysed as a process in which the player is addressed as and takes on the role of a hero or heroine of a preset narrative – fighter-pilot, kickboxer, racing car driver, monster-annihilating space marine, whimsical explorer, etc.

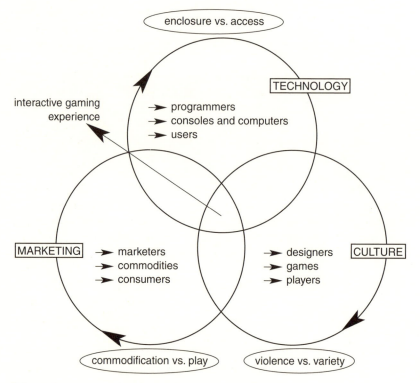

Diagram 5
Contradictions in the Three Circuits of Interactivity

In this cultural context, "interactivity" is understood as the allowance of varying degrees of openness or closure, option, and limitation, within the given scenario. The players choose their characters, teams, and roles; the course of the game follows branching paths according to the players' choices and skills; the outcome of the game depends on these factors; indeed, with the increasingly common inclusion of editing capacities, players can design their own scenarios and share them with others. To this degree, video game play seems like a consummation of the theory of the "active" reader or audience, with the power to refuse, subvert, or alter meanings implanted or intended by the artificer or author of a text. But a moment's reflection shows that most interactivity is a matter of tactical choices and issues that arise within scenarios whose strategic parameters are preset by a design practice: an invitation to comply or collude in the construction of a particular universe rather than in the deconstruction of its boundaries. For example, in the superlative military computer game *Combat Mission*, set in 1944 Normandy, we can elect to fight as Germans, Americans, British, or numerous other armies; we can command tanks or infantry; we can change the mix of our units and the terrain of battle. But we cannot be a civilian or a nurse or a conscientious objector. Coding sets limits. Different games are designed with varying degrees of openness or closure, option and limitation. But the issue of dominant or preferred meanings inscribed in the game code and gamer communities and the associated issues of hegemonic values and constrained roles and identities is by no means surpassed just because the technology enables the players' choices to be made in different ways.

From this perspective we can begin to discuss how interactive games construct or subvert social meanings and identities within the larger intertextuality of popular culture, conforming to or deviating from a broad repertoire of myths, stereotypes, and social practices. For example, we can broach the issues of violence and gender representation that have been so central to discussions about gaming. From our point of view, a key issue in the circuit of culture is the ascendancy, within the "flow" of digital game design, of a player identity based on the positions of what we term "militarized masculinity." The primary question that the industry faces is whether to stay within the tried and true formulas of digital death, destruction, and dominion, or risk experiment with more diverse game models – a choice between "violence or variety."

In the technological circuit, the subject of the interactive experience is now positioned not only as a player but also, and simultaneously, as a "user" of computers and consoles that are increasingly linked to a networked telecommunications environment. We examine the ways in which the video game medium is the message, interactively inculcating the skills, rhythms, speeds, and textures of the computerized environment; cultivating digital aptitudes; squeezing out or devaluing other nonelectronic capabilities; socializing players as subjects of and for a high-technology society; building cyborg identities of human/machine identity as gaming pleasure drives successively more sophisticated levels of virtual experience, involving new expectations about verisimilitude and complexity of interaction (e.g., solo/multiple/networked play).

But the technological circuit also involves the process of technological innovation and diffusion – the way inventions pass into popular use. Here too recent theory has seen a move from linear to circular models. Accounts of visionary geniuses realizing technological options that are then passively adopted by a recipient population have been discredited.[72] Recent studies in the "social construction of technology" have shown how for new media, from the telephone to the radio and television, potentialities are only gradually determined and crystallized into specific forms through a play of institutional and organizational forces. Yet the unanticipated improvisations of users may be as or more important than the intentions of inventors in establishing the lines along which large-scale, usually commercial, development eventually takes place.[73]

Interactive games are the prime instance of this process – spun off by hackers and hobbyists from military research, subsequently commercialized and domesticated as a popular entertainment technology that has become the basis for vast commercial empires. Thus, a major part of our account of the "technology circuit" of gaming is about the complex path by which inventions and technological possibilities pass from initial experimentations through the market and into mass consumption. But as Innis and Williams remind us, recognizing the dynamic nature of this design and adoption process also means allowing for escapes, departures, and divergences from this commercial logic. Today, the digital games industry confronts the paradox that the very technologies it has created are running beyond its control, in an epidemic of game "piracy" that drains profits on a global basis. As with many other branches of information capitalism, the games business today faces in the circuit of technology an acutely uncomfortable contradiction between "enclosure and access."

At the marketing level, we examine the interaction between marketers, commodities, and consumers. Cultural studies scholar Angela McRobbie notes that marketing professionals have a privileged status in cultural industries today; consequently, cultural analysis must not be limited to the final "cultural product" but must encompass the whole "commercial process": "attention must be drawn to the various levels of activity and the relations of power played out in the decision-making processes which produce the marketing campaign and the product itself."[74] We agree with McRobbie's conclusion that the "business of culture" is "a missing dimension in cultural debate and until we know more about it we cannot speak with much authority on anything other than the cultural meaning, significance and consumption of these forms once they are already in circulation."[75] In our view, marketing "negotiations" between producers and consumers assume a vital role in information capitalism, especially in youth cultural markets where trends change so rapidly.

Moreover, although video game companies at first relied on the novelty of the new technology to attract interest, they rapidly realized that high-intensity advertising was critical to continued growth and aimed at beckoning consumers into being through an array of marketing communication practices, surveillance, prediction, solicitation, and elaborate feedback relations. Nintendo, Sega, and other games companies became pioneers of branding on the electronic frontier, enveloping game play in a branded ambience of custom, myth, status, and craft-lore that involved not only television promotions but also games-tips phonelines, magazines, films, merchandising tie-ins, virtual tournaments, sponsorships, Web sites, game rentals and trials, and a host of other marketing synergies and public relations strategies. The scope and depth of this promotional web is one reason why our account gives special attention to marketing.

To the immediate fictional identities stipulated by the game itself these promotional discourses add on a whole metalevel of signifying practices about what it is to be an ideal game player in general (whether cool, combative, youthful, boylike, girlish, technosavvy, popular, etc.), as well as a whole series of brand-specific identifications (i.e., what distinguishes a Nintendo fan from a Sega enthusiast, or a Sony player from a Microsoft one). This marketing apparatus not only transmits commodity-promoting messages but simultaneously gathers data about customers, culling information with each transaction about their preferences and habits to provide highly detailed demo- and psycho-graphic maps of the market terrain. This in turn shapes the design of future games and of future advertising and promotional strategies. The aim is to "close the loop" between corporation and customer, reinscribing the consumer into the production process by feeding information about his or her preferences and predilections back into the design and marketing of new game commodities.

Many celebrants of the game industry see this as a democratization of corporate practice and a triumph of consumer sovereignty. But when we consider the relationship of the game producers to the video game audience, we see that the corporate "responsiveness" is constrained by marketing imperatives and specific research practices that tend to lead to the distribution only of the kinds of gaming experiences that can maximize profitable growth. Indeed, as video game development technology gets more complex, it also escalates the risks of publishing genres for markets that have not been proven. This can often lead to a narrowing of the potential diversity of interactive entertainment to the games preferred by the most loyal and frequent buyers. Ultimately, we suggest that in the marketing circuit there is a deep tension between the calculated, organized, and oligopolistic marketing of game culture and the experience of freedom, adventure, and transgression its imaginary worlds promise. There is at the heart of the gaming industry a contradiction between "commodification and play," a tension that paradoxically drives its frenzied creativity and subverts its own success.

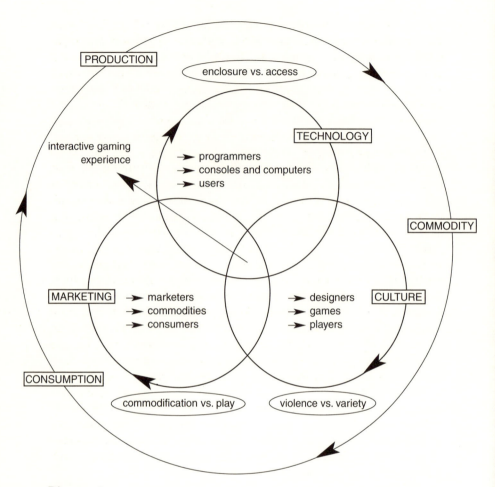

Diagram 6
The Three Circuits of Interactivity in the Mediatized Global Marketplace

CONCLUSION:
COMPLEX PROCESS, MODEST PROPOSAL

Our map of the mediatized global marketplace situates interactive gaming at the intersection of dynamic processes in the spheres of technology, culture, and marketing (diagram 6). The subcircuits are mutually constitutive.[76] In theory, they can be abstracted and described as semiautonomous moments. In practice, they interpenetrate and dynamize each other. Our diagram therefore shows the subcircuits passing through each other, to suggest this imbrication and interplay:

they are wheels within wheels, reverberations within feedback loops, and synergistic relations constructed within a cybernetic system of production and consumption. We use the term "circuits" because it invokes homologies and affinities between the cultural "circulation of meanings" flowing amongst authors, texts, and audiences, the cybernetic "circuitry" of computer technologies programmed by systems engineers and adopted and adapted by users, and the "coevolving" information loops involving advertisers, commodities, and consumers.

There are undoubtedly dimensions of the gaming industry and of the gamer's experience that escape our geometry. However, this map is intended to offer a productive way of looking at the processes of contemporary media industries in general. Intertwining cycles of cultural, technological, and economic activity have always been constitutive of social existence. But today, the interdependence of these processes is acutely visible. Information industries based on computers and telecommunications – such as the video game business – are predicated on innovations that are at once immensely profitable commodities, astounding technological innovations, and radical cultural departures. Our model conjures a historical moment when cultural processes, market growth, and technological innovation have been assimilated into the ensemble of management practices that are focused on fostering and exploiting the dynamism that is created *between* these circuits in a wired marketplace that is beset with instabilities in meaning and identity. So far we have unpacked the interactive game as a complex composite of technological, cultural, and marketing forces. But to understand the extraordinary dynamism of this convergence of myths, machines, and markets we also need to look at some accounts of recent dramatic changes in digital capitalism and its culture, a task we turn to in our next chapter.

3

An Ideal Commodity?
The Interactive Game in Post-Fordist/
Postmodern/Promotional Capitalism

INTRODUCTION: SEA CHANGE

We have presented a model of the technological, cultural, and marketing forces that mutually constitute the experience of interactive play. The previous chapter reviewed some of the theoretical influences on our thinking. Now we shall say more about the broader set of historical circumstances in which the three circuits are situated. For the way the circuits interact with one another is, remembering Raymond Williams, both "an intention and an effect of a particular social order."[1] We have to set this model in motion by providing some account of the dramatic social changes that have occurred in all three spheres since the inception of the digital game some thirty years ago. Earlier, we rejected the digerati's claim that the information revolution marks a completely new moment of civilization, making all previous understandings of media obsolete. Nonetheless, the end of the twentieth century has seen what David Harvey terms a significant "sea change" in the way capitalist societies operate, a change in which the emergence of new digital technologies – from automated production lines to new media of communication like video and computer games – plays a very large part.[2] Critical media analysis must take account of these transformations, without lapsing into either utopian millenarianism or apocalyptic doomsaying.

While we situate our investigation of the interactive game in the tradition of Williams, we recognize that in several respects the theoretical perspectives Williams devised for his examination of television need to be radically revised and updated. Three points require elaboration to construct an analysis, following the example of his work, that is applicable to new media in general and the video game in particular. First, because key developments in the information economy had yet to unfold at the time he was writing, Williams's analysis was innocent of the new digital media technologies that have restructured both capitalist markets and entertainment. Second, though Williams would scratch its theoretical surface in later work, his analysis did not address the postmodernization of culture, with its seeming triumph of image over substance and its dazzling cascade of simulations, virtualities, and hyperrealities. Third, Williams did not fully examine the changing context of commercialization in the mediated marketplace, particularly how marketing practices and cultural intermediaries were influencing popular culture and communication in profound ways. So in this chapter we expand our framework for analyzing interactive media by drawing on the contributions of contemporary theorists whose ideas emerge from debates about the relations between digital technologies, information capitalism, and postmodern culture. This discussion enables us to situate our three-circuits model within a broader set of developments in the technological, cultural, and marketing spheres of the contemporary era.

To frame these discussions historically we start in the overall circuit of capital with the distinctions between "Fordism" and "post-Fordism" – or "industrial capitalism" and "information capitalism" – now widely current in political economy circles. These terms, we suggest, offer a way of acknowledging the rapidity of recent change in capitalist economies, while still recognizing the continuities in a system that remains based on profit, commodification, wage labour, commercial media, and consumerism. We then go on to show how this shift is manifested in our three subcircuits of technology, culture, and marketing. Discussions about the increasing role of high-technology innovation and digital networks in "post-Fordist" capitalism connect with debates about the nature of a postmodern culture characterized by ever-escalating difficulties in separating illusion from reality, image from substance, and virtual from actual. Amongst theorists of the postmodern, Jean Baudrillard's work on simulation and hyperreality is central to any discussion of interactive gaming. But we argue that Baudrillard's ideas should not only be seen as describing the hallucinatory power of virtual technologies. They also address an undertheorized aspect of postmodern culture and post-Fordist economics – the high-intensity marketing and promotional practices of contemporary consumerism that are so strongly displayed in the operations of interactive game empires such as Nintendo, Sony, and Microsoft. We then bring together these several theoretical threads by suggesting that interactive gaming might be seen as what Martyn Lee intriguingly terms an "ideal commodity" of post-Fordist, postmodern promotional capitalism.

FORDISM AND POST-FORDISM

The concepts of Fordism and post-Fordism originate with theorists of the French Regulation School.[3] These writers draw on Marx's heritage but develop the aspect of his work that looks at capitalism not just as a matter of the workplace but as a system that organizes society as a whole. Capital's managers, so the Regulation School theorists argue, are plagued by the problem of balancing production and consumption. In production, individual corporations increase profits by pushing down workers' wages and social benefits. Yet all corporations need these same workers to act as consumers for the goods they produce, a role that requires ample purchasing power. Market society's success therefore depends on its ability to find a "regime of accumulation" – a set of economic, political, social, and cultural arrangements that walks the tightrope between these contradictory requirements, holding production and consumption in equilibrium.[4] Reaching this delicate balance is essential if the "circuit of capital" is to maintain its momentum.

"Fordism" – named after the pioneering industrial experiments of Henry Ford – is the term coined to describe the "regime of accumulation" in advanced capitalist economies during the first three-quarters of the century.[5] It involved three crucial elements. At the point of production, it depended on assembly-line factories, which united routine deskilled monotonous work with increasingly high levels of mechanization. In consumption, it involved the creation of mass markets for standardized manufactured goods. In the broader social sphere, it departed from *laissez-faire* policies in favour of greater state management of the economy through public spending and the welfare state. The relatively high wages of mass-production workers (Ford's famous "five-dollar day") provided the purchasing power to absorb the goods rolling off the assembly lines. Government spending ironed out the up and downs of the business cycle. This created a virtuous circle of economic growth, allowing business to afford steady rises in wages and finance the welfare state that in turn supported the relatively healthy and educated "human capital" necessary to support an increasingly technological production system.

Fordism was not just an economic arrangement. It was a whole way of life, integrating factory work discipline, the cultivation of consumerism, and an increasing state presence in the management of society. It reshaped work and leisure time around the industrial nine-to-five job, and space around the new configurations of factory centres and cheap suburban housing. It codified gender roles around the nuclear family and the division between the male "breadwinner," devoting his life to productive activity, and the female housewife, who was given over to domestic care, shopping, and child raising.

Fordism brought with it a massive development not just of workplace machinery but also of domestic technology. A special kind of industrial technology – the mass media – was developing alongside the production line. Radio and television carried into living-rooms astounding new cultural experiences and unprecedented exposure to advertising messages. In both ads and programming, commercial broadcasting was transmitting to mass audiences the cultural norms that would help to bring consumption patterns into balance with the flood of cars, canned goods, domestic appliances, and clothing fashions that were streaming off the mechanized production lines. Consumer goods and a host of domestic technologies – from vacuum cleaners to radios – became central to the very practices of everyday life. New discourses on taste and style were opened. In the consumer culture of "getting and spending," leisure, pleasure, and necessity were taking on new meanings. Perceiving the pivotal role played by the commercial mass media in the social, economic, and cultural arrangements of Fordism reveals an important moment in the consolidation of a distinctly "mediatized marketplace" as a site of social communication – a system not only for communication between producers and consumers but, in the process, for circulating cultural discourses on the "way of life" in a Fordist mass society.[6]

The various elements of Fordism did not arise at once or as part of a master plan. They came together through a process of protracted trial and error spanning several decades. But in its full-fledged form, Fordism seemed to be the answer to capitalism's cyclical crises. For some twenty-five years after the Second World War, market societies in North America and Europe seemed to enjoy a golden age of affluence and profit.

In the late 1960s and early 1970s, however, the good times came to an end. A variety of factors, including the saturation of mass markets, the discontent of assembly-line workers organized in powerful trade unions, new sources of competition (such as Japan), and the oil shocks of the 1970s, threw Fordism into crisis.[7] The high wages and social expenditures it required became increasingly burdensome to business. As the social contract was revoked both in the workplace and the larger social arena, strikes and social unrest multiplied, productivity fell yet further, and the Fordist circle of growth began to reverse itself into a downward spiral of falling profits and economic recession.

If we accept the outline of the Regulation School analysis, the last third of the twentieth century has been occupied by the attempts of capitalism's managers, including business and political leaders, to escape from the crisis by building a new "regime of accumulation." The precise complexion of this post-Fordist regime is a matter of some debate. But broadly speaking post-Fordism, just like Fordism, involves changes in the workplace, in patterns of consumption, in media of communication, and in the role of government. Its tendency has been to promote economic growth not so much on the basis of "mass" production and consumption but rather by connecting business more intensively with a thinner stratum of high-income consumers, while cutting down on the expenses of supporting the less fortunate.

In the post-Fordist workplace, the mechanical mass-production assembly lines have been replaced with new modes of computerized labour-shedding "flexible production" or sent to cheap labour zones in Asia, Latin America, or Eastern Europe. In the field of consumption, mass markets have been increasingly broken down into more custom-ized and segmented "micromarkets," while at the same time these niches are cultivated to a far greater extent on an international basis. In the area of government, the Fordist programs of social benefits have been eroded by privatization and deregulation and the creation of a state that is far more oriented towards promoting business interests and policing public order.

New technology has been central to all these developments. More than twenty years ago Michel Aglietta, one of the first Regulation School theorists, speculated that a "neo-Fordist" regime would replace the "mechanical principle" of the assembly line with computerized systems based on the "informational principle."[8] Since then other theorists have gone much further in describing the emergence of a new post-Fordist "industrial paradigm" centred on information and com-munication technologies, or ICTs.[9] Computers and telecommunications have been crucial to the breaking up of Fordist factories of mass production by automation and globalization.

Many commentators saw the computer ushering in a post-industrial age, celebrating its potential for the demassification of society. It promised to reverse the homogenizing aspects of Fordism.[10] Enhance-ments in information storage, processing, and transmission would allegedly allow the cultural industries to challenge rather than extend the rigidities of mass society. Some observers, mindful perhaps of the seeming responsiveness of the cultural industries to changing cultural preferences, believed that digitization and commodification were democratizing culture. For example, audience choice might have been limited in the Fordist years of television broadcasting, but by the 1980s it looked as though networks were responding to diverse audience tastes and lifestyles, a process enhanced by cable technology as it fragmented the mass audience into demographically distinct niches. (By the late 1990s, of course, these proliferating niche channels were housed in a narrowing number of media conglomerates.) The complexion of the media environment seems not only to register but also to facilitate and intensify other transformations associated with post-Fordist cultures and markets.

These new digital technologies also create new information products – from personal computers to Internet services to palm pilots – to sell to the affluent stratum of consumers. They enable new marketing techniques, using narrowcast media and new means of surveillance, from swipe cards to Web "cookies," to advertise to and monitor complex niche markets. New digital weaponry and security systems have been vital to the construction of a lean and mean post-Fordist state that is capable of fighting wars in the Persian Gulf and Kosovo while cracking down on welfare recipients at home. Post-Fordism is, in short, a regime of high-technology capitalism – the sort of capitalism in which video and computer games are right at home.

The notion of a Fordist/post-Fordist divide has proved controversial.[11] Theorists who use it have been criticized for overstating the extent of change, lapsing into technological determinism, and, in some cases, falling victim to a futuristic optimism that ignores persistently ugly aspects of capitalism.[12] Despite these criticisms, we find the idea of a post-Fordist regime useful. There is now an impressive body of analysis that accepts the idea of a transition away from Fordism without ignoring either the unevenness of the process or the scale of conflict and disruption it entails.[13] As Lee remarks, while we must beware not to overstate the extent of recent technological and social transformations, to reduce those changes to mere "superficial adjustments" denies "the very dynamism and revolutionary nature" that many social theorists, including Marx, saw as the core of capitalism.[14] In some ways, then, the post-Fordist concept doesn't go far enough. The Fordist reference suggests a hungover attachment to the primacy of industrial manufacturing.[15] Some theorists have proposed other names such as "Toyotism," "Sonyism," or "Gatesism."[16] This sequence, moving from the automobile industry to consumer electronics to computer software, shows a growing awareness that the epicentre of the newest phase of capitalist development lies in media, information, and digitization.

At the same time the ambiguous nature of the term "post-Fordism," as it teeters between the old (post-*Fordism*) and the new (*post*-Fordism), has its merits. It emphasizes the paradoxical nature of change. Yes, computerization does bring new technological innovations, forms of work, media, and styles of consumption, departing very significantly from the assembly-line world of industrial corporations, mass production, and mass consumption. But no, it does not – as the silicon futurists proclaim – leave all that behind for some brave new virtual world in which people are freed from corporate power, exploitative work relations, mass-mediated culture, and commodified desires. The problems of the market reappear, mutated but intractable. It is precisely in this equivocal space between continuity and discontinuity, between *post*-Fordist and post-*Fordist*, that we make our analysis of the circuits of interactive games.

PERPETUAL INNOVATION

The ideas of the Regulation School about post-Fordism are valuable, but they need some addition. We use another term that covers much of the same ground but from a different angle: "information capitalism." This was coined by Tessa Morris-Suzuki in her studies of Japan's high-technology economy – the very heart of the video game business – but it has been more widely applied.[17] The central characteristic of information capitalism, Morris-Suzuki says, is that "the centre of economic gravity shifts from the production of goods to the production of innovation – that is, of new knowledge for the making of goods."[18] Computers and robotics and other automating technologies enormously increase the speed and reduce the cost of production for all kinds of commodities – so much so that there is a risk that demand will actually be satisfied. From the point of view of capital, this would mean "market saturation and stagnation."[19]

To avoid this, corporations "devote a growing share of their resources to the continual alteration and upgrading of their products."[20] The result is the creation of what Morris-Suzuki calls the "perpetual innovation economy."[21] Business seeks to maintain continual expansion by generating a ceaseless stream of new commodities with ever-shortening product cycles and life spans. Arun Kundnani elaborates:

In high-technology fields, perpetual innovation is characterized by short product cycles – the time it takes from the launch of a new product to the point where it becomes obsolete and production ceases. It is now common for many computer software and microchip products to have a product cycle of six months. Perpetual innovation is characterized by the need for constant creativity in finding new ways to build audiences. The constant reworking of genres and styles found in the music, film and television industries derives from this.[22]

Kundnani notes that this speed of innovation is particularly marked in media industries such as film, music, television, and, we would add, interactive games, which are characterized by "the need for constant creativity in finding new ways to build audiences" and hence by a "constant reworking of genres and styles."

In perpetual-innovation capitalism, the strength of corporations depends on their knowledge resources – research teams, accumulated know-how, data collections, and so on – and on their ability to legally protect innovations from competitors and consumers by means of patents, copyrights, and trademarks. Intellectual property rights become critical in enabling the corporation to fence off the new corner of knowledge from the public and make a profit from its application.[23]

Perpetual innovation also has major consequences for work. As hands-on labour is increasingly replaced by sophisticated automation systems, living labour is channelled into "the incessant generation of new products and new methods of production."[24] There is a "softening" of the economy whereby "non-material elements such as research, planning and design come to constitute an ever larger share of the total value of output."[25] Corporations rely on a core workforce of skilled researchers, scientists, product designers, marketers, and consultants to create and sell the stream of new commodities. Experts in information technologies – programmers, software engineers, and systems analysts – have an obvious centrality.

Shortening product cycles, accelerated by "the increased cycles of information flowing from producers to consumers and back," promise information capital a higher rate of profit than the slower cycles of traditional manufacturing.[26] But this dynamism also brings greater risk "as the investment needed for innovation is high and the window of opportunity to realize the investment is ever smaller."[27] Morris-Suzuki's emphasis on "perpetual innovation," short product cycles, and high risks in some ways does better justice to the frantic boom-and-bust cycles that we see in the interactive game industry than the more static Regulation School language of "regimes of accumulation." In this sense her analysis of information capitalism complements that of post-Fordism. Both remind us that behind the wonders of the PlayStation, the GameCube, or the xbox lies a whole complex of technological and cultural production and organization, involving the creation of new work forces, managerial strategies, mobilizations of expertise, institutions of ownership and legal enclosure, marketing strategies, and – a point that from our perspective warrants special attention – the organization of new forms of consumption.

EXPERIENTIAL GOODS
AND FLUIDIFIED CONSUMPTION

One aspect of the transition to post-Fordism that several commentators have remarked on is what Lee calls a shift in emphasis "from material to experiential commodities."[28] By this he means a switch towards goods "either used up during the act of consumption, or, alternatively,

based upon the consumption of a period of time, as opposed to a material artifact."[29] One aspect of this is the "enormous increase in the
commodification and 'capitalization' of cultural events"; another is the
growing emphasis on the production of software and informational
products.[30] By turning towards increasingly fleeting and ephemeral
goods (music, films, games), post-Fordist capital has sought to overcome
the sluggishness of markets for expensive and long-lasting Fordist consumer durables (washing machines, vacuum cleaners, automobiles).
While material-goods markets eventually reach a point of saturation,
the "act of exchange associated with the commodification of experiential goods ... is potentially always renewable, and the market far less
prone to exhaustion."[31] These tendencies are enhanced by the use of
new technologies – VCRs, Walkmans, or cellular phones – to free up the
spatial and temporal dimensions of social life, allowing "previously dead
time and empty space to be animated by purchasable experience."[32]

In its march from the laboratory to the arcade and into the house,
the interactive game exemplifies this process. As Brian Winston
remarks, video games are a spin-off of the computer that in a way
happened "in advance of the parent device," since in their early days
the diffusion of consoles outstripped that of personal computers, which
are usually considered the centrepiece of the information revolution.[33]
By penetrating deep into domestic space and time, digital games built
on and extended older Fordist generations of broadcast media; console
games depended on physical connection to the ubiquitous television
set for their entry into the home. Interactive games altered the economics of this penetration in that they did not depend on indirect
commodification by advertising revenues but on direct sale of a media
technology (although product placement and advertising are increasingly being introduced into game content). But digital games also
depart from the logic of broadcast programming since they can be
played whenever one wants, and in a variety of locations. The intensification of this process is charted by countless benchmarks in video
game history, from Nintendo's introduction of the hand-held Game
Boy to the edicts issued by North American office managers against
the playing of *Doom* on the job, to the development of game-playing
capacities for cellphones and personal digital assistants, or PDA's.

THE POSTMODERNIZATION OF CULTURE

From our point of view, one of the most important points about the
concept of post-Fordism is that it points beyond political economic
analysis towards debates about cultural change. Many theorists have
pointed to a correlation between the post-Fordist economy and post-

modern culture.[34] Harvey's examination of the "sea change" in contemporary capitalism, for example, notes that postmodern cultural
styles in architecture, fiction, and the visual arts began their ascendancy
in the 1970s, at the very time of the demise of Fordism. Fordism,
Harvey says, was associated with certain distinct modernist cultural
styles – solidity, progress, and standardization. The shimmering sur-
faces, disposable forms, and dazzling simulations of postmodernism
corroborate and reinforce the post-Fordist regime of "flexible accumulation" with its reprogrammable production systems, just-in-time
inventory, and contingent workforce. "The more flexible motion of
capital," says Harvey, "emphasizes the new, the fleeting, the ephemeral,
the fugitive, and the contingent in modern life, rather than the more
solid values implanted under Fordism."[35] In this sense, recent cultural
change mimics the new productive forces. But it is also itself part of
those forces, providing the new images, fashions, and styles that are
so central to cybernetic production cycles, niche marketing, the new
culture of consumption, and the growth of the media industries.

Postmodernism is, of course, one of the most difficult of contemporary critical concepts. We might try to sum it up by saying that the
term designates a situation in which image overpowers reality in the
accelerated circuitry of the mediated marketplace. In the postmodern
condition the implosion of illusion and authenticity creates, it has been
argued, a floating, foundationless world.[36] Everything is surface and
nothing depth. Increasingly, the playful sign-world of media culture
disconnects history, needs, social roles, and rationality from any defi-
nite grounding. Reality disintegrates into an ever-shifting, recursive,
and cross-referencing kaleidoscope world of lifestyles, language games,
and entertainment. Identities and meaning are ceaselessly subverted by
the apprehension – corrosively nihilistic or whimsical and playful –
not so much that nothing is as it seems as that seeming is all there is.

This of course is the burden of the French cultural theorist Jean
Baudrillard's famous theory of the "simulacrum."[37] According to
Baudrillard, the omnipresence of powerful electronic media increasingly confounds our ability (always somewhat dubious) to distinguish
fiction from fact. Disneyland's sanitized representations of North American life convince us more than the cities we actually inhabit; digital
photographs effortlessly add or erase loved ones in the family album;
virtual sex partners alter genders and species at will to ignite actual
orgasms. We enter a "hyperreality," where models are as efficacious or
more so than the reality they supposedly emulate.

It is precisely this hyperreal experience that the video gaming industry promises. Interactive play is simulated experience par excellence.
Finding examples of Baudrillard's hyperreality in the world of video

and computer games is like shooting fish in a barrel.[38] One thinks of the digitally created actors in the film of Nintendo's smash hit *Final Fantasy*, whose computer-animated facial expressions look no more wooden than those of scores of real Hollywood stars. Or of *Marine Doom* in which real US soldiers practise for real combat on modified versions of a commercial "first-person shooter" style of video game. Or of Electronic Art's *Majesty*, an *X-Files* type of conspiracy game in which players will be supplied with clues through their "real-life" e-mail, fax, or cellphone and may thus be contacted by game characters during a work meeting, over dinner, or in bed. It is this apprehension of the total engulfing of society in gamelike simulations that is – in another ironic Baudrillardian twist – thematized by Hollywood in films such as *The Matrix* and *Existenz*.

In a seminal essay, Fredric Jameson has suggested that the postmodernist sense of being wrapped in a seamless but ever-shifting envelope of imagery and spectacle, so forcefully articulated by Baudrillard in his theory of the simulacra, is the product of an historical moment in which culture itself has become fully, and technologically, commodified.[39] If this is so, video gaming is a pre-eminent culprit. "Hyperreality" is precisely what the industry promises in selling its "experiential" media commodities: "game play so real it's hard to tell where your living-room ends and the software begins"; or "a surrealistic adventure that will *become* your world."[40] Jameson's famous description of postmodern culture as one where "the world ... momentarily loses its depth and threatens to become a glossy skin, a stereoscopic illusion, a rush of filmic images without density" – an experience alternatively or simultaneously "terrifying" or "exhilarating" – could well be an account of playing a level of *Unreal Tournament*.

MARKETING MANAGEMENT
AND CULTURAL INTERMEDIARIES

There is more to Baudrillard's theory, however, than just an observation about the illusion-creating powers of digital technology and the dizzying surge of images in hypermediated culture. Other aspects of Baudrillard's work open up another, somewhat neglected front in the investigation of post-Fordism: the analysis of the cultural symbolics of mediated marketing. His theory about hyperreality is not only a theory about the creation of virtual "realities" but also about the intensification of "hype" in a promotionally saturated society. Postmodernism is not just about digital illusions but also about supercharged cultural commodification. For this reason our account of video games draws attention to the importance of the marketing circuit of the interactive game.

Published in the early days of post-Fordism, Baudrillard's book *The Consumer Society* (1970) argues that it is no longer possible to believe in a fixed and universal set of human needs, nor in the traditional social relations that political economists claimed organize those needs. Consumer culture, he argues, is motivated less by the "need for a particular object as the 'need' for difference (*the desire for the social meaning*)."[41] The perceived instabilities of postmodern culture arise not only from digital technologies but also, according to Baudrillard, from the chaotic and self-consuming "commodity-sign system" of the mediatized marketplace. This means that commodities are not just objects but sources of meaning. Jeans, cosmetics, cars, food choices, games – all communicate messages about who we are, where we stand, or what we aspire to be. But these goods are also invested with meaning in advertising discourses and through an ever-expanding apparatus of promoting commodities through media.

Baudrillard gives us a portrait of the mass-mediated global culture as a proliferation of promotional signs, and of the dizzying semiosis of a culture – and the circuit of capital itself – as driven by consumer advertising. The system has become propelled by an unanticipated set of symbolic relations: a new circuitry established in media and markets that act in a ceaseless process of the creation and destruction of meaning. In Baudrillard's view, the instabilities at the level of cultural sign systems are bound up with the ceaseless "differentiation" of consumer products. Owing to its role as an engine of both a commercial media system and capitalist growth, Baudrillard says "advertising is perhaps the most remarkable mass medium of our age."[42] The mediated marketplace is central to the growth dynamic of capitalism because advertising constantly empties and refills commodities of their cultural meaning. Baudrillard therefore calls consumption itself a "communication system."[43] Consumer culture is a treadmill whose machinery of desire is a marketing communication system based on the creative destruction of meaning in media, and a perpetual innovation and exhaustion of signs.

But as the cultural theorist Andrew Wernick points out, "missing from Baudrillard's account ... is an appreciation of how the whole normative apparatus of the sign-commodity, publicity and consumer culture is mobilized."[44] This requires a study of marketing communication practice. Several historians and theorists have noted that marketing is what synchronizes economic imperatives and cultural change.[45] Producers realized that to sell they had to learn to communicate with consumers about their desires and aspirations. Marketers embraced consumer research and the mass media to conduct this communication. The promotional practice of "selling through media" is what underwrites

Baudrillard's ever-accelerating circulation of "commodity-signs" and means that, as Wernick says, "the very distinction between the symbolic and material economies, between regimes of accumulation and the regime of signification, cannot be clearly drawn."[46]

Represented by our marketing circuit, this apparatus for market management is related to another point our analysis emphasizes – the role of cultural intermediaries. As early as the 1960s, theorists of postindustrialism such as Daniel Bell were suggesting that the growth of this cultural marketplace involved a new cadre of workers that included product designers, graphic artists, researchers, and marketers.[47] As aesthetics and "trendiness" became more entrenched aspects of the consumer culture, industries had to maintain a squad of cultural intermediaries capable of keeping apace with fast-changing fads and fashions. Although such activities have always been important, they assumed a new importance in the post-Fordist regime that emerged in the 1970s, with its emphasis on cultural markets, experiential goods, and ever-shortening product cycles.[48]

Here, once again, the concept of a Fordism/post-Fordism shift identifies both continuities and discontinuities. Lee and others recognize that Fordism was predicated on "demand management and the manipulation of consumption" and made a "huge investment into consumer motivation, analysis and advertising." But, they suggest, such efforts largely consisted of attempts "simply to clear the continuous stream of commodities being produced in the first instance," with little evidence that the "qualitative composition or indeed the quantitative extent" of demand was being systematically investigated.[49]

In post-Fordism, by contrast, markets are "intersected by a wide variety of consumer targeting strategies" that combine "hard" demographic indices, such as income, occupation, and residential location, with analysis of "softer" variables such as consumer taste, social attitudes, psychology, and lifestyle.[50] This fragmentation of the mass audience into a "bewildering matrix of interconnected market segments" manifests itself in a diversification of style, design, and aesthetics in consumer goods that is made possible by flexible production techniques, sales strategies geared to increasingly finely sliced market segments, a huge increase in expenditures on advertising, sales promotions, and direct marketing, and employment of the most sophisticated and brilliant postmodern cultural idioms. At the same time these new promotional narratives, with their emphasis on difference and customization, in turn help to prop up flexible production techniques.[51] Lee emphasizes how these marketing practices result in a new constellation of cultural sensibilities and behaviours. We would equally emphasize that they represent an intensification, rather than transcendence, of capital's long-standing practices of market management to balance production and consumption.[52]

Interactive gaming gives an extraordinary demonstration of the new forms and intensification of marketing practices. As we have already remarked, computer and video games, unlike television, do not depend mainly on advertising-based revenue (although various forms of "advergaming" and product placement are becoming increasingly important). In that sense they represent a break from the Fordist broadcast mass media model. But interactive gaming has been one of the most intensively and creatively marketed of new industries. The need first to establish a new product and then incessantly to renew the appeal in a perpetual-innovation product based on constant techno-logical upgrades and the exhaustion of short-lived entertainment values has propelled it to extraordinary lavishness and ingenuity of promo-tional effort. Moreover, to profit in this circumstance requires cultural intermediaries who are concerned with aesthetics and popular taste, a legion of designers and artists, and advertising and promotion wizards engaged in engineering or orchestrating the incessant changes in demand on which the absorption of perpetual innovation depends.

The circuits of the video game industry exemplify this dynamic. We can start with the technology circuit. Every three years or so, for exam-ple, the industry is convulsed by the appearance of a new cycle of video game consoles. This pace of innovation requires the efforts of interme-diaries in the marketing circuit to "build" a consumer audience for the new systems. In 2001 the latest generation of video game consoles was being launched in North America with the support of marketing bud-gets that soared as high as two hundred million dollars. The industry was intensifying and broadening its base of loyal consumers with care-fully targeted communication; consolidating brands within core mar-kets through TV advertising that featured game characters such as Joanna Dark and Crash Bandicoot; and saturating and integrating pro-motional communication across the media environment. Video game marketers and cultural intermediaries strove to infiltrate every cultural space where young consumers might hang out, from multiplex cinemas showing *Tomb Raider* movies to Sega theme parks to Nintendo-sponsored snowboarding terrain parks. Linking the marketing and cul-tural circuits, the industry's "cool hunters" tracked trends from skate-boarding to electronic music to antiglobalization "riots" and fed them back into game content. Its designers entered into elaborate negotia-tions with their hard-core base of fans and *aficionados*. Its viral mar-keters diffused out through the World Wide Web, masquerading as enthusiastic fans spreading the word on the latest hot game. It is thus not only the "fluidification" of consumption time or the "hyperreal" nature of the digital worlds that makes the interactive game industry such an exemplary manifestation of information capital. It is also the degree to which its "perpetual upgrades," saturation marketing, and creative game design render it a cynosure of promotional communication in the wired mediascape. In this scenario, the industry's profitability rests significantly on trying to synchronize technological innovation, cultural trends, and marketing strategy. Managing this interplay is required for the video game industry's profitability.

IDEAL COMMODITY?

In his study of changing cultures of consumption, Lee makes the intriguing suggestion that for each regime of accumulation, such as Fordism or post-Fordism, it is possible to identify an *"ideal-type* commodity form."[53] The "ideal commodity form" embodies the most powerful economic, technological, social, and cultural forces at work in a regime. Such a commodity tends to "reflect the whole social organization of capitalism at any historical and geographical point in its development."[54]

Thus, for Fordism, the ideal type of commodity form was "standardized housing and the car."[55] Cars and suburban housing did not merely sustain core industrial sectors of the Fordist economy; they were goods around which a whole set of social practices and values that were vital to the regime were arrayed. Beyond this, they were an "unambiguous objectification of the practices and requirements of the production processes themselves."[56] Houses, the appliances inside them, and the automobiles in their garages were all imprinted with the stamp of a mechanical production process that emphasized structure, solidity, and reliability. These concerns were objectified in Fordist consumer goods in their "sense of fixity, permanence and sheer physical presence."[57]

If post-Fordism is a major shift within capitalism, Lee says, we should expect to see that shift reflected in a changing commodity form.[58] Fordist commodities were governed by a "metalogic" of massification, durability, solidity, structure, standardization, fixity, longevity, and utility. Post-Fordism's "metalogic," in contrast, is one of intensification and innovation; its typical commodities are instantaneous, experiential, fluid, flexible, heterogeneous, customized, portable, and permeated by a fashion with form and style.[59] Lee's examples, however, are somewhat vague: "high-tech commodities," services such as "information, data and access to means of communication," and other goods "geared to the deeper penetration of existing markets and creation of new needs via the compression of the times and spaces of consumption."[60]

We propose that the interactive game fulfils Lee's prescription for an ideal type of commodity for post-Fordism. It is a child of the computer technologies that lie at the heart of the post-Fordist reorganization of work. In production, game development, with its youthful workforce of digital artisans and netslaves, typifies the new forms of post-Fordist enterprise and labour. In consumption, the video game brilliantly exemplifies post-Fordism's tendency to fill domestic space and time with fluidified, experiential, and electronic commodities. Video and computer games, moreover, are perhaps the most compelling manifestation of the simulatory hyperreal postmodern ambience that Lee and other theorists see as the cultural correlative to the post-Fordist economy. The interactive gaming business also powerfully demonstrates the increasingly intense advertising, promotional, and surveillance strategies practised by post-Fordist marketers in an era of niche markets. In all these aspects the interactive game industry displays the global logic of an increasingly transnational capitalism whose production capacities and market strategies are now incessantly calculated and recalculated on a planetary basis.

But why should video and computer games be considered the "ideal commodities" of post-Fordism when so many other industries, goods, and services are participating in the same digital dynamic? Why games rather than, say, the Internet, or office software, or the digitally transformed worlds of film and music? There are features of interactive gaming that make it peculiarly revelatory of the processes of informational capital. Although gaming and the Internet both emerged from the same hacker culture, the Net for several decades grew on the basis of public funding and free access for hundreds and thousands of users in academia. Video games, by contrast, were very quickly selected for commercial development. Unlike office software – which is mainly a business application – games are popular entertainment products that tap into the volatile dynamics of the consumer marketplace that is a driving force of economic growth and cultural change. And unlike Hollywood or the music industry, which long predated computer technologies and whose established economic and cultural power is in some ways threatened by digitalization, video and computer games are a new media whose *sine qua non* is the logic of chips and bits. We repeat Garnham's observation that "videogame industries like Nintendo and Sega were in fact the first companies which could be said to have created a successful and global multimedia product market."[61] In nominating interactive games as the ideal commodity of post-Fordist information capitalism, we are suggesting the importance of understanding the processes from which this global achievement emerged.

Video and computer games embody the new forces of production, consumption, and communication with which capital is once again attempting to force itself beyond its own limits to commodify life with new scope and intensity; they play a crucial role in a digital transformation of the texture and processes of everyday life; they typify the strategies and imperatives of the new regime of accumulation marked by increased reliance on simulations both as work tools and as consumer commodities. Indeed, as so many information-age pundits have suggested, video game play can be seen as a sort of low-level domestic socialization for high-tech work practices. This observation is trenchantly put by J.C. Herz in her book *Joystick Nation* (1997): "Video games are perfect training for life in fin de siecle America, where daily existence demands the ability to parse sixteen kinds of information being fired at you simultaneously ... kids weaned on videogames are not attention-deficient, morally stunted, illiterate little zombies ... They're simply acclimated to a world that increasingly resembles some kind of arcade experience."[62] It is the centrality of interactive gaming to the new range of work, recreational, cultural, and social practices emerging in contemporary capitalism that the concept of the "ideal commodity" allows us to unpack in an analytic and systematic way.

THE LIMITS OF POST-FORDISM: RIDING CHAOS

The concepts of "ideal commodity" and "post-Fordism" are not without their difficulties, however, and in naming interactive games as an ideal commodity for contemporary capital, we are not suggesting that they represent a trouble-free answer to the problems and controversies of digital capitalism. That would just be repeating the euphoria of the "new economy" and "friction-free capitalism" crowd. Our position, on the contrary, is that high-technology capitalism continues to be full of frictions, tensions, and unresolved problems. Interactive games are "ideal" for post-Fordism not because they are a problem-free slam-dunk success story but because they are representative of the social forces of an age. As we will see later in this book, that includes being a point of convergence for a whole series of conflicts and crises characteristic of post-Fordist capitalism: crises arising from the difficulty of managing the blistering speed of perpetual innovation, and the relentless exhaustion of the entertainment values of experiential goods; conflicts that revolve around labour issues, piracy and hacking, militarism and violence in games, the exclusion of women from the industry, and the high-intensity commodification of young people's play. In some respects interactive games are a "dream" commodity for post-Fordist capital, providing the basis for whole new industries and markets. But in others they are a commodity "nightmare" that manifests many of the most acute instabilities and uncertainties of the new regime.

Indeed, the Regulation School's discussion of successive "regimes of accumulation" is rather misleading. It has a tendency to emphasize closure, stability, and inevitability – as if, after a brief period of restructuring and upheaval, the managers of capitalism always find the right balance between production and consumption, get things sorted, and then settle down sedately to business as usual. Our examination of the interactive game industry, with its dramatic cycles of boom and bust, suggests that, on the contrary, managing accumulation on the shifting sands of virtualization is continually problematic, a process shot through with recurrent dangers and unexpected risks. We are impressed by the post-Fordist writers' interest in flexibility, but more attention should be drawn to the dynamism and sheer chaos involved in creating, sustaining, and sometimes destroying, media industries. A closer look at the digital play business reveals post-Fordism as something less like a fully fledged "regime of accumulation," and more a matter of "riding chaos" – a constant attempt to strategize responses to a highly unstable, fluid, crisis-ridden conjunction in which managing markets, workers, consumers, and commodities proceeds by incessant improvisation, and where today's solution becomes, overnight, part of tomorrow's problem. Paradoxically, it is in "riding chaos" that the interactive game industry finds both its profits and its perils.

But to get to the heart of this turbulence we need to zoom in more closely on our object of study and look in detail at the story of the interactive game industry. In this chapter, we suggested that the concepts of post-Fordism, postmodern culture, and high-intensity marketing provide a way of locating the appearance of the interactive game industry within larger changes in contemporary capitalism. Lee's notion of an "ideal commodity" affords a conceptual strategy for tracing outward from the specific commodity to the broader array of economic, social, and aesthetic connections that bind together production and consumption in a given historical moment. Our aim now is to test the analytic concepts laid out in these theoretical discussions within specific institutional contexts, social settings, and historical moments crucial to the emergence of interactive games as a major form of popular entertainment. We therefore restate our agreement with Williams that it is only through a materialist history of a new medium that we can uncover the dimensions of intentionality and conflict that ground both critical evaluation and progressive advocacy with regard to culture and technology. It is to this task that we turn in the next section.

PART TWO

Histories:
The Making of a New Medium

The story of the emergence of interactive play and of its uncertain crisis-filled transformation into one of the premier industries of digital globalized capital is both exciting and revelatory. Historical perspective is vital to critical understanding. We strongly agree with Williams that it is impossible to diagnose the cultural impact of a new medium until the specific institutional circumstances of its development are understood. Moreover, critical media analysis requires historical perspective in order to argue against the deterministic view that technology "is a self-acting force which creates new ways of life."[1]

Historical perspective also prevents us from isolating a media "text" – a single video game, for example – from its grounding in specific material conditions and human practices. In our historical narrative we have therefore tried to heed Williams's suggestion and examine the character of the various practices that produce the text in the first place (e.g., hacking, designing, branding) while keeping our eye on the wider social "conditions of a practice."[2] It is on this last point that epochal concepts such as "post-Fordism" or "information capitalism" provide a conceptual orientation for understanding changes in the mediascape and therefore serve as continuing reference points in our account. But we also need concrete accounts of particular industries in order to see how such transformations are built up from specific clusters of technological innovation, sequences of cultural change, and lines of profit-accumulation strategy. We hear calls for this sort of research coming from many quarters in communication studies. As Angela McRobbie put it: "What is needed now is a better, more reliable set of cultural maps. We need to be able to do more than analyse the texts, we need data, graphs, ethnographies, facts and figures."[3]

Our purpose is to unearth the institutional contexts in which the technological possibility of interactive media was transformed into the most profitable global cultural form of entertainment. Legend has it that the first interactive video game, *Spacewar*, was conceived in 1962 in idle off-hours by MIT graduate student Steve Russell in what seems like a creative act of pure technological innovation. Yet the long march to interactive entertainment profitability was neither smooth nor inevitable. This is why Williams promoted a method of media analysis that would try to "restore *intention* to the process of research and development," which means studying technologies "as being looked for and developed with certain purposes and practices already in mind."[4] As Williams showed in the case of television, the transformation of a new media technology into a domestic entertainment industry only unfolds through a process of institutional decision-making and intentions, social choices, paths selected and abandoned, crises survived or succumbed to by particular agents, connections made or ignored, all of which can crucially swerve and shape the evolving directions of a new medium of communication before congealing to invest the medium as it exists now with a spurious aura of fixity and inevitability. Our history of interactive gaming traces the interplay of the three circuits of interactive gaming theorized in chapter 2 – technology, culture, and marketing – within the wider context of the competitive and expansionary dynamics of the post-Fordist global cultural marketplace described in chapter 3. But now our narrative concentrates on the changing, contingent configurations of technological innovation, game design, and marketing strategy that, coupled in what we term "digital design practices," propelled the ascent of this cultural industry to its place as a preferred source of entertainment for postmodern youth.

In chapter 4 we show how the apparently serendipitous invention of *Spacewar* (and other ur-video games) was the outcome of a conjuncture of military-industrial funding, hacker experimentation, and science fiction subcultures. We also show the uncertain path towards commercialization and the place of the first great video game company, Atari; how the particular route taken by the video game imprinted it with a series of crucial cultural templates of lasting influence; and how in the mid-1980s inadequate marketing and managerial strategies contributed to a catastrophic industry crash that threatened the extinction of the new medium.

Chapter 5 shows how the Japanese company Nintendo, with extraordinary transnational audacity, revived the digital play business, rationalizing its design, marketing, and intellectual-property practices. As we shall see, the creation of a "Nintendo generation" familiar with digital play was the consequence not only of techno-logical advances and consummate artistry in game design but also of an ensemble of marketing practices copied from earlier genera-tions of mass media. Nintendo's controversial intervention in the United States market established a transpacific flow of game culture and technology that made the industry an example of corporate globalization.

We then trace the interweaving processes of corporate restruc-turing, game design, and youth marketing strategies through the continued boom-bust cycles of the 1980s and 1990s, during which the video game industry grew by leaps and sputters. Chapter 6 shows how Nintendo was challenged in the 1990s both by other video game console makers, such as Sega, and by games developed for the personal computer. This era saw the creation of ever better performing "generations" of game consoles; the creation of ever more extreme and violent games; and a series of brand wars that drove game marketing in spirals of escalating symbolic investment aimed at an increasingly mature and media-savvy niche of youthful male players. This period of internecine warfare brought disaster to many individual companies and threatened new crises of technolog-ical and symbolic overproduction. But its overall effect was to deepen the transformation of the industry from a technology-driven to a consumer-driven sector, further enlarging its transnational scope and making it a hothouse for experiment in the management of digital design for youth markets.

Chapter 7 tells how in the later 1990s the growth of the game industry attracted the established electronic empires. The multi-national giant Sony advanced on the video game business with its enormously successful PlayStation console, while Microsoft gradu-ally leaned its massive weight on the growing computer game market before joining the console fray with its xbox. The appear-ance of these behemoths marks the recognition of interactive gaming as a key competitive arena in the struggle between the largest of the global multimedia corporations. We give particular attention to the growing importance of the Internet as a meeting place for the gaming community and as a distribution system for games.

Our narrative concludes in chapter 8 with a portrait of the digital play industry as it enters the third millennium. Caught up in the torrid quest for digital productivity, the interactive game industry's short history reveals why garage inventiveness quickly mutated into a set of oligopolistic corporate alliances where an apparent diversity of game development companies is increasingly dominated by a handful of publishers and multimedia giants, whose most recent oligopolistic conflict is currently being fought out with supersophisticated video game consoles – Microsoft's xbox, Sony's PlayStation 2, and Nintendo's GameCube. We note too that simulation gaming is a field of high-tech know-how that has had firm, if hidden, government backing. Gestated in the remarkable conjuncture of American culture where corporate, military, and technological interests support experiments in simulation, cybernetics, and networking, the game business reveals why "military entertainment complex" is not just a cute phrase. On the consumption side of the business, we examine the changing composition of the game market – rapidly growing, changing in age and gender composition but also subtly stratified and complexly segmented. The game business is now central to the commercial colonization of cyberspace, where online gaming and chat-rooms became the fulcrum of youth marketing on the Web, and the problems and possibilities of massively multiplayer gamesites are studied as trailblazing e-commerce models. The whole interactive game complex now operates within a globalized context characterized by both international expansion of markets and deep exclusionary digital divides. Our overview closes with a brief evaluation of the industry's position in the wake of the dot.com crash and Internet meltdown of 2000.

This tale of management savvy turning an arcade fad into a domestic entertainment medium that competes with television and movies for the affections of its youthful primary audience has been told before. But our account attempts something other than a chronology of technological marvels, or an anecdotal celebration of entrepreneurial smarts, or nostalgic-ironic reflections on changing generations of popular culture. It is a thematic history that exposes how the logic of capital sets limits, exerts pressures, and manages the unfolding possibility of video gaming. This logic is not wholly determinative; there is, as we show, a process of "negotiation" between consumers and producers in a cultural marketplace. Our history is about how interactive gaming became a mutating matrix of experimentation in the practices of managed innovation, cultural commodification, digital entrepreneurialism, and intellectual property control critical to the emergence of a post-Fordist world market. It investigates the commodification of play, and the role of marketing in making visible and targeting consumer segments through strategies of branding, targeting, and synergy. It identifies the growing intersection of innovation, design, and promotion as practices managed through cultural intermediaries and the integration of audience feedback into cyclical loops of production and consumption. It is about the instabilities generated by perpetual technological innovation, and reveals the symbolic exhaustion that arises from the constant renewal of a market-driven entertainment form. It maps the growing transnationalization of an industry that originated in the United States, was matured by Japanese corporations, took hold in Europe, and now operates in culturally hybridized networks spread unevenly across Asia, Russia, Latin America, and throughout the planet. It illustrates the importance of online gaming to burgeoning and imploding e-business models. It is about the linkages of digital play to a massive simulation-based military-entertainment complex central to the survival of advanced capitalism. In this process, interactive gaming has come to be situated at the centre of a series of imminent and uncertain but, for strategically positioned enterprises, immensely profitable changes taking place in the mediatized global market.

4

Origins of an Industry:
Cold Warriors, Hackers, and Suits
1960–1984

INTRODUCTION:
DREAD, DISTRACTION, AND INNOVATION

In 1962, as President Kennedy confronted the Soviets over the missiles in Cuba, Americans stared apprehensively into the night sky for the signs of atomic attack, then turned their gaze back to science fiction films on the screens of drive-in cinemas. Rocket fins sprouted on everything from Cadillacs to hotdog wrappers. Technological development was a source of both dread and distraction. In this ambiguous context of nuclear angst and consumer confidence, a prototype video game saw the light of day. *Spacewar* was a serendipitous digital doodle. It enabled two players to steer vapour-trailing rocketships and fire torpedoes at each other by twiddling knobs to control blips on an oscilloscope screen. Jointly conceived by the self-styled Hingham Institute Study Group on Space Warfare, the game was the brainchild of a group of engineering graduate students working in the basement lab at the Massachusetts Institute of Technology.[1] The chief architect, Steve Russell, had no idea his creation would lay the foundations of an industry.[2]

Creative adaptation of previous ideas and the destructive creation of new ones have always been crucial to the long march of America's technological innovations.[3] But innovation – especially radical innovations of entirely new products – is hard to achieve. Most innovation proceeds along trajectories of incremental improvement in technical components or processes. The introduction of radically new technological directions and the creation of organizational forms capable of taking advantage of these technologies is rare.[4] Conventional wisdom sees the capitalist entrepreneur as the main source of such innovation. Many accounts of Silicon Valley business celebrities and dot.com millionaires as makers of the information revolution reflect that view.

But as Richard Barbrook and Andy Cameron point out, ascribing digital innovation to the dynamic powers of entrepreneurs and the unfettered market – what they scathingly term "the Californian Ideology" – is deeply unhistorical.[5] It overlooks two crucial contributors, both of which are outside of the market and in some ways antithetical to it, as well as to each other. These two contributors are publicly funded military-space research, and the playful "gift economy" of hackers. It was only by building on and appropriating the technological foundations of these other agencies that industrial capital could launch itself from a Fordist to a post-Fordist regime. The genealogy of the video game is a prime example of this process, for it is at the intersection of warfare state and hacker culture that we find the point of departure for the digital play industry.

PENTAGON PLAY

That war in space provided the topic of the first video game is no coincidence. The Russian Sputnik launch of 1957 had shocked the US military establishment by showing the precariousness of its lead in science and technology, which were key to a cold war victory. Funds poured into missilery, ballistics, and space defence, largely funnelled through the Pentagon's Advanced Research Projects Agency (ARPA), an institution that was to become a matrix of digital development. At the same time, the National Aeronautics and Space Administration (NASA) launched its attack on the final frontier of outer space in a program that was conceived partly as a civilian alternative to military development, yet that was also deeply and inextricably intertwined with strategic designs to command the "high frontier" of the heavens.[6]

Nuclear mobilization and space exploration both required prodigious achievements in the nascent science of computing. This pushed America's space warrior towards deeper interaction with two other sectors – universities and business. Military funding supported academic inquiry in institutions such as Harvard, Stanford, the University of California at Los Angeles, Johns Hopkins, and the Massachusetts Institute of Technology, whose cutting-edge "Artificial Intelligence" department, founded in 1959, became a primary beneficiary of ARPA's largesse.[7] Defence contracts underpinned the corporations developing computer equipment, such as IBM, General Electric, Sperry Rand, Raytheon, and Digital Equipment Corporation. The military-industrial-academic complex provided the triangular base from which the information age would be launched.

Computers were still lumbering through the era of giant mainframe machines.[8] Card readers, accumulator toggles, and machine language

were the idioms of the day, algorithms, subprogram loops, decision
trees, and assembly language the basic design tools. But the ambitions
of nuclear warriors and moon-landing planners pushed the envelope
of archaic computing capacity. By the 1960s, MIT's artificial intelligence
wizards were starting to depart from the logic of mainframes and batch
computing, exploring the possibilities of real-time computers, and
smaller machines such as the Programmable Data Processor-1 (PDP-1)
"mini-computer" (which was "the size of about three refrigerators")
created by the defence contractor Digital Equipment Corporation.[9] The
invention of the video game could not have taken place without these
foundational developments in the computer industry and at university
research institutes – all subsidized by the military-space complex. As
we shall see, this symbiosis with military research was ongoing, not
only assisting the commercial industry throughout its early years but
also continuing with unabated force into the present day.

HACKER GAMES

Interactive games would never have emerged without the activity of
another group: the first hackers. Today the term "hacking" is associated with digital theft and delinquency. But this usage marks the
criminalization of activities that were once considered legitimate and
vital for the development of computer networks. "Hacker" originally
meant simply a computer virtuoso, someone "who enjoys exploring
the details of programmable systems and how to stretch their capabilities; one who programs enthusiastically, even obsessively,"[10] while
"hack" referred to "a stylish technical innovation undertaken for the
intrinsic pleasure of experimenting – not necessarily fulfilling any more
constructive goal."[11] The ur-hackers were young male programming
wizards whose unauthorized but accepted experimental computer play
was crucial to the explorative work of digital centres such as MIT. The
hackers were at once indispensable to cold war research, yet in many
ways profoundly contrary to its spirit. Élite employees of the military-
industrial complex, many were disillusioned by Vietnam and Watergate
and committed to the idea that computers could be an empowering
democratic technology.

     Their idealism was encapsulated in the early "hacker ethic," described
by Steven Levy. The most famous article of this credo was that "information wants to be free," a view that would eventually lead hacker
culture into collision with commercial empires.[12] But other slogans were
even more important to game development, such as "Always yield to
the Hands-On Imperative," a command Levy explains as meaning that
"essential lessons can be learned about the systems – about the world

– from taking things apart, seeing how they work, and using this knowledge to create new and even more interesting things."[13] Other hacker assertions were that "you can create art and beauty on a computer," and, simply, "Computers can change your life for the better."[14]

Hacking was a way of dealing with the tedium of long hours spent in the presence of unforgiving machines and the mind-numbing programming problems of massive mainframe computers. This is where creative puzzle solving and programming – which have a distinctive cognitive challenge in common – come into play. As Allucquère Rosanne Stone notes, studies in programming subcultures and online working groups revealed that a significant part of the time that people spend developing interactive skills is devoted not to work but to what, in our common understanding, would be called "play."[15] It is in fact the promotion of creativity and the "*play* ethic" that engineering designers now celebrate when they promote "divergent thinking" as a means of accelerating radical innovation through creativity.[16] Divergent thinking is a particular kind of playfulness – a constructive exploration with strategic value that comes to be very important in the programmer subculture. It promotes radical innovation by substituting quick intuition for the slow process of deductive trial-and-error testing. Regression into a "childlike" frame of mind in which possibility is less restricted enables a less-linear flow of ideas.

Computer engineering for military science had hitherto pursued a systemic trajectory, testing out known decision trees and designs against past results; incrementally optimizing the strategy that seemed most likely to produce successful outcomes. The hacker trajectory was strategic and intuitive in design, imagining radical alternatives, seeing the system of relations as a pattern that could benefit from the rearrangement of its components.[17] What made institutions such as the MIT laboratory special was that this whimsical activity was not merely tolerated but actually encouraged.

Spacewar grew out of this delirious nocturnal technoculture. Up until their emergence computers, because of their cost and size, had been designed for complex and abstract calculations. *Spacewar* was a radical innovation because it used interface controls for navigation and made the screen a graphic input to the player, so that, for example, a visual vapour trail behind the accelerating space blips accentuated the sense of movement. These two crucial operations of *Spacewar*, navigation and display, are the foundation of digital interactive entertainment – the crucial "core design" subsequent hardware and software designers would work up and sophisticate through generations of games.

It took playfulness to create digital play. Using algorithms and oscilloscopes to make a game was a radical interpretation of the possibility of those computers that went beyond their designed uses as advanced calculators and modelling machines. Russell's invention did not derive from computer technology per se but from imaginative speculation about how it could be used. It introduced the possibility of pleasure from the ability to navigate moving objects through a simulated environment represented on a screen. His demonstration revealed the potential of a new way of thinking about computers: video games broke the fun barrier.[18]

Here lies the seemingly paradoxical quality of innovation: it arises in a predeterminate institutional and (sub)cultural context along a sanctioned and funded research trajectory; and yet it is radical, formative, original, and unique to the group that develops it and comes to influence the future evolution of both technology and culture. Like many experimental programs, *Spacewar* was quickly installed on other university machines as a demonstration of what computers could do. Gradually, it dawned on many people that computers need not remain associated with scientific calculations, drudgery and boredom, long hours of concentration and mechanical problem solving but could be a source of entertainment and diversion. With that realization we move from the moment of creativity to the problem of how innovations diffuse through organizations and society.[19]

SCIENCE FICTION AND CYBERNETIC CULTURE

Spacewar wasn't a very elegant game and perhaps not even properly the first video game. Russell himself says that if he hadn't created it, someone else would have because it was so much a part of the "moment" in early computer circles. Many programmers, steeped in simulation and gaming theories, were rethinking the application of computational capacities for modelling. Some had created computer versions of chess, noughts and crosses, and solitaire. Even more importantly, a convergence of cybernetics theory, artificial intelligence research, simulation, and science fiction was having a deep and powerful impact on North American culture. Indeed, as Nick Heffernan has recently argued, this cybernetic culture was laying the social and technological foundations for post-Fordist developments that would within a decade transform work and consumption.[20]

Here we confront a little-understood aspect of the circuitry of technological innovation, especially within the digital disciplines: the role of cultural contexts and subcultural practices in the dynamics of innovation and design. Breakthroughs in interactive gaming were accelerated significantly by fringe activities favoured by programming subcultures: puzzles, Lego, board games, and science fiction. Sci-fi in particular was central to the cultural milieu of applied cybernetics. *Spacewar* had been inspired by "Doc" Smith's "Lensman" series of intergalactic adventure novels.[21] Throughout the 1950s, writers like Isaac Asimov had offered glimpses of a cybernetic future awaiting the baby-boom generation. His "Foundation" series, conjuring a ten-millennium empire predicted by computer modelling, was perhaps the first novel to counter the dystopian themes of George Orwell's *1984* (1949) with a more optimistic reading of the potential of computers to anticipate and control the future.

Real-life experiments in this direction were becoming widespread. With the mantle of space exploration increasingly assumed by the civilian NASA, Pentagon agencies such as ARPA focused their resources on computers and networks as high-prestige high-technology competitors to the headline-grabbing attempts to land men on the moon. "Game theory" was riding high amongst the technocratic planners of both the Vietnam War and nuclear deterrence, and computers were indispensable to ultrasophisticated versions of Kriegspiel in which Reds and Blues endlessly battled each other for global dominance.

The central theories of systems-and-information sciences began to circulate more widely through the social sciences, as computers were used for modelling and teaching about complex systems like the economy, city planning, and biological evolution. In 1967 a computer model of a whole social system, called "Simsoc," was tested by real people playing roles in university classrooms, creating the ancestor of *SimCity*. MIT bioscience researchers were creating *The Game of Life* to emulate evolutionary processes; *Lunar Landing* continued the space-program origins of video gaming; *Hammurabi* offered the opportunity to administer an ancient kingdom.

So just as science fiction was beginning to predict social reality, virtual reality was being inlaid with playful cybernetic visions that were widely circulated for free, through communities of academics and hackers who were increasingly interconnected by computer networks. Many of these games reflected the technocratic preoccupations of researchers at the cutting edge of the physical and social sciences. But there were also more whimsical strains in simulation culture. Commercialized boardgame versions of large-scale social experiments – *Blacks and Whites*, *Diplomacy*, and *Risk* – had already become popular pastimes in the 1960s for university students. Perhaps the most popular of all, however, was *Dungeons and Dragons*, published in 1972. It made Role Play Games (RPG) a new category of entertainment that was very attractive in programming circles – one that rapidly found translation into the academic and hacker culture.[22] Programmers of this era remember the quest game *Find the Wumpus*, which came with the UNIX operating systems installed on their machines. *Dungeons and Dragons* in turn inspired its programmed equivalent, *Adventure,* which debuted on university computers in the 1970s.[23] In the growing underground culture of role-playing games, Stone observes, there was an "unalloyed nostalgia for a time when roles were clearly defined, folks lived closer to nature, life was simpler, magic was afoot, and adventure was still possible."[24] This is not the least of the paradoxes of the emergent hacker game culture: while many of the creations of this ultramodern technoculture looked forward to a science-fiction future where digital capacities for command and control attained maximum development, others expressed a fascination with misty premodern fantasy in which the political and ethical dilemmas of the nuclear era military-industrial complex were sublimated, simplified, and made playable.

ENTREPRENEURIAL PATHS

Digital play initially flowed from military infrastructures into subcultures of hacker play, sci-fi speculation, and cybernetic simulation. But it was not yet a for–profit business operating in the consumer marketplace. The interactive game awaited metamorphosis into a commodity. This crucial transformation was the work of a handful of programmer-entrepreneurs, many of whom were connected to the military-industrial complex, all of whom had a keen eye to the commercial potential of spin-off technologies. Their search for profit from interactive games took several different routes during the 1970s. As Leslie Haddon observes, what was involved was not the emergence of a single device but rather "a family of related technological forms."[25] Only gradually were these tentative experiments winnowed and synthesized to create what we recognize as the commercial video game industry.[26] Following Haddon, we will look briefly at three of these entrepreneurial paths: the arcades, the home video game console, and the personal computer.

The Arcades

Nolan Bushnell, a research-design engineer, had encountered *Spacewar* while studying at the University of Utah and doing stints of holiday work at amusement parks. In 1970 he made a clone version dubbed *Computer Space* and added a crucial ingredient that Russell never had – a marketing vision. A "successful commercial video game," Bushnell discovered, would need to be "a fixed-purpose game-playing machine."[27] *Computer Space* was the first coin-operated arcade video game. It should be no surprise that video games first became available to the public in arcades, malls, and bars whose noisy traditions of public gaming dated back to the seaside entertainments, amusement parks, peep-shows, coin-op phonograph booths, and fair grounds of the previous century. These were sites for dynamic and sexualized active entertainment: gambling, shooting, betting, racing, and contests of might and skill. Pinball parlours had become especially popular hangouts for young males, mostly from the working classes. Electronic devices had already been integrated into this milieu: the flashing lights and beeps of pinball and gambling machines punctuated the action, heightened the carnival atmosphere, and beckoned to new players with dramatic announcements of success and top scores.

Table models of several digital games began appearing in pubs and pinball arcades throughout the early 1970s. Companies that were soon to become household names, such as Atari, cut their digital teeth in this environment. Transition to the arcades democratized the arcane games of computer hackers. But it also reinforced social biases that were already implanted in the technology. Arcades were primarily male venues. The themes of shooting, violence, and intense competition, present at the very origins of gaming's Pentagon-sponsored inception, at once made the arcades a natural setting for the video game's commercial placement and were amplified in this environment. From the arcades, video gaming also inherited an aura of danger. The darkened arcades were socially suspect, sites of young male delinquency and corruption, and the domestic video game industry would soon try to decide whether to capitalize on this aura so as to sell to boys, or repudiate it to reassure parents.

The arcades were also important in establishing the interactive game industry's reliance on a feedback loop of detailed information about customer preferences. In the arcade business, one gets an instant sense of game popularity by emptying coins from machines. When Atari persuaded Andy Capp's, a well-known Sunnyvale bar, to install the first coin-operated version of *Pong* in 1972, success was signalled by an emergency phone call announcing that enthusiastic players had jammed the coin slot.[28] Similarly, when Nintendo brought *Donkey Kong* to the US in 1983, its American managers knew immediately by assessing one day's take from a selected bar that, contrary to expectations, this story of an urban ape hunt would be a smash hit.[29]

Arcade success rapidly became evident outside America. David Rosen, an American who had served in the US Air Force and was stationed in Japan, started a company in Japan named Rosen Enterprises that imported pinball machines as entertainment from the US.[30] In 1965 Rosen's company merged with a Japanese electronics firm, Service Games. The merged company, Sega Enterprises Ltd., released in 1966 an electronic shooting gallery game, *Periscope*, that succeeded amongst

Japanese businessmen who were fans of the pinball-like game
pachinko.[31] Sega was soon followed into the world of electronic play
by Nintendo, a hundred-year-old card company whose director perceived that the future of entertainment lay in an expanding postwar
economy and branched out into arcade gaming. The seeds of video
gaming's international future were thus sown in the arcades.

The Home Console

In 1966 Ralph H. Baer, another engineer working for a defence
contractor, approached the emergence of interactive computing from
another direction, seeing in it a new use for surplus television sets.[32]
Baer had seen a tennis simulation game invented in 1958 by the
engineer William A. Higinbotham at the US Department of Nuclear
Energy's Brookhaven National Laboratory. Higinbotham programmed
the game, *Tennis for Two*, on an analogue computer with two control
boxes, each with a direction knob and serve button, controlling a five-
inch oscilloscope screen. The game was demonstrated during Visitors'
Day at the facility. Higinbotham never planned to market his tennis
simulator and chose not to patent the game.

Eight years later, Baer, now manager of consumer product development for Sanders Associates, a military electronics consulting firm, set
out to develop a game system. By linking the screen display of a TV
set to electronic input devices he made an interactive game device that
required eye-hand coordination. By 1967 Baer's interactive television
gadget – first called the "Television Gaming Apparatus" and eventually
the "Brown Box" prototype – had taken shape as a "primitive" game
called *Fox and Hounds.*[33] When he demonstrated it to his bosses at
Sanders, they were impressed and gave him $2,500 in funding.[34] But
it remained classified as a military training device until 1968, when
Baer was permitted to continue commercial development and applied
for exclusive patent rights to the "Television Gaming Apparatus." It
wasn't until 1971 that Magnavox, a consumer electronics company –
that manufactured televisions, of course – purchased the technology
from Sanders Associates and developed the "Brown Box" as the
"Magnavox Odyssey," a TV plug-in device that played simple games,
such as *Ping-Pong*. The unit could not keep score and had only black-
and-white display and minimal graphics; games came packaged with
colour overlays that could be taped onto one's television.

Unpromising though it seemed, this was a momentous début. By linking console to television, Baer had connected digital gaming with the
most pervasive mass media technology of the era, enabling the convergence of the twentieth century's two most important communications

media – television (and the popular entertainment industry that had developed around it) and computers (with their ability to design more complex interaction with technology). If video games can be described as a post-Fordist media, the console's tie to television exemplifies the ambiguity of this description: *post*-Fordist in so far as it was a pioneer of the digital technologies that would overturn the regime of industrial mass production and consumption; post-*Fordist* in so far as this mini-computer depended for its acceptance on its tie to the very media most strongly associated with the same mass-consumption regime.

Home Computers

Video game consoles, in the arcade and in the home, emerged as part of the hacker innovation that converted computers from a mainframe tool of military and corporate bureaucracy to a household media. The consoles made by Bushnell and Baer were in essence just simplified computers with all their power dedicated to the single function of gaming. Another obvious route for the marketing of digital games was through the home computer industry that Commodore started in 1977. The conversion of games from mainframe and minicomputers to the far more limited capacity of "microcomputers" was a chief attraction of the hacker/hobbyist culture – again, predominantly youthful and male – that sustained the early personal computer market. But despite their common origin, video gaming and personal computing followed divergent paths. In North America, home computers were widely dismissed as a platform for commercial games because of their high cost (two thousand dollars and up, compared with consoles that sold for less than two hundred dollars) and poor graphics and sound quality. As Haddon points out, many early home computer makers played down the gaming capacities of their machines for fear it would make them seem frivolous or toylike.[35] In 1982 *Time* magazine declared the PC its "Man of the Year" but declared ironically that the "most visible aspect of the computer revolution, the video game, is its least significant" and speculated that "the buzz and clang of the arcades" might be "largely a teen-age fad, doomed to go the way of Rubik's Cube and the Hula Hoop."[36] The universe of early domestic digital devices thus tended to be roughly divided between the "serious" computer – for work or education – and the "playful" console.

The split was not absolute or universal. In Britain the rapid take-up of cheap home computers such as the Sinclair created a computing-gaming culture that at times rivalled console play in popularity. Even in North America, a vigorous subculture of computer players, circulating games on floppy disks, persisted.[37] This was in many ways a hard-core gaming culture. In spite of limited graphics and slower chips, computer games had the advantage of being programmable. They often appealed to an older or at least more technically literate group of players. Moreover, these games, unlike their console cousins, were part of the exciting innovations in computer networking. Electronic bulletin boards and online connection provided a semiclandestine arena for those interested in playing, programming, and hacking games. In a community that tended to regard digital networks as electronic commons, piracy was frequent. Free-release, or "brown bag," versions of beta versions of game software were a common method of getting the gaming community involved in a new product. Thus, although the world of computer gaming was relatively small in North America, it was experimental, interconnected, and technologically sophisticated.

ATARI ERA

The company that was to organize and dominate this swirl of trends has become a legend in gaming history – Bushnell's Atari, founded in 1972. That Bushnell, rather than Baer, should emerge as the industry's ur-entrepreneur at first seems surprising. His arcade game, *Computer Space,* was a failure, while *Ping-Pong* succeeded beyond Baer's dreams. McLuhan's idea that media are "rear-view mirrors" – new technological forms relying for content on old social contexts – is relevant here; while *Computer Space'*s inspiration was futuristic space-weapons logistics, *Ping-Pong* was based on an already familiar pastime: anyone could figure it out.

But despite Bushnell's initial setback he proved to be both a clever engineer and a good businessman. Atari was quick to produce a rip-off of *Ping-Pong* – *Pong,* which, licensed to the arcade game company Midway, proved a triumph. In 1974 "several Atari executives mysteriously defected" to another company, Kee Games, and released *Tank,* which played as it sounds and became the top game of the year, outselling Atari.[38] Competition for the limited arcade spaces across America seemed to be heating up and other companies were leaping into the fray. It was later revealed that Kee was a wholly owned subsidiary of Atari created by Bushnell to evade exclusivity agreements with distributors. Eventually it merged back with its parent company, having both widened Atari's reach and stimulated the whole arcade game industry – which by 1982 generated more than three billion dollars in North American revenue.

From the arcades, Atari began to move on the home market, with *Home Pong,* a one-game dedicated console box with two mounted "paddle controller" dials that attached to a television. Atari secured a contract with Sears for 150,000 units, and venture capital to finance the expansion of its productive capacity. *Home Pong* was a best-selling item. Its runaway success confounded many in the entertainment industries. But there were problems with home systems. Chief among them was the inflexibility of the circuit boards and the limited power of computer chips. Not only were early home game systems rudimentary in terms of graphics, sound, and navigation but they could play only one or a few games. Big coin-op arcade systems had better sound and graphics and could amortize the cost of expensive technology over the life of a machine. But the dedicated console was a deterrent to consumer investment in a relatively expensive piece of equipment. The high price of microprocessors stalled the development and marketability of home game systems.

Fortunately, manufacturers such as Fairchild Camera and Instrument and National Semi-conductor were looking to branch out from their initial focus on automated machine tools to make components for consumer goods, and on a wider basis than just adding digital chips to watches and toys.[39] It wasn't until 1976 when Fairchild released "the first programmable home game console" that players could "actually insert ... cartridges into the console and change games."[40] The release in the following year of the Atari Video Computer System (VCS) 2600, a console capable of playing and displaying a wide variety of games on cartridges, was a giant step towards creating a viable home market for video games. At least thirty other electronics firms released comparable home video game consoles over the next few years, many based on the same chip, showing that the technology circuits of the industry were becoming clearly established.

With software programming, games could easily be transferred from arcade to domestic systems, so that "coin-ops were used ... as a testing ground for products which might then be cross-licensed to the home market."[41] *Space Invaders*, released by Midway in 1978, enjoyed huge success as an arcade game: so Atari purchased licenses for domestic use in 1980, dramatically boosting its home sales. "By the end of 1979, *Space Invaders* sold a record-breaking 350,000 units worldwide; 55,000 of them in North America."[42] Bushnell's company thus built a successful interaction between two of the main routes to the commodification of the digital game: the arcade coin-op and the home console. The third route – the personal computer – was for the moment largely ignored. But there too Atari's influence was felt, albeit indirectly. The VCS 2600 and its many clones provided millions of American households with their first exposure to digital consumer items, thereby laying the ground for an emergent "micro" computer market.[43] And this market was in large part the creation of one-time employees of

Bushnell's: in 1976 two Atari employees, Steve Wozniak and Steve Jobs, having completed the successful game *Breakout*, left the company to found Apple Computers, which pioneered many of the interface features that a decade or so later would return the home computer to the forefront of game technology.

GOLDEN EGGS AND EASTER EGGS

By the late 1970s not only the technology of interactive gaming but also its cultural content was beginning to blossom. These were exciting times to be a game designer. Game genres were being invented. Some continued the legacy of war themes: in 1980, for example, Atari designer Ed Rotberg created *Battlezone*, lauded as the first three-dimensional game in a "first-person perspective as seen through a periscope that simulated the interior of a tank," the players rolling around on a virtual battlefield.[44] Others followed up on the sport, puzzle, and role-playing themes first established in the hacker underground, embellishing the original formats with better sound and colour into a more complex entertainment terrain, adding narrative, humour, and cute graphics to enhance play.

In 1980 Namco released *Pac-Man*, an amalgam of the skill-and-action, puzzle-and-maze adventure games already available. Pac-Man navigated through a labyrinth environment eating energy capsules, winning prizes, and avoiding monsters to become the most popular arcade game to date. In 1981 "250 million games of *Pac-Man* were being played on over 100,000 *Pac-Man* machines in arcades every week ... "[45] The game produced one billion dollars in revenue in 1980 alone, energizing innovation in game development, while its successor, *Ms Pac-Man* released by the American company Bally/Midway in 1981, in turn became the biggest arcade game in America. Atari had just released *Atari Football*, a game that was overtaken by Midway's *Space Invaders*, which was so popular in Japan that it "caused a nation-wide yen coin shortage that would momentarily cripple the Japanese economy."[46] Staying a little closer to the tradition of the arcade, Sega released *Monaco GP*, which set the standard for racing games. Interactive games were no longer just a technological novelty; they were becoming a cultural industry.

While hardware designers expanded the speed, display, and storage capacities of systems, the leading edge of game development had shifted towards software engineers and designers preoccupied with game play dynamism and entertainment value. Companies learned to value these new techno-artists. Atari's Bushnell claimed, "We treated engineers like mini-gods."[47] Stone describes Atari's mixed bag of "young men in their late teens and early twenties, the first generation of their kind, who lived their lives perpetually in semidarkened rooms, sleeping at their terminals or under tables, and seemingly subsisting entirely on nothing more than Fritos, Coca-Cola, and wild determination ... [yet] pulled down salaries in the $50,000 to $60,000 range for what looked like nothing more than *playing games*."[48] These "disreputable" software programmers transgressed conventional norms of workplace discipline but were untouchable because they laid the golden egg for Atari.[49] As the game industry boomed, holding on to programming talent became a managerial priority.

One clear sign of the increasing importance of software designers, and of the cultural circuits of the industry, was the emergence of game developers independent from hardware companies. Once cartridge-based consoles had detached games from consoles, there was no reason why both should be made by the same company. Independent developers licensed the right to make games playable on other firms' home and arcade machines, relying on the artistry and inventiveness of their game designers to bring in revenues that would make such deals profitable. In 1979, a short three years after Atari was taken over by corporate giant Warner Communications, key programmers defected to set up the first such third-party developer, Activision. They reportedly left Atari because they were "frustrated with the lack of recognition and compensation. Their company" would be "the first to prominently credit its programmers."[50]

But in this organizational reformation, it was Trip Hawkins, a former Apple employee and founder in 1982 of an upstart software design company called Electronic Arts, who was most closely identified with bringing a unique new media cultural aura to the development and marketing of video games. Hawkins brought the attitudes and borrowed lessons from the youth music industry. While Activision had introduced "an artist-oriented environment and approach to game development," Electronic Arts went one step further "by incorporating an artist-oriented marketing effort," packaging its games with album-like artwork and liner notes, and promoting its developers like rock stars in game magazines.[51]

Signs of unrest amongst the new digital labour force showed the importance of this recognition. In 1978, irked by Atari's "policy of crediting the design of a game to the company as a whole, rather than to individual employees," Warren Robinett, a designer of *Adventure*, placed "the first 'Easter Egg' or hidden feature within a game." Robinett designed "a hidden room that [had] his name in bright rainbow letters. To access the room, players [had] to find a grey pixel and carry it back to the beginning of the game."[52] "Easter eggs" still abound, either as a subversive expression of programming boredom or disaffection (one of the most notorious was a gay hot-tub scene smuggled into *Sim*

Copter) or, increasingly, as a planned design feature intended to intensify player interest – a striking illustration of how management often
gets its best ideas from the resistance it provokes.

THE VIDEO GAME AESTHETIC

One of the most influential and innovative game designers was Chris
Crawford, who worked at Atari's Research Lab, where he authored
such classic games as *Eastern Front (1941)* and *Balance of Power*. But
Crawford also wrote the first treatise on video game design, *The Art
of Computer Game Design* (1984). Its central premise was that "interactive games were a new and poorly developed art form."[53] To survive,
they had to be lifted out of commercial mediocrity and the hands of
"technologists" (as opposed to "artists") by the creation of an "aesthetic of the computer game."[54] Crawford wrote of how the computer,
originally developed as a "number cruncher," was now transformed
by "graphics and sound capabilities."[55] "With this capability came a
new, previously undreamed of possibility" – that of "using the computer artistically as a medium for emotional communication art." The
computer game, Crawford claimed, "has emerged as the prime vehicle
for this medium."[56]

Crawford set out a credo of video game design that took the programmer well beyond technology. While technological improvements
would continue, game design was no longer hampered primarily by
hardware limitations. Rather, "our primary problem is that we have
little theory on which to base our efforts": "We don't really know what
a game is or why people play games or what makes a game great. Real
art through computer games is achievable, but it will never be achieved
as long as we have no path to understanding. We need to establish our
principles of aesthetics, a framework for criticism, and a model for
development. New and better hardware will improve our games, but it
will not guarantee our artistic successes ... "[57] Designers needed to wres-
tle with psychological issues like frustration and reward, levels of challenge and duration of play, and they had to understand what motivated
players to play. The challenge facing the new cadre of programmers was
not to emulate reality or to facilitate the transfer of skills and competencies, as it might be in a military or social science simulation, but
rather to structure the quality of the player's involvement with an imaginary universe – that is, to understand what we today call "virtuality."

Enumerating the appeals of virtual fantasy, Crawford noted "nose-
thumbing" as "a means of overcoming social restrictions."[58] Many
games, he observes, "place the player in a role that would not be
socially acceptable in real life."[59] He also lists "proving oneself" where
people play games "as a means of demonstrating prowess"; "social

lubrication" – the "focus around which an evening of socializing will be built"; "exercise," including the development of cognitive skills; and "acknowledgment," that is, "the need to be recognized by other people."[60] This wide range of appeals explained why "a truly excellent game allows us to imprint a greater portion of our personalities into our game-playing."[61]

The task of the game designer, according to Crawford, was to sculpt the "play value" into the graphics, interface, and software, so that a game worked seamlessly as a virtual interactive environment, making the window display and joystick give the feel of, say, flying. Most game designers, Crawford bemoaned, were mere "amateurs with no further preparation than their own experience as game players."[62] But most were aware that, for players, "the most fascinating thing about reality is not that it is, or even that it changes, but *how* it changes, the intricate webwork of cause and effect by which all things are tied together."[63] The only proper way to represent this webwork is to "allow the audience to explore its nooks and crannies to let them generate causes and observe effects ... Games provide this interactive element, and it is a crucial factor of their appeal."[64] Designing with these dimensions of interactivity was, in Crawford's view, the central art of the game designer. Realizing that interactive gaming was part of the much broader changes sweeping through America, he predicted that "the relationship between society and the computer will be one of reciprocal transformation."[65] This transformation would see a shift "from the pragmatic toward the recreational, from the functional to the frivolous," in which, Crawford speculated boldly, games could turn out to be "the primary vehicle for society to work its will on computers.[66]

Crawford's treatise, published in 1984, was a sign of the growing confidence and sophistication of multimedia artists who were coming to recognize their essential role in a new and booming cultural industry. But there were other interests shaping the business whose perspectives on interactive play were far more utilitarian, and which did not share the aesthetic and social concerns broached in Crawford's manifesto. Principal among these interests were the military and, increasingly, large corporate conglomerates.

IN THE BATTLEZONE

The early years of the interactive game industry's development continued to be heartily subsidized by military research. Game developers and war planners had overlapping interests in multimedia simulation and virtual experience. *Spacewar*, multiplayer networked systems, the protocols for the Internet, and the three-dimensional navigation of virtual environments had all emerged from a context of Pentagon-funded research.

Early computer war games had been fairly abstract affairs. But the rapidly improving powers of simulation technologies promised to alter this.

Realistic and engaging training for tactical action was of great interest to the US forces. Their interest was heightened by Israel's successful rescue of airline passengers from Palestinian guerrillas at Entebbe airport in 1976. The Entebbe raid had been rehearsed with an exact replica of the airport built in the Negev desert. Observing the obvious advantages of such scaled models but realizing their expense, the Pentagon approached Nicholas Negroponte (who later become the silicon utopian discussed in chapter 1) to develop a simulation trainer with videodisk displays. Negroponte solved the problem of the massive storage of visual information necessary for real-time tactical simulators by linking the computer with a videodisk player. He set out to map Aspen, Colorado, one foot at a time with a camera, storing the images on a videodisk, and recreating the city to give the virtual experience of it in three-dimensional space. The same technological and programming strategies were, of course, perfectly suited to adventure games, which also began to use side-scrolling screens as a way of mapping fantasy space. Although the videodisk technologies Negroponte experimented with were not generally available until the 1990s, his research demonstrated the role of military funding in driving forward the digital development on which interactive play depended.

The networking of computers dramatically accelerated this confluence of war and games. The origins of the Internet lay in the same defence strategy and funding that subsidized *Spacewar*. It was within the military-sponsored ARPANET – originally devised as a distributed command-and-control system capable of operating in nuclear war conditions – that the hardware and cabling linkage of computers, the multi-user multitasking operating systems that run on them, and the data-exchange protocols (TCP/IP) that allow for two-way data flows emerged. And it was in this environment that the hacker-students, who played such a crucial role in defence research programs, conducted early experimentation in networked play. With the spread of multitasking, multiplayer games became formative experiences in programming subcultures, emulated, learned from, stolen, and avidly talked about through e-mail, file transfers, and, later, news nets.

Eventually, the hackers' cold war masters caught on to the potential of online games. In the early 1980s the Atari arcade game *Battlezone*, featuring an army tank, caught the attention of the US Army Training Support Centre, which saw in it the basis for a simulation-and-training tool.[67] Defense Advanced Research Project Agency (DARPA) planners also saw in the new games potential training tools. One DARPA project gave birth in the early 1980s to the concept of SIMNET (or simulator networking), which is described as "a large-scale tactical training environment." Says Warren Katz, one of the designers, "SIMNET training facilities typically consist of several dozen vehicle simulators, data logging systems, after-action review stations, intelligent automated adversary simulators, and offline data analysis devices. When trainees close the doors of their simulators they are transported to a time and place where they can rehearse tactics, re-fight historical battles, test hypothetical weapon systems, become familiar with enemy terrain, and so on."[68] The system "could so drastically reduce the fidelity," hence the expense, "of simulators and still maintain 'suspension of disbelief' ... [because] for the first time, manned crews were fighting against other manned crews. Even with cartoonish displays, the fact that both friendly and hostile players were controlled by other humans made the system believable and engrossing."[69]

SIMNET incorporated lessons the military learned from Atari.[70] But the loop of influence also moved in the other direction. Within a decade, software engineers who had worked on SIMNET were adapting its networked play and artificial intelligence components to cutting-edge war games such as *SpecOps* that were intended for the general market.[71] The idea of real-time interconnected immersive environments was being put to work in commercial ventures such as the BattleTech Center, which opened in Chicago in 1990, a networked "virtual reality" arcade described by its owners as "the first entertainment use of military-type multiuser networked simulation technology for out-of-home entertainment," where paying customers shot it out against each other while commanding on-screen thirty-foot-high walking mechanical fighting devices.[72] From ARPANET and *Spacewar* to Atari, and then back again from *Battlezone* to SIMNET to *BattleTech*, the interactive game industry and the military had developed a circular, self-reinforcing, synergistic dynamic. As in all symbiotic relationships, the benefits flowed both ways: the relationship between military research and video gaming, although at first a classic case of civilian spin-offs from war preparations, was also becoming a sophisticated way of getting the entertainment sector to subsidize the costs of military innovation and training. As we shall see later, the complexity of this interplay would fundamentally shape the directions of virtual creativity and profoundly influence both the technological and the cultural circuits of the video and computer game industry.

THE SUITS STEP IN

The initial success of interactive games was created outside the parameters of established corporations, by small bootstrapped enterprises.

But the meteoric success inevitably attracted the attention of business giants. The consequence was a flurry of corporate repositioning to take advantage of this emerging market. The most dramatic manifestation was the purchase of Atari by Warner Communications in 1976.

Warner was one of the largest media conglomerates of the era. It ran one of the Big Five Hollywood studios, cinema chains, television production, and a huge record and music publishing division. During the past two decades, which saw its 1989 merger with Time and then the acquisition of Time Warner by America On Line in 2000, it has become one of the building blocks out of which the biggest entertainment complex of the twenty-first century was constructed. But in the mid-1970s Warner was experiencing difficulties. Expensive deals with superstar artists and a dip in record sales in the mid-1970s were cutting profits in the music division, and the film studio was not doing as well as some of its Hollywood rivals. Warner was looking for a new product line and found it in Atari's games.[73] Atari in turn was looking for a cash injection to manufacture a home-based system.[74] On the strength of *Home Pong* sales, Atari was sold to Warner Communications for $28 million: Bushnell personally made $15 million on the deal.[75]

The other corporate sector that was keen to get in on video gaming was the toy business. During the late 1970s, gross sales of toys and games began to increase rapidly, providing toy companies with a flow of capital. The 1977 launch of the *Star Wars* movie marked a watershed both in the history of children's culture and in the development of toy marketing. Its enormous popularity not only consolidated science fantasy as a prime youth genre but also helped accelerate the steady postwar growth in "character" toy sales and other ancillary goods. The worldwide licensing of *Star Wars* playthings dramatically demonstrated the profits awaiting those toy makers who could effectively manage the synergistic linkages between media, play, and popular culture. Cultural saturation, advertising, and global distribution arrangements were being reorganized as a new generation of toy marketers emerged as the "software specialists" of the entertainment industry. Promotional marketing transformed all aspects of corporate strategy from product design and production schedules to advertising targeting and distribution channels in the playthings industries.[76]

Since electronic and learning games had been a minor part of the toy industry from the early 1970s, the growth of interactive games was watched carefully by the toy majors. Their first attempt to venture into the electronic gaming market was Coleco's Telstar, a console launched in the late 1970s. Mattel, already heavily involved in hand-held electronic toys, followed in 1980 with its Intellivision home game console, which became Atari's biggest competitor, while Coleco leap-frogged both Atari and Mattel with its cartridge-based game console, Coleco-Vision, released in 1982, which had better graphics capability, more colour, better sounds, and better interfaces than its rivals and popular games such as *Pac-Man*, *Frogger*, *Galaxian*, and *Donkey Kong*. By the early 1980s, then, any doubts in the toy trade about the effects of electronic games on other play industries completely disappeared. Arnold Greenberg of Coleco proclaimed electronics "the greatest thing for our industry since the development of plastic."[77] A Milton Bradley executive stated simply: "In the future all games will probably be electronic."[78]

The toy companies decided that, with their foot already in the door of children's lives, they and not the upstart electronics firms were the heirs apparent to the video game market. To justify this claim, they trumpeted their own keen understanding of product innovation and marketing as the key advantage. It was the toy industry, one pundit noted, that had a management tradition best suited to the fast-paced consumer entertainment sector. In contrast with the "obese and immobilized bureaucracies" found elsewhere in corporate America, toy industry representatives liked to depict their sector as "a model of success through innovation."[79] The essence of the industry, after all, "is about having fun, being clever, taking risks, trusting intuition, being first, beating the competition and winning. Toy people are on their toes sensing the air for the next move of the enemy; figuring out strategies. Market share is tenuous at best and liable to slip or turn in a matter of weeks or months ... the toy industry is about taking advantage of the latest advances in technology – using them to stimulate new ideas for new products." The same article pointed to "the phenomenal growth of the video game segment of the industry" as showing the need for "serious studies into the creative processes in management."[80] And the toy companies believed they were the exemplars of innovative management who could properly handle the new media. The truth, however, was to prove a little different.

MELTDOWN

For anyone impressed by the contemporary interactive game industry, it is sobering to look back to the early 1980s, for relative to other media of the period, video and computer games were as successful then as they are now. By 1982 worldwide home sales of video games were about three billion dollars.[81] The arcade game business was even larger, grossing eight billion dollars. By comparison, pop music had international sales of four billion, and Hollywood brought in three billion. Revenue from the game *Pac-Man* alone probably exceeded the box-office success of *Star Wars*. Consoles had penetrated seven million North American homes;[82] predictions of fifty percent penetration by 1985 were common. In 1980 Atari accounted for over thirty percent of Warner Communications' operating income.[83] Yet within a couple of years the industry was little more than a smoking economic crater.

The crisis was a consequence of the boom cycle itself. As we have seen, Atari and many other companies were competing with very similar consoles in what remained quite a "thin" market. As demand for components heated up, supply and delivery problems occurred. Merchants became wary of ordering systems in advance. As early as 1977 a minor crash occurred because machines and software were not available. Many companies got out of the business. Atari survived because it had Warner's deep pockets and an established relationship with distributors and suppliers; the big toy companies had similar advantages of scale.

By 1982 Atari, under the aegis of Time Warner, controlled eighty percent of the US video game industry. But it too was beginning to experience problems. Maintaining the pace of innovation was difficult. The programmers in their scruffy clothes and tennis shoes were valued but not understood by the "suits" that held the real power in the organization. Project managers were often brought in from military and corporate environments: "Management was thinking *product*; the coders and gamers were thinking *fun*," notes Stone. "While Time Warner itself was acutely aware that fun had to be packaged, Atari management simply didn't get it."[84]

The video game sector lacked a strategy for managing the creativity of game designers. Differences in understanding were caused by differences in age, class, and aspiration. It was difficult to maintain communication across groups with divergent knowledge systems and aesthetic sensibilities. The company environment soured, and game designers became less innovative or jumped ship as the industry was afflicted with an undercurrent of unrest. Bushnell said, "We were smitten with the Hollywood thing and saw the potential synergies with Warner films and music. We didn't realize that the sale introduced something we were completely unaware of – politics."[85] After two years of losses and growing tension among Warner executives, the wily Bushnell left Atari in 1978, signing a lucrative five-year agreement not to compete with the company he had bootstrapped, and went on to develop the Chuck E. Cheese complex.

Back in the industry he founded, rushed product-development cycles became the norm amidst a digital gold rush atmosphere. By 1983 there were more than two hundred different game cartridges on the market, manufactured by about forty companies. New games were constantly being developed. But the games in many cases were mediocre and shoddily put together, offering nothing new to the player. They were too repetitive and their intrigues were solved too easily to give long-standing play appeal. Arcade game revenues were falling as rejigged versions of *Space Invaders*, *Asteroids*, and *Pac-Man* failed to generate the excitement of the originals. But while it was one thing to spend a quarter at the arcade to find out the games were no good, the price-point of home game systems deepened the frustration. When parents saw kids retire systems to their closets after playing for five minutes they were unlikely to hand out money for new cartridges. Nor were the anticipated synergies with Hollywood a surefire recipe for success. In 1982 Atari released its E.T. game. Hoping to cash in on the success of Steven Spielberg's film of the same name, for whose license they paid a premium price, the company again rushed development. Massive shipments to eager shops were followed by equally massive consumer indifference.

In spite of their vaunted expertise in domestic entertainment, the toy companies did no better than the media conglomerate. As Stern and Schoenhaus note, the costs and timeline for product development in electronic toy lines is very different from that of other playthings.[86] They require longer lead times and they cost more. Microelectronic chip shortages occurred periodically, so the toy giants had to stock up early, carry larger inventories, and commit to longer production runs. As competition heated up, the costs of promotion increased dramatically, cutting into profits. Pushing electronic lines nearly tripled Mattel's advertising budget between 1981 and 1982. These difficulties were compounded when the toy giants unwisely attempted to compete with the emerging home computer industry, creating machines such as Coleco's Adam Computer, which was meant to fight against Apple and Commodore machines. But above all, none of the advantages of the toy industry mattered if their games couldn't beat the "suck factor." Software pumped out without quality control failed because the experiences it offered were simply not worth the investment of time or money.

The reckoning was brutal. What began as general slow-down in demand careened over the brink into a vertiginous crash that all but wiped out the industry in North America. Time Warner/Atari sales of two billion dollars in 1982 dropped forty percent the following year and the division lost $539 million. The crisis worsened because companies had leveraged capital in anticipation of constantly escalating sales. By 1984 revenues had dropped to less than half of what they had been two years before. Coleco was heading into receivership and Mattel had to change management. Warner laid off several thousand

workers in 1984 and then sold off Atari, which split into two companies – Atari Games, which continued to make video games under
various owners until 2000, and Atari Corporation, which attempted
to make its way in the personal computer business. An image that was
to haunt the interactive game business for decades was that of millions
of unsold and unwanted game cartridges being bulldozed into New
Mexico landfill sites like the contaminated residues from some
unspeakable industrial accident.

CONCLUSION: BRIGHT, BRIEF FLARE?

The interactive game emerged from the social vectors of its time. It
was transformed from geek invention to billion-dollar industry over
the very decades – from the 1960s to the 1970s – that are usually
identified as crucial for the transition from a Fordist regime of mass
industrial production to an increasingly digitally centred post-Fordist
capitalism. Scholars such as David Harvey have gone so far as to
nominate 1973 as the year post-Fordist economics and culture began
to crystallize:[87] 1972 was the year Magnavox released the first commercial home video game console, and the year Atari's *Pong* became
an arcade sensation. The position of interactive gaming at the crest of
a wave of techno-cultural-economic change arose from a unique conjunction of institutional factors. The military subsidies for computer
research, the problem-solving playfulness of the hacker subculture, and
the risk-taking entrepreneurship of Silicon Valley whiz kids all coalesced in a moment of innovation. Once brought together through the
act of design, however, they defined the new trajectory of interactive
media for the rest of the century.

The actual shape taken by the industry appeared only through
uneven combined developments in all three of its emergent circuits of
technology, culture, and marketing. Although hacker-inventors such as
Russell had devised the core navigation, display, and control technologies in the early 1960s, it took a series of successive breakthroughs
to allow them to become the basis of a popular entertainment device.
Only rapid improvements in the processing power of computers
allowed the game to pass from the laboratory into the arcade, and
then into the home. Yet further increases in microchip performance
were necessary to permit home consoles that could play a variety of
cartridge-carried games. These speedy technological developments
were a spectacular early example of "Moore's Law," named after
Gordon Moore of Intel, which predicted that computer processing power,
or the level of data density, approximately doubles every eighteen
months – the digital propulsion of a perpetual-innovation economy.

The interactive game industry was one of the first beneficiaries of this process. But it is also important to note that the popularization of the interactive game only really occurred with the linkage of consoles to the display screen of the domestic television set. That step allowed the video game to enter the home on a big scale. But it did so as a hybrid technology, straddling the new post-Fordist world of digital devices and the old Fordist era of mass entertainment.

These technological dynamics reverberated with the nascent cultural logic of the industry. The themes and preoccupations of interactive entertainment were indelibly stamped by their emergence within a context of, first, military and space research, and then, more broadly, technocratic social and scientific planning. The hacker subcultures of those environments brought play to computer technology. But it was the play of an overwhelmingly masculine world, centred around themes of abstract puzzle solving, exploration, sport, and, centrally, war, a preoccupation constantly reinforced by the military's ongoing support for the new simulation technologies. The bias of virtual experience was reinforced by the video game's passage through the arcades, a popular entertainment site with a very different class composition from the laboratory but an equally strong masculine ethos. Such factors meant that when the video game did appear in the living-room – linked to the television and appealing strongly to a culture already habituated to screen amusement for children, with a powerful tradition of action-adventure genres – it would be immediately constructed as a "toy for the boys."

Within these gendered boundaries, however, the cultural circuit was still marked by considerable creativity. The initial game making of hackers established game genres that were energetically diversified and elaborated in commercial production. The appearance of independent software houses and the growing recognition of game making as a digital art – signalled by Crawford's aesthetic manifesto – were manifestations of a cultural exuberance inspired by the possibilities of interactivity. It is not for nothing that the late 1970s and early 1980s are known as electronic gaming's "Golden Age."

The marketing circuits of the new industry, however, were rudimentary – fatally so. Initially, electronic games companies, flush with the excitement of digital innovation, relied on the sheer novelty appeal of their product. For a while this worked triumphantly, generating what seemed a classic American success story in the rocketlike rise of Atari. But that success was self-destructive, as the market became saturated with competitors and mediocre product. When major toy and media conglomerates entered the field, producing the first major concentration of ownership in the business, they only worsened the problem.

Game development was subjected to the logic of corporate speed-up in a way that soured relations with the maverick software artists who were the industry's foundation, and games were marketed according to models that showed little or no understanding of the new digital subculture at which they were directed. In the great crash of 1983–84 the industry imploded thanks to its own perpetual-innovation dynamic, collapsing under a self-generated product glut that seemed to destroy all the dazzling promise of a few years before.

Our examination of the early years of interactive gaming shows how uncertain and tumultuous the moment of origin was: digital play might have taken many other paths. What if computer research had been funded not primarily by the Pentagon but by educational or medical institutions? What if digital games had not passed into the home through the masculinized ambience of the arcades? What if they had not blossomed until the home computer market matured; in other words, what if they had not been linked to television and built on the basis of a pre-existing mass entertainment market? Many of the features and themes of interactive gaming today might be very different from what we know. Above all, digital play came close to failing completely. We close this chapter at a point where the interactive game industry seems to have gone up and come down like a brilliant but quickly spent flare. It would take new agencies, a new approach, and considerable entrepreneurial nerve to try again.

5
Electronic Frontiers:
Branding the "Nintendo Generation"
1985–1990

INTRODUCTION: PLAYING WITH POWER

The revival of the North American market was the achievement of a company whose name has become practically synonymous with video games: Nintendo. No other corporation has so firmly put its stamp on the interactive gaming business; very few have as decisively altered the domestic and cultural habits of millions of children and families. In a series of bold gambles, Nintendo brought the video game back from the verge of extinction and then drove it to a new pinnacle of popularity and profitability. The phrase "Nintendo kids" designates an entire generation familiar with console and joystick.[1] The company did it with much help from brand marketing, and by learning how to manage global youth markets.

Nintendo's success was the result of a ruthlessly rationalized business strategy. They integrated innovation in hardware, creative software development, rigorous control in the quality and quantity of games, litigious protection of intellectual property rights, and a sophisticated marketing and monitoring apparatus. From 1985–90, this system gave it a quasi-monopolistic control over the video game market. The company's slogan, "Now you're playing with power," explicitly spoke to the technological mastery it promised youthful players but tacitly conveyed the corporate confidence of an enterprise that had remade and dominated the interactive games sector through much of the 1980s.

FROM THE EAST UNTO THE WEST … AND VICE VERSA

Nintendo was a Japanese company whose path to the home video game lay through traditional Japanese playing cards, novelty toys, and arcades.[2] In 1907 it began making western-style playing cards, adopting the name Nintendo (literally "leave luck to heaven") at the same time, and enjoyed an early success by adding Disney characters to them.[3] At its very beginnings, therefore, the company that was eventually to "invade" the US entertainment market had strong links to North American cultural industries and displayed a peculiarly globalized hybridization of Western and Eastern influences.

Nintendo and its president, Hiroshi Yamauchi, were something of outsiders to Japan's business world, with few connections to the great *keiretsu* (a network of businesses that own stakes in one another) that dominated the economy. But Nintendo's entry into the electronics age occurred at the very moment when the upper echelons of Japanese business and political leaders were looking to the concept of *joho shakai* – the information society – as a key to future economic growth, and when promises of "computopia" were vivid in the Japanese public imagination.[4] In the 1970s it released a hand-held computer game, then licensed Magnavox consoles to sell in Japan, and in the early 1980s began making arcade games.[5] In Japan in 1984 it released the machine on which the fortunes of the company would be built – a home video game console, the Nintendo Famicom (family computer).[6]

The Famicom was built around the same eight-bit microprocessor used by Atari games. But it had exceptionally good graphics and colour and was modestly priced. Its launch coincided with the opening of Tokyo Disneyland, an event many saw as marking a shift in Japanese popular culture from the compulsive work ethic of postwar reconstruction towards a greater interest in entertainment and leisure activities.[7] Very swiftly, Nintendo came to dominate the Japanese video game market. But despite this achievement, Nintendo doubted whether it could compete in North America. It was in the process of licensing the Famicom to Atari when the North American video game market plummeted towards zero.[8]

The crash annihilated Nintendo's major competitors but also destroyed all North American confidence in the viability of its product. Nintendo took an extraordinarily bold step and set out simultaneously to resurrect and capture the US market. In 1985 its US subsidiary, Nintendo of America (NOA), launched a modified version of the Famicom, the Nintendo Entertainment System, or NES. Retailers and market analysts were unanimously sceptical about Nintendo's prospects: "It would be easier," a toy industry executive reportedly told NOA president Minoru Arakawa, "to sell Popsicles in the Arctic."[9]

But determination paid off. Despite the apparent obliteration of electronic gaming, "the potential was still good, for a very high proportion of American youngsters had already played video games and so were prime targets for the revival. Indeed, within five years," the video game market was back up to "the 1982 peak level of $3 billion" – with Nintendo controlling eighty percent of it, and "a 20 percent share of the entire US toy market."[10] In 1988 the NES was the best-selling toy in North America; by 1990 one-third of American homes had one, and by 1992 the video game industry was flourishing again, passing five billion dollars in retail sales.[11] As the company made its vertiginous climb to success, however, Nintendo's managers were acutely aware of the lessons of the Atari disaster. The commercial system they developed was calculated to prevent repetition of such a catastrophe by using technological advantage to consolidate a near-monopolistic power, wielding that power rigidly to control the quantity and quality of game production, and unleashing an unprecedented intensity of promotional practice on interactive game culture.

PERFORMANCE PLAY: A WAR OF STANDARDS

Today, Nintendo's North American launch of the NES is cited by gurus of the networked economy as a model "performance play" – a strategy for breaking into a high-technology market by "introduction of a new, incompatible technology over which the vendor retains strong proprietary control."[12] Nintendo could not rely, as Atari had, on the sheer technological novelty of the game-playing experience. It had to offer a better virtual experience. The key factor was cheap microchips. The improved colour and graphics quality of the Famicom and NES was the product of special chip design. Usually, such enhanced performance means higher costs. But Nintendo executives determined that in order to meet their price target, the designer chips would have to be supplied at "a rock bottom price."[13] The only way to do this was to contract for an enormous quantity. Nintendo guaranteed the semiconductor giant Ricoh that it would buy three million chips over the space of two years, a promise that most competitors regarded as preposterous and suicidal. But the price/performance edge Nintendo won with this deal yielded staggering sales, first in Japan and later in North America. On the basis of its designer chips, Nintendo could offer a machine cheaper and better than anything left in the stable of the US video game industry.

The superior playing power of the NES was not its only important feature. Equally significant was its digital "lock-and-key" device. This consisted of two patented chips, one of which, the lock, resided on the console, while the key was on the cartridge. A copyrighted program known as "10NES" made the two chips communicate by singing a digital "song" to one another.[14] Only if the right song were sung would the key open the lock. It was thus impossible to play cartridges that had not been approved by Nintendo on the NES. Nintendo justified this device as a block to counterfeiters. However, "it stopped more than counterfeiters because no one could manufacture their own games for the NES without Nintendo's approval."[15] The "lock-out chip," as it was referred to in the industry, effectively enforced Nintendo's ability to control the software side of the video game business.[16]

In 1988 an Atari subsidiary, Tengen, tried to break the lock-and-key system. It resorted to a classic exercise in what hackers term "social engineering," relying not so much on technological ingenuity as human duplicity. Nintendo had deposited a copy of the program with the US copyright office. One of Atari's lawyers applied to the office for a copy, stating – falsely – that Atari needed the code only as evidence in an intellectual property lawsuit.[17] Using the information it obtained, Atari reverse-engineered the lock and key and produced unauthorized games for the NES. Nintendo sued. Atari rested its case on the copyright-law distinction between "ideas," which cannot be restricted, and "expression," which can, arguing that it had only copied the idea of the 10NES, not its specific expression. This disingenuous defence was soiled by the dishonest way the code had been cracked. In a decision celebrated as a victory for proponents of rigorous software protection, the judge ruled against Atari. Because Nintendo games were only playable on Nintendo hardware, the remaining pockets of US console production were inexorably wiped out as the NES won a standards war that left only one gaming hardware system standing.

RAZOR AND BLADES

The video game industry is a "razor and razor-blade" business.[18] The "razor" is the hardware – the console, the game playing system; the "blade" is the software – the cartridge containing the digitized game experience. Marsha Kinder suggests that Mattel introduced the "razor marketing theory" into the toy industry in 1959 with the endlessly renewed outfits for its Barbie doll.[19] But the approach was raised to a new height by the video game business. For a company such as Nintendo that produces an entire system comprising both hardware and software, profitability depends on innovation in, and complex interaction between, both the "razor" and the "blade."

Profits come primarily from software sales. Hardware is sold sometimes at a loss, sometimes at break-even point, at best with profit margins that are "razor-thin."[20] But although the immediate source of profits is from software sales, these sales in turn depend on the successful selling of hardware. To play Nintendo software presupposed owning a Nintendo hardware system. A company that dominates the console market stands to reap a steady harvest of software profits. A good hardware system, however, is not enough in and of itself to ensure competitive success, for if it does not support attractive games it will not sell.

Since the software "blade" constitutes the sharp edge of profit taking, it would seem logical for Nintendo to make all the software for its own machines. But in practice this was impossible. Software making is extremely expensive. Game development represents a huge investment. Even at the time of *Mario,* the process from conception to programming could cost well over $100,000. Moreover, software is a hit-and-miss affair. As in many other sectors of the entertainment industry, such as films and recorded music, a handful of successful products must pay for a large number of flops. Writing of the video games industry, Michael Hayes and Stuart Dinsey note that there are "chaotic conditions surrounding the software market and it is certainly true that the 80:20 rule applies to video games software. (This is where 20 per cent of all game titles published represents 80 per cent of all sales.)"[21] To cover itself against frequent failures, a software company must develop several games – all the more so because video game software has a short lifecycle: even successful games are reckoned to be obsolete within about six months.[22]

Without sufficient variety of high-quality games to attract consumers, the video game industry would die. But "no one company has the resources, in terms of either money or personnel, to monopolize software creation."[23] Therefore, even Nintendo, with the technical capacity to shut out rival manufacturers' games from its consoles, found it "prudent ... to encourage developers and publishers of software, who are not hardware competitive, to flourish."[24] Thus, the revived industry included a vital sector of software licensees who bought rights from Nintendo to make games playable on its machines. Nintendo produced only a handful of titles itself, including its famous "platform games," such as the *Mario* and *Zelda* series, which accounted for a huge part of its commercial success and cultural influence. But it relied on third-party suppliers – us companies such as Electronic Arts, Acclaim, and Data East and Japanese software makers like Konami, Capcom, and Namco – to fill in the volume. By the early 1990s there were more than one hundred companies legally making home video game software.

Relations between the hardware producers and software licensees have been one of the most contentious elements of the interactive games industry. For hardware companies, the ideal is to ensure that licensees produce games only for *their* systems. The aim is to create a momentum whereby a large customer base attracts many developers, whose games encourage console sales, thereby further building the customer base in a spiral of "increasing returns."[25] But multiple games carry with them the danger of Atari-style overproduction. Nintendo therefore steered between the Scylla of too few games and too few sales, and the Charibdis of too many games of low quality at too cheap a price.

To realize its goal, Nintendo imposed rigid conditions on licensees. It required all suppliers to submit games, packaging, artwork, and commercials for its seal of approval, limited the number of game titles a licensee could develop in any one year, and prohibited licensees from making games for other hardware systems for two years. The lock-and-key device functioned as a technological enforcement of near-monopolistic position, backed up by a formidable litigation apparatus that could be turned against competitors like Atari who dared to try and break the system.[26] Informal pressures backed up the contract arrangements; during microchip shortages, for example, Nintendo allocated manufacturing capacity to favoured suppliers.

This system allowed Nintendo to control the number and content of games circulating for play on its machines. It argued that such control was in the interests of the industry as a whole. But software licensees could hardly fail to notice that it enabled Nintendo to extract impressive profits, making about five dollars in royalties on each cartridge sale. And it did this regardless of whether the games sold, because the contracts stipulated that a minimum order of ten thousand cartridges be bought from Nintendo and paid for in advance.[27] Nintendo could afford to charge game developers for the right to put their games on its system, because "no individual game created by these developers was crucial to Nintendo, but access to the Nintendo installed base was soon critical to each of them."[28] With revenues streaming in from royalties and from sales of cartridges to licensees, as well as from sales of its in-house games and hardware, the company profited from every aspect of the game business, fulfilling president Hiroshi Yamauchi's declared business strategy: "one strong company and the rest weak."[29]

THE CORPORATE SOFT-WARS

As gaming technology became commercial, intellectual property rapidly became a major industry concern. Companies sought copyrights and patents to establish monopoly rights over their software and hardware innovations. Licensing rights to and from other companies – for programming tools, game codes, design features, characters, stories, and "just about everything else that could go into a compelling game" – became a crucial activity.[30] So did testing the strength of the competitors' control over their critical knowledge assets. Such testing might take the form of a court challenge or, more likely, an on-the-ground encroachment that would inevitably provoke a legal response. So lively was this aspect of the industry that in the early 1990s one Nintendo employee reportedly described his company's business as "games and litigation."[31]

One of the landmark cases involved the game *Donkey Kong*. The creation of Nintendo's foremost game maker, Shigeru Miyamoto, it presents a scenario in which the hero has to recapture his girlfriend from the clutches of a marauding ape. The name was intended to suggest a renegade primate, like King Kong, who was obstinate and foolish, like a donkey.[32] So successful was the game that in 1982 its sales in the US were worth one hundred million dollars. In the midst of this wave of revenue, company president Yamauchi received a telex from MCA Universal declaring that Nintendo "had forty-eight hours to turn over all profits from 'Donkey Kong' and to destroy any unsold games."[33] The Hollywood studio claimed the game infringed its copyright for the film *King Kong*. Nintendo fought the case. Miyamoto testified that he called the character "Kong" because "'King Kong' in Japanese was a generic term for any menacing ape."[34] The trial revealed that MCA Universal's claim was a bold bluff. Despite its aggressive stance, the studio did not own the copyright to *King Kong*, had not trademarked the name, and in the past had even argued successfully in court that the name was in the public domain. The case was appealed all the way to US Supreme Court, but eventually Nintendo won $1.8 million in damages.[35]

Nintendo was no champion of collective ownership of cultural icons. It had quickly secured copyright over its primate production. Even as it was being sued for copyright infringement by Universal, Nintendo was in the midst of a campaign against *Donkey Kong* pirates, hiring private detectives to track down bootleggers, pressuring US marshals to raid the offenders, and filing thirty-five copyright-infringement suits against individuals and companies. Despite these efforts, the company claimed in 1982 that it "lost at least $100 million in potential sales because of counterfeiters."[36] Nonetheless, *Donkey Kong* was the starting point for what was probably Nintendo's most successful intellectual property rights venture ever, since the game's hero proved to be the first incarnation of *Mario*, later to become the company's signature character. No one got away with any unlicensed films starring diminutive Italian-American plumbers.

DESIGNS ON CONSUMERS

The design of the games themselves was a crucial problem for Nintendo's launch into the US market. Would the Japanese games appeal to the US players? Would there be market resistance to a Japanese company? The Atari crash had taught the importance of design – good graphics, dynamic structure, exciting symbols – in a rough-and-tumble market. Nintendo was among the first video game companies to realize the commercial implications of Chris Crawford's earlier insights into software artistry, grasping that they were not selling just a new technology but a new experience and that their platform provides access to an imaginary world whose attractiveness was dependent on qualities that could be measured, evaluated, and, at least to a degree, predicted.

Nintendo set out to understand the "play value" of the games. In doing so it drew on the experience of arcade games. With greater storage and graphics capacity the arcade game makers had earlier been in a position to feature cool themes and characters and better backstories. Moreover, they had garnered invaluable experience from managing the profitability of their games. In the arcade, games that were too challenging were not attractive, because players had to invest too much money up front before they could achieve competence. But games that were not complex or challenging enough would either be conquered too quickly or become boring. These considerations taught arcade game makers to design games that could attract novices and sustain their interest but also deliver fresh experiences and appropriate goals to the *aficionado*. All these ideas were adopted by Nintendo and later by its archrival Sega, which grew its business back from the arcades into the domestic environment.

As it established itself in the US, Nintendo also developed a research-and-intelligence network that rivalled any in the consumer marketing industries. Its research program not only vetted games through expert panel evaluations but also tried them out on children. By 1993 1,200 kids played premarket games every week and rated them on the Nintendo evaluation instrument. The research extended previous taxonomies for game evaluation (i.e., challenge, fantasy, curiosity) by measuring player engagement along eight dimensions of design (graphics, sound, initial feel, play control, concept/story, excitement/thrills, lasting interest and challenge, overall engagement). The company then developed four evaluative dimensions that were critical to successful game development – "production values," such as good visuals; "fun themes," with strong characters, intriguing storylines, and attractive fantasy environments; "play control," that is, easy responsive interaction with console and screen; and "challenge," the ability to produce excitement and repeat-playability.

Although Nintendo only made a few of the games played on its machines, this "in-house" software was of the very highest quality. Competing development teams, working with ample time and technology and carefully recruited personnel, subjected their products to a rigorous vetting, testing, and evaluation process, the culmination of which was an assessment by an evaluation committee (the "Mario Club") made up of staff members and outside recruits. Games that got less than thirty out of a possible forty points were sent back for reworking or scrapped entirely. This screening reflected Yamauchi's creed that "boring software will be the death of both Nintendo and the entire video game market."[37]

The result was a series of spectacular successes, including the company's famous "platform games." This ambiguous term sometimes refers to a genre of games in which characters jump, spring, fly, or move through a series of levels, or "platforms," as they battle enemies and collect items, and sometimes to a game that is closely associated with a new console, launching the technological "platform" that will in turn support further software-title sales.[38] In practice, however, the two meanings often coincide: Nintendo, and later Sega and Sony, all used games in which whimsical protagonists overcome obstacles and puzzling situations as they progress through various levels of adventure as a sort of signature software strongly associated with their hardware systems. This is partly because such titles contain a wide mix of game-play elements – speed, dexterity, puzzle solving, fighting, identifiable characters and heroes – but also because "cutesy jump 'n run games" are relatively benign and escape the controversy associated with more violent and disturbing game genres.[39]

The development of high-calibre games that are strongly associated with hardware platforms such as the NES and frequently "bundled" together with them proved a recipe for success for Nintendo. In the late 1980s its in-house games accounted for twenty to thirty percent of the entire software market, making it the world's largest single game supplier.[40] The most famous of Nintendo's platform games was the series built around the character Mario, first seen in *Donkey Kong*, then reappearing, with his brother Luigi, in his own game, *Mario Bros.*, in which the pair battle attacking turtles in a sewer system. Like *Donkey Kong*, this was "essentially a jumping and climbing game as Mario (or Luigi in two-player games) had to jump over moving creatures and knock them out."[41] Nintendo followed up with *Super Mario Bros.*, which "took the basic jumping and climbing theme ... and transplanted the action to an entire scrolling world."[42] In 1985 Nintendo had packaged *Super Mario Bros.* with the Famicom in Japan, with great success. It then repeated the ploy in the US: "As the company had anticipated, many people bought the NES just to get the game."[43]

It was through the *Mario* games that Nintendo put its unique stamp on video game culture. While many earlier and later games – from *Spacewar* to *Doom* – obviously display their deep affiliation with military-industrial culture, *Mario* appears to be made of different stuff, a stuff of purer playfulness, wit, and humour. The designer of *Mario*, Miyamoto, described the game as an attempt to recreate childhood experiences, hiking and "stumbling on amazing things as I went." For Miyamoto, the most important goal in designing a game was to "create a kind of livability that makes people want to go back to it." One reason why kids "immediately head for the game machine when they come home from school is that the game provides them with an enjoyable, livable world. The music and tempo are likeable: you feel what fun the game is with your entire body."[44]

Emphasizing this quality of "livability," Katayama says, "The pace of the game, the rhythm of the music, the movement on the screen and the breathing of the player must all match."[45] He attributes the success of *Mario* to three factors. First, it introduced a sideways scrolling screen. Second, it "replaced distant, aerial long shots with a camera that was trained in on the main character. Third, it introduced a new sense of dynamic, thrilling, and flexible movement when Mario jumped, fell or leapt." These three elements, Katayama says, enabled the developers to create "a new 'game space,' a new 'information space.'"[46] It was clearly a space many children, not to mention adults, found extraordinarily pleasurable and exciting. Of the billion Nintendo cartridges sold by 1997, *Mario* games accounted for 120 million.

PESTER POWER: CLOSING THE LOOP

Nintendo's control of the video game industry depended on careful management not only of production but also of consumption. Though video games had already been introduced to consumers during the Atari boom, the subsequent bust meant the task had to be taken up again. In 1985 Nintendo spent "$30 million in advertising to convince retailers and consumers that their games were different."[47] It was under Nintendo's hegemony that the video game industry began to see the systematic development of a high-intensity marketing apparatus, involving "massive media budgets, ingenious event marketing, ground breaking advertising and spin-off merchandising."[48] According to one competitor, "Nintendo still would never have gained its enormous sales without phenomenal marketing – 'the kind that America had never seen before.'"[49]

Video game companies face the dilemma, common to other toy producers, that the main purchasers of their product are not the main consumers. Consumer profiles show that most video games are bought by parents, for their children.[50] In the 1980s the industry gradually realigned itself around the notion that its core market was youth. Boys between eight and fourteen years old were the regular and faithful users of NES. Nintendo refined their analysis of those loyal consumers and learned to communicate with them effectively. Hayes and Dinsey summarize Nintendo's tactics of mobilizing "pester-power": "there's little point in attempting to convince the parent to purchase something they neither understand nor approve of. Delicate negotiations such as these are best left to be expertly executed by the kids themselves. If a parent had been sold into really 'wanting' their kid to have a games console, the child would probably have at once assumed he was getting a video games equivalent to Clark's shoes or a 'sensible' school bag."[51] But since they actually paid for them, parents had to be reassured about the nature of interactive games. Nintendo was very careful, at least at first, to avoid the most violent or provocative games. Promoting the Nintendo brand as a family-oriented entertainment industry was central to the company's thinking about product and market development, from the Famicom name to the resources dedicated to *Donkey Kong, Mario*, and later *Zelda*. But it couldn't allow caution to negate its appeal to children's rebellion and independence. Creating the Nintendo image was thus an exquisite balancing act.

Nintendo eventually brought its marketing to a high art, but it took time. The in-house TV advertisements that accompanied Nintendo's invasion of the North American market in 1985 were amateurish: "Nintendo executives didn't know much about advertising a video-game system."[52] Later, Nintendo gave its account to major agencies, and the ads became more sophisticated. Because Nintendo produced relatively few products, it could lavish money on a handful of commercials, budgeting up to five million dollars for advertising, four or five times more than other companies spent. But it was only later, in the period of the Sega-Nintendo wars, that the advertising aspect of Nintendo's marketing was truly driven into high gear. Nintendo marketers focused initially on other avenues. They developed in-store "World of Nintendo" merchandising displays; sponsored video game competitions; established cosponsorships and cross-licensing arrangements with Pepsi, Tide, and McDonald's; and set up a network of over 250 fan clubs.[53]

As they developed, these promotional activities assumed an additional function – that of gathering information about the tastes and preferences of players. Such information could then be used to shape the development and presentation of new games, thereby "closing the loop" between production and consumption. One crucial lesson Nintendo learned from the arcades was the need for quick feedback on which games would have repeat play appeal. In the arcade business, as we noted earlier, one can get a strong sense of a game's popularity by emptying coins from the machines. When *Donkey Kong* was brought to the US, Nintendo managers knew immediately from assessing the "take" from a day's play in a selected location that, contrary to industry expectations, it would be a smash hit. This rudimentary exercise in market research proved to be only the first in a series of experiments by which Nintendo kept its diagnostic finger profitably positioned on the game player's pounding pulse. Two of the most important steps in this process were the creation of Nintendo's own magazine, *Nintendo Power*, and the establishment of its game-counselling phonelines.

One of the most striking signs of the growing presence of video gaming in North American culture was the explosion of magazines devoted to it. By far the most important of these was *Nintendo Power*, which by 1990 had become the biggest-selling magazine for children, with a paid circulation of two million in the US. The journal was essentially "one long Nintendo advertisement."[54] Children's culture critic, Henry Jenkins, describes it as "a kind of techno-porn: children spend hours ogling the fascinating places they might visit in a new game, eyeing the magazine's uncloaking of those secret sites to which they have so far failed to gain access."[55] Even Sheff, who is usually celebratory about Nintendo's achievements, says, "There was something bordering on the insidious about *Nintendo Power*."[56]

Subscribers paid fifteen dollars a year, a price that covered most costs. Any advertising revenues from Nintendo licensees were almost pure profit. All mention of games from potential rival companies could be excluded from what had become the main organ of gaming culture. Most importantly, the magazine "guaranteed that Nintendo was in touch with millions of its most dedicated customers."[57] This sort of access meant that Nintendo "didn't have to waste money developing hundreds of games. It could develop a select few each year and be all but guaranteed that the games would sell at least a set minimum amount."[58]

But *Nintendo Power* became more than a covert marketing arm. It provided not only advice but clues to "Easter eggs" that were planted in hidden rooms or levels. The practice of offering tips, clues, and "cheat codes" – that is, ways of "hacking" into the game and changing some of its properties – became especially important to the more advanced players. By fostering this sort of practice, *Nintendo Power* played a critical role in nurturing a gaming subculture with its own semiclandestine lore of codes, secrets, and expertise, a subculture that was at once a reliable market for Nintendo products and a source of information about the changing tastes of gamers.

Nintendo's other major marketing innovation, its phoneline help system, catered to players who were experiencing difficulty in operating their systems, or who were unable to crack the secrets necessary to progress to the next level of play. "Hundreds of game counsellors huddled in partitioned work spaces, each equipped with a Nintendo system and stacks of games, a computer terminal, notes, and 'green bibles,' bound volumes of game maps and secrets."[59] They handled fifty thousand calls a week, giving out esoteric game knowledge to kids over toll-free lines.

These services were costly. Nintendo of America reportedly spent fifteen to twenty million dollars annually on games counselling. But it was considered money well spent. Not only did Nintendo make money by charging for help calls but also, and more importantly, the phonelines, along with the magazines and other promotional activities, built a game culture. They "bonded players to the company."[60] As industry analyst Andrew Pargh put it: "If you call and get help, you feel good about your buying decision and you go out and buy more games."[61] According to Peter Main, Nintendo of America's vice-president of marketing, Nintendo had discovered that "more and better information whets players' appetites for more games and accessories."[62] But that was only part of the story. The other part was that the magazines and phone tips that grew out of the game culture enabled Nintendo to gather information about customers and then incorporate it into its development process.

This evaluation process became enormously sophisticated as magazine subscriptions and phone calls gave Nintendo a window into the minds of its customers. Main declared, "The phone system is really the closing of the loop in a fashion no other consumer company in this country has been able to do."[63] Counsellors not only answered questions but also asked them, finding out from callers their gaming likes and dislikes. "We used those calls," says Main, "as market research."[64] Sheff explains: "The information about consumers – not from dated market research studies but from the daily input of diehard customers – gave Nintendo a living, breathing line to its customers every day, seven days a week, twelve hours a day. The feedback helped steer the company's product development and marketing strategies; the information went right back into the development process."[65]

"JAPAN PANIC" AND "TECHNO-ORIENTALISM"

Through the technological marvels of its hardware, the creativity of its software design, the intensity of its marketing techniques, and the aggressiveness of its lawyers, Nintendo conquered North American children's culture. Indeed, in the late 1980s and early 1990s the image of occupation by a foreign power was likely to be understood literally – as part of a looming threat to the US from "Japan Inc." The growing economic power of Japan, the inroads made by its automobile and electronic companies on US domestic markets, and the size of the US trade deficit stimulated a wave of North American anxiety about the superior performance of Japanese workers, Japanese managers, Japanese industrial planning, and, in particular, Japanese high technology. Capturing first television sets, then VCRs, then semiconductors, and then the video game industry, the "Silicon Samurai" were marching relentlessly towards global domination.[66]

As David Morley and Kevin Robins point out, "Japan panic" reversed the traditional contrast between "advanced" West and "ancient" East. In an inversion of long-standing stereotypes, "technological East" was now contrasted with the "human West." Japan became "synonymous with the technologies of the future – with screens, networks, cybernetics, robotics, artificial intelligence, simulation."[67] Such "technomythology" mobilizes anxieties about "some kind of postmodern mutation of human experience," so that "within the political and cultural unconscious of the West, Japan [came] to exist as the figure of empty and dehumanised technological power ... This provokes both resentment and envy. The Japanese are unfeeling aliens; they are cyborgs and replicants. But there is also the sense that these mutants are now better adapted to survive the future."[68] These concerns were occasioned by real shifts in global patterns of production and trade, but they were charged with racism and paranoia, fuelled by media metaphors harking back to Pearl Harbor and the War in the Pacific, stirred by xenophobic films such as Ridley Scott's *Rising Sun* (1993) and *Black Rain* (1989), based on a Michael Crichton novel about corruption in the Japanese corporate world. The sentiments stirred by this heady brew coalesced around certain key moments that epitomized the United States' vulnerability to the threat from the East. Several of these moments involved the media sector – notably the Japanese "invasion" of Hollywood with Sony's purchase of Columbia Pictures in 1989, followed the next year by Matsushita's purchase of MCA Universal, events that crystallized anxieties that Japan was "stealing America's soul."[69]

Nintendo's revival of an industry that had originated in American innovation fed into these anxieties. For those who liked to complain about how Japan stole American ideas, improved them, and then sold them back, video gaming seemed an exemplary case. In 1992 Nintendo found itself at the centre of "Japan Panic." Company president Hiroshi Yamauchi purchased the Seattle Mariners baseball team. The sale was announced during the same week that a prominent Japanese legislator publicly suggested that US workers were "lazy."[70] It evoked a wave of protest. The baseball commissioner, Fay Vincent, rejected the deal on the grounds that baseball should not allow foreign – or at least non-North American – ownership. The issue was further inflamed when it resulted in the revival of earlier accusations that Nintendo, along with other Japanese companies operating in America, discriminated against blacks in hiring and promotion. These charges in turn fed into the discovery that "in the video-game industry – Nintendo's industry – many games imported from Japan had to be modified so that they would be acceptable in America; the villains in these games were frequently black or otherwise dark-skinned."[71] Such claims were sufficient for the State of Washington to suspend its practice of referring job seekers to Nintendo until it instituted an affirmative-action program.[72]

Such incidents created a political climate that was hostile to Nintendo. As Sheff observes, "Tactics that would have been called aggressive if used by an American company were viewed in some quarters as unscrupulous when the company was Japanese."[73] Japan bashing at least partially fuelled the investigations by congressional antitrust investigators and the Federal Trade Commission into licensing and price-fixing practices that plagued Nintendo from 1989 to 1992, and that eventually played a part in undermining its dominant position in the video game market.[74] But within a decade "Japan Panic" had dissipated. The abrupt decline in Japan's economy in the 1990s, the flourishing of American software companies, and the integration of companies such as Nintendo, Sega, and Sony with the multinational media giants based in the US rendered it obsolete. Today, the story of the video game industry appears not as a narrative of Japanese success over America but as the creation of a global media complex in which issues of national ownership are blurred by alliances, links, collaborations, and interdependencies at the production end, and by the global scope of the marketing campaigns at the consumption end.

CONCLUSION: MARIO AS COLONIZER

Nintendo's impact on children's culture in North America was immense, generating amazement and anxiety among educators, psychologists, and parents. To Henry Jenkins, his son's Nintendo console seemed like an extraterrestrial artefact surreptitiously inculcating an alien mode of thought – "x-logic." He wondered whether he should allow the machine such centrality in his son's life. "What did we really know about Nintendo anyway? Where did it come from and why do I have such difficulty understanding its particular version of 'x-logic'?"[75] As Nintendo games insinuated themselves into more and more households, controversy raged amongst adults about their negative and positive effects.

Most of the negative assessments centred on violence and sexism. In his book-length study of Nintendo, *Video Kids: Making Sense of Nintendo* (1991), Eugene Provenzo castigated the company for its games' obsession with narratives of conquest and their stereotypic representation of gender.[76] Other critics, however, said that focusing on violent plots and stereotyped characters missed a major element of children's gaming experience – that of simply investigating a fascinating virtual environment. Thus, Janet Murray argued that Nintendo gaming is substantially about "the pleasures of navigation" and "exploring an infinitely expandable space," pleasures akin to the sport of "orienteering" where players follow clues across a large and complex terrain.[77]

In our view, however, the most useful concept is neither "conquest" nor "exploration" but "colonization," a notion introduced by Jenkins and Fuller. Like Murray, they argue that analysis of violence and gender stereotypes misses the "central feature" of games like *Mario*, the "constant presentation of spectacular spaces" in landscapes that "dwarf characters who serve, in turn, primarily as vehicles for players to move through these remarkable places."[78] But they give this argument a new twist by looking at the frequent constructions of cyberspace or virtual worlds as an "electronic frontier" – a term that is redolent of the colonial exploration and conquest of the "New World."

There are similarities, say Jenkins and Fuller, between "the physical space navigated, mapped, and mastered by European voyagers and traders in the sixteenth and seventeenth centuries and the fictional, digitally projected space traversed, mapped and mastered by players of Nintendo video games."[79] This analogy reveals itself in the resemblance between the "travel writing" of the New World colonists, with their rambling narratives of endless travels, ceaseless wonders, and episodic combats, and the wandering, labyrinthine and open-ended experience of a game like *Mario*. Focusing on such platform games, Jenkins writes: "When I watch my son playing Nintendo, I watch him play the part of an explorer and a colonist, taking a harsh new world and bringing it under his symbolic control, and that story is strangely familiar ..."[80]

The crucial question Fuller and Jenkins raise is, "In this rediscovery of the New World, who is the colonizer and who the colonized?"[81] One answer is that the colonizer is Nintendo itself, and the colonized the children who play the games.[82] To understand this we need to turn for a moment to the issue of "branding." A term whose origins lie in the ranching and slavery practices of the terrestrial frontier days, branding now refers to the marketing processes by which corporations win the allegiance and identification of customers. These range through high-intensity media advertising campaigns, to the creation of logo-inscribed products, to the weaving of an enveloping cultural ambience based on complex networks of sponsorships, endorsements, and synergistic alliances. The aim is to implant in consumers an ongoing awareness of and identification with the branding corporation.

Some marketing specialists say frankly that the intention of branding is to "own" consumers: "in truth, mainstream customers like to be 'owned' – it simplifies their buying decisions, improves the quality and lowers the cost of whole product ownership, and provides security that the vendor is here to stay."[83] It is generally recognized that branding is particularly powerful, and profitable, in its effects on young people. As Steven Miles writes, "their experience as consumers provides young people with the only meaningful role available to them, during a period in their lives in which they are struggling to come to terms with exactly who it is they are."[84]

Nintendo was the first company to "brand" the video game market. Indeed, the creation of the "Nintendo Generation" was one of the earliest and most successful extensions of branding practices, which were already familiar in the world of clothing, toys, and cigarettes, to the world of high-technology digital entertainment. Bill Green and his coauthors argue that the branding of a Nintendo Generation "is a register of the power of advertising in popular culture, as well as evoking the corporate contexts in and through which such a formulation comes into common usage."[85] The creation of the Nintendo brand represented the conjunction of innovative technology, a carefully structured system of game design, development, and licensing, and an enormously sophisticated marketing and intelligence apparatus. For some five years, Nintendo was synonymous with video gaming, just as Xerox had once meant photocopiers and Hoover had meant vacuum cleaners. It achieved this recognition on the basis of an undoubted technological and organizational superiority over its few competitors. It then created a net of cultural events, allusions, slogans, and spin-offs that as completely as possible absorbed the attention, and the purchasing power, of game players.

Nowhere was this clearer than in the character marketing of *Mario*. Here Nintendo's strategy was to "build the videogame character into a celebrity à la Disney, and license, license, license."[86] In the early 1990s Mario had greater name recognition amongst US children than Mickey

Mouse. He became the basis of a full-length film, *The Wizard*. By the early 1990s there were two successful *Super Mario Brothers* TV shows, one produced for the Fox channel that came to rank fourth on nationally syndicated children's programs, and an NBC program that became the top-rated Saturday morning cartoon for six- to eleven-year-olds.

This transition to mainstream media is hardly surprising, given Nintendo's historical connections to Disney. J.C. Herz suggests that when it launched its own "cute, round, colorful arcade hero," it had "one eye constantly trained on the Magic Kingdom."[87] Herz quotes Peter Main, Nintendo's vice-president of marketing, as saying: "Well, Disney has a lot more experience at this and that's why I unabashedly say we don't care where we borrow our smarts from. Disney's been through it. We try and learn everything we can from publications, conversations, licensing arrangements which we've been involved with them on in the past. We look to them for learning."[88]

Where this was most apparent was in the realm of "spin-offs." Mario became the basis for scores of agreements for T-shirts, sweatshirts, shoes, bicycle helmets, comic books, novels, stickers, and peel-off tattoos. The agreements also cover adult clothing, ceramics – lamps, mugs, cookie jars – and foodstuffs: candy, frozen pudding, soft drinks, cake decorations, and snacks. To preserve the value of such arrangements, Nintendo made elaborate regulations as to how Mario and his fellow characters could be represented in public appearances. These defined what costumes they could appear in, what they could and could not be associated with (no alcohol or tobacco products, no direct sale), and what they could do (no speaking or touching); the regulations also stipulated a specific range of authorized poses: as Herz observes, these characters were Nintendo's "crown jewels" and had to be defended accordingly.[89]

With these branding practices in mind, we can return to Fuller and Jenkins's "colonization" metaphor. What does it mean to say that Nintendo colonizes its child players? It means that the child's attention, time, desires, ambitions, and fantasies become attached to the Nintendo world, from which he or she derives not only the immediate pleasure (and frustration) of gameplay but also an array of metaphors, narratives, and codes for the interpretation of life, and often a whole range of social activities – contests, conversations, clubs, etc. Minds, bodies, and social interaction are thus increasingly "occupied" by Nintendo activities and purchases.

This process may be developmental, fostering certain digital aptitudes: colonialism often develops the territories it occupies, building roads or mines, implanting into the inhabitants new skills and literacies while suppressing or obliterating others. But it is above all profitable for the colonizer. In the business of gaming, sensational virtual aggression and simple navigational pleasures are both only means to an end, which is that of converting children's attention and time into a resource that fuels a capitalist enterprise, and of claiming that attention and time over any competing or alternative use. Thus, the microworld colonizer – the player who enjoys exploring the virtual territory of the game – is also the real-world colonized whose imagination is increasingly occupied and shaped into a source of profits for Nintendo. Through identification with Mario the conqueror/explorer/colonist, the player is annexed/invaded/occupied/ by Nintendo, the company for whom Mario has become conquistador.

Fuller and Jenkins describe the experience of playing Nintendo games as "a constant struggle for possession of desirable spaces, the ever shifting and unstable frontier between controlled and uncontrolled space, the need to venture onto unmapped terrain and to confront its primitive inhabitants."[90] This could well be an account of how a video game business is made. As we saw, Nintendo's success was carved out in fluctuating and uncertain market conditions, where every device of magic (read technology) had to be used to "confront primitive inhabitants" (read child consumers), and every sort of digital charm and tricky entrepreneurial move deployed to dominate the markets of the New World. The result, for a while, was powerful brand hegemony over new territories of digital culture. But this dominance was not immutable. Eventually the dynamics of technological innovation, globalization, and high-intensity marketing that Nintendo itself had largely perfected exploded its supremacy and brought a new burst of feverish turbulence to the industry.

6

Mortal Kombats: Console Wars and Computer Revolutions 1990–1995

INTRODUCTION: MARTIAL MOVES

For some five years, Nintendo's command of the home video game business seemed unassailable. But in the early 1990s a challenger armed with faster technology, riskier games, and more aggressive marketing appeared. The contest between Sega and Nintendo revolutionized video gaming, propelling new extremes of technical innovation, marketing intensity, and cultural audacity and opening the way for other contenders. The suddenly competitive interactive entertainment market of the mid-1990s was often compared to one of the era's most controversial games, *Mortal Kombat*, in which a group of martial arts experts strive to finish their adversaries with gruesome "fatality moves." In the corporate warfare of the digital game business some contenders suffered nasty ends, others triumphed spectacularly. Sega, after teetering on the verge of triumph, fell to the very forces of innovation it had unleashed. Nintendo, after reeling back in disarray, managed a remarkable recovery. Several other would-be video game console makers perished. Meanwhile, behind the console wars the video games' long-lost digital cousin, the personal computer, made an aggressive reappearance as a medium for interactive gaming. Regardless of the successes and failures of individual companies, however, the net effect was intensification and acceleration in the technological, cultural, and marketing circuits of the game business.

SONIC GOES HUNTING: "THE REVOLUTION STRATEGY"

Sega, like Nintendo, was a Japanese company, but very Westernized – globalized, in fact, before the term became fashionable. In 1954 an

American, David Rosen, started Rosen Enterprises in Japan and then merged in 1965 with its only rival, Service Games, to become Sega Enterprises, a Japanese company focused on arcade games. In 1969 it was bought by the US communication conglomerate Gulf and Western Industries. In 1984, however, following the Atari collapse, Gulf and Western divested itself of its electronic games holdings, and Sega returned to Japanese hands in a management buy-out.[1]

Building out from its arcade business, Sega ventured into home video games in the early 1980s in Japan, where it was eclipsed by Nintendo, though it enjoyed significant success in Europe. Full hostilities did not break out until Sega entered the North American market in 1989. The immediate focus of its attack was Nintendo's obvious weakness – technological stagnation. Having so heavily invested in the 8-bit NES system and built such an extensive library of games for it, Nintendo was reluctant to innovate and vulnerable to being technologically leapfrogged. Sega followed the logic of the perpetual-innovation economy and took advantage of this opening by producing the 16-bit "Sega Genesis," with microprocessors that were superior to the 8-bit NES, creating bigger animated characters, better backgrounds, faster play, and higher-quality sound.

This was a prime example of what Carl Shapiro and Hall Varian call the "revolution strategy" of technological innovation, an exercise in "brute force" whose principle is to "offer a product so much better than what people are using that enough users will bear the pain of switching to it."[2] The strategy is risky. It needs a new technology whose superiority to existing products is absolutely manifest. Since it aims at establishing a new industry standard, it cannot work on a small scale and usually requires powerful allies. To break into the market quickly, Sega sold its console cheaply, at a price comparable to the "obsolete" but established Nintendo system. Sega's strategy was aided by the many potential young customers who had not yet invested in video game technology; as Shapiro and Varian note, "there is a new crop of ten-year-old boys every year who are skilled in convincing Mom and Dad that they just have to get the system with the coolest new games or graphics."[3] Nonetheless, Sega had made a big gamble: "It is devilishly difficult to tell early on whether your technology will take off or crash and burn."[4]

The Genesis was not an instant success, largely because of a shortage of interesting games. Nintendo's draconian licensing agreements gave it a grip over the software supply. Few developers were prepared to support a rival and risk retaliation that could shut them out of an established console base and push them off store shelves. Breaking this embargo was a major problem for Sega, but it got its opening from the US government. The US-Japanese trade wars and the intellectual property battles between Nintendo and Atari led to antitrust investigations into Nintendo's monopolistic practices. In 1990 Nintendo was compelled to relax its licensing agreements.

Sega leaped at this opportunity to recruit developers who were already resentful of Nintendo's strong-arm practices. It charged only a "nominal fee to create titles for their new platform."[5] By 1992 nearly all Nintendo's third-party suppliers in the US had gone over to Sega.[6] One important defector was Electronic Arts, the leading independent game software publisher. It helped make the Genesis a success with a new genre of sports games – *John Madden Football* and NHL *Hockey* – based on actual players and teams, using the voices of well-known commentators and including features such as slow-motion replays. But the breakthrough for Sega was the discovery of an appropriate platform game, one that would give it brand identity and a signature character who could take on Mario. *Sonic the Hedgehog*, featuring the adventures of a speedy blue hedgehog "with attitude," was introduced in 1991. It featured animation that was faster and more richly detailed than anything put out by Nintendo. With the arrival of *Sonic*, the Genesis had a family hit and, crucially, a character-marketing device with which to fight Nintendo. The console wars had truly begun.

"IN YER FACE" MARKETING: THE ENGINEERING OF HYPE

Sega not only made good games and consoles; it also galvanized the gaming industry's marketing circuit. The head of Sega of America was Thomas J. Kalinske, an advertising *aficionado* who had previously worked in the toy industry – as the president of Mattel.[7] In 1991 Kalinske convinced Japanese headquarters of the importance of "aggressive marketing" in the US.[8] As one of his marketing researchers put it: "We knew that we would have to make Sega a cultural phenomenon if we were going to beat Nintendo."[9] Sega's marketers made a crucial decision: target older children.[10] Nintendo had defined the primary video game market as made up of eight- to fourteen-year-old boys. Sega pitched to fifteen- to seventeen-year-old boys. This opened a previously untouched demographic sector. Just as important, selling to this "mature" market would draw younger boys who were keen to emulate their elders, persuading the Nintendo Generation to switch loyalty. In this way, Sega aimed to "captured the high ground" that gave command over the entire gaming market.[11]

To do this, Sega launched an "in yer face" advertising campaign to take on Nintendo, smearing its rival as infantile and boring and building its own "cool" image. This campaign marked an extraordinary acceleration in symbolic production. Sega spent ten million dollars to launch Genesis under the slogan "Sega Genesis Does What Nintendon't." Over Christmas 1992 it invested forty-five million dollars in an unprecedented serial advertising campaign, producing thirty-five commercials in less than four months.[12] The new advertisements "debuted not during Saturday morning cartoons, but on the MTV Music Awards."[13] In format they were like music videos, capturing the "mores of a culture hooked on visual images, an impatient culture that absorbs and processes information literally in four-frame riffs."[14] They included baseball players getting hit in the groin, hockey games turning into ballets, and elves turning on Santa Claus. They made advertising video games entertaining, and they turned the problem of media-savvy youth into its own solution through an ad style based on irony and twisted humour, often at the expense of its rival, Nintendo.

Sega's advertising, aimed at male gamers, was "all about testosterone."[15] It made rebellious attitude and adolescent sexuality constant reference points. Its 1991 "Jimmy" campaign in the UK starred a hero who had "a life" as well as remarkable games skills, "even to the point of being able to 'get the girl.'" Other ads were darker but wittier. In a 1994 *Street Fighter II* commercial, a geeky teenage male sits, stooped over the edge of his bed, unable to keep still, groaning and grunting as he frenetically shakes his body. Listening outside his locked bedroom door, Mom is getting worried. Holding the laundry in her arms, she tries to lure him out, saying in mildly seductive and anxious tone: "The new lingerie catalogue is here." Then she threatens, "Mommy's got a gun." Immersed in the fighting on his game screen, the boy shouts back: "Go away!" Mom gives up, murmuring as she walks away, "I think I saw a chainsaw in the garage?" The advertisement gives an exaggerated visualization of the oft-made remark that "video games provide teenage boys with a form of electronic masturbation."[16] But the ad ironically transforms this familiar disparagement into a source of pride and self-identity, using twisted humour to vindicate the intense isolated involvement of the compulsive gamer.

The "trademark yell" of "SEGA" made brand competition the central dynamic of the video game industry. Faced with a promotional onslaught, Nintendo was compelled to respond. Both parties poured vast sums into advertising campaigns, and "costs spiralled expensively upward as each side vied with the other in the volume and originality of their sales pitch."[17]

Sega and Nintendo each defined a characteristic corporate "style" for themselves. Sega's was "flamboyant, off the wall, even anarchic," Nintendo "rounded, reliable, safe, and, perhaps most damning of all in

such a lively business, predictable."[18] Sega self-consciously developed a
brand personality, while Nintendo presented itself as "product focused,"
less concerned with form than content – a stance that was in itself an
image-conscious selection of corporate personae.[19] But there was also
"common marketing ground" between the competitors. Both used their
"famous, immensely marketable, mascots – Mario and Sonic"; both
focused heavily on TV but also used many other promotional tactics;
and both targeted boys and adolescent males in their advertising, using
ninety-five percent of their marketing funds to influence teenagers.

In the short run, Sega was the winner. It set the trend in marketing
budgets and symbolic values, bringing absurd scenarios, intense emo-
tions, mordant humour and pop culture icons into video game advertising. Even today, "it's almost impossible to find a videogame ad that
doesn't owe something to Sega's shock-tactics marketing innovations."[20] But being an advertising trendsetter also brings problems. Sega
milked the marketing approach it started with the Genesis even further
with its subsequent consoles but did not account for how quickly its
strategy would be exhausted. Marketers didn't plot a symbolic amor-
tization schedule, or factor the relationship between product scheduling
and symbolic obsolescence. The level of intensity and humour in Sega's
early ads set a standard that was hard to beat – even by Sega itself.
The accelerated style of video game marketing demands that codes for
difference are always fresh. As competition intensified and as other
companies emulated "shock" advertising, Sega lost its ace in the hole.
Moreover, it flooded the market with so many of its own systems –
Game Gear, Genesis, 32-x, and later, the Saturn – that it had difficulty
not only differentiating itself from competitors but also making distinctions between its own commodities. It out-trumped itself. Once the
initial impact of its cultural blitzkrieg was exhausted, Sega had no way
to prevent emulation of its tactics by competitors with deeper pockets
and greater organizational resources, competitors who would prove
perfectly capable of retro-engineering its stylistic flair.

BLOOD CODES: INTENSIFYING GAME VIOLENCE

In the meantime, Sega reinforced its appeal to teenage males with another
crucial innovation, this one in the cultural circuit of game development:
it introduced a new level of violence into mainstream video game experience. As the journalist Bill Moyers observes, what drives representations of violence in television, videos, and other media is "the bottom
line ... if you want juice, money, power, ratings, use violence because
violence sells."[21] But violence does not simply sell; it sells to specific
audiences with an acculturated preference for such fare – specifically,

teenage boys and young men. Upping game violence was integral to Sega's repositioning in the market, and to the product-differentiation strategy it deployed against Nintendo, rejecting the "kiddie" orientation of its rival and marking its own games as excitingly transgressive.

Sega's roots in the arcade business placed it closer to some of the more violent game genres than its competitor. Two Sega games in particular, *Night Trap* and *Mortal Kombat*, became a focus for controversy about the "video nasties." *Night Trap* involved a vampire hunt in a sorority house. Failure to defeat the vampires resulted in young women being hung from meat hooks and drilled to extract blood – scenes recorded in "live action" video footage. The misogyny of these images and execrable acting in the videos made *Night Trap* notorious – though but for its role in public debates about video game violence, it would probably have sunk without a trace in the marketplace.

Mortal Kombat, a fighting game that brought arcade-quality graphics, two-person combat controls, and luridly detailed lethality to the home game system, was a different proposition. Its reputation rode on notorious "fatality moves" – ripping an opponent's still-beating heart from his chest or tearing off his head and holding it aloft, spinal cord dangling. In 1993 the developer Acclaim released home editions of the game for both Nintendo and Sega platforms. The Nintendo versions were significantly sanitized – sweat substituted for blood, the most gruesome moves removed. At first sight, the versions shipped for the Sega Genesis and Game Gear appeared to have been even more effectively neutered, "so innocent that the warriors couldn't even sweat, let alone bleed."[22] But it was soon discovered that players could enter "secret" codes that restored the gore of the arcade version, including the notorious spine-ripping sequence. Game magazines and newspaper articles all printed the "blood codes," which were soon "in the hands of every potential buyer."[23] The Sega version outsold Nintendo's two to one.

This result was what Marsha Kinder terms a "spiralling escalation of competitive violence."[24] Sega's advertisements made the point that its *Mortal Kombat* was more violent than Nintendo's *Streetfighter* series. Nintendo responded by condemning Sega but released later versions of *Kombat* for its own consoles that restored most of the bloodletting. Sega replied by releasing *Mortal Kombat 2* with its new "fatality moves" – ripping the torso in two, bisecting it vertically with a buzz saw, sucking in an opponent's body and spitting out bones, or a lethal kiss that explodes the body (these last for *Kombat*'s female warriors). It grossed a record fifty million dollars in its first week in the stores.[25]

Producing violent games involves juggling two simultaneous concerns: consolidating a stable target market while staving off public criticism that could damage sales. In playing the violence card, Sega gambled that it could ride out public outcries, turning them to its advantage to build an outlaw pitch to teenagers, while maintaining an official face that was sufficiently responsible to avoid complete stigmatization by the adults who, in the last resort, controlled the purse strings. Nintendo's response was equally ambivalent. Overtly, it condemned Sega's violent games and emphasized its own "family values" image. Tacitly it boosted the violence content of its own software. Issues of violence, rating, and censorship became inextricably entwined in the struggle between the two companies and were exploited by both as weapons for strategic advantage.

The early 1980s had seen a spate of concern by parents, educators, and journalists about the effects of arcade video games, particularly violent games, on children and adolescents. Suspicion about the "addictive" nature of video games and their violent content were mixed with very traditional anxieties about arcades as uncontrolled spaces in which young people congregated – concerns that, as some sceptics pointed out, mirrored earlier "moral panics" about the effects of, say, jukeboxes in the 1950s.[26] Some US cities banned video arcades altogether, while others "passed laws restricting arcade hours for minors."[27]

As the games migrated from the arcade into the home, the concern about violence followed and intensified. On 27 September 1993 the cover story in *Time* magazine was "Attack of the Video Games." The violence-inducing qualities of games became themes in arenas ranging from television detective shows to public defenders' offices.[28] In California state legislation required that violent games be stored out of reach of minors and would allow the victims of violent crimes to sue developers "if it was proven that the game directly led to the violent attack."[29] In 1993 two US Senators, Joseph Lieberman of Connecticut and Herbert Kohl of Wisconsin, launched an investigation and proposed a bill that would impose a rating system unless the industry developed its own within a year.[30]

As the controversy intensified, Sega and Nintendo both tried to spin it to their advantage. Sega's CEO admitted that his company was developing games to appeal to different age groups, announced that it would rate its own games for violent content, and suggested that other video game manufacturers adopt similar codes. Nintendo replied that it didn't need a rating system since "every one of its games was rated G and appropriate for all ages," and the company strove to maximize public criticism of Sega over the violence issue.[31] At a 1993 Senate press conference about violent video games, the crowd was shown an exceptionally gruesome scene allegedly taken from Sega's home edition of *Mortal Kombat:* the clip actually came from the arcade version and was supplied, it was later revealed, by Nintendo. But as it insisted on the innocence of its own products, Nintendo sent new guidelines to developers relaxing its proscriptions of violent content, and it partnered with WMS, the parent company of the arcade game company Midway, to produce a gory futuristic fighting game, *Killer Instinct*, which, once released for the home platform, became one of the fastest-selling games of 1995.

By now, however, the video game industry had enough corporate players to manifest a certain collective self-interest in transcending the strategies of its two duelling giants. The prospect of the Lieberman–Kohl bill spurred game makers to organize voluntary rating codes. Indeed, the industry came up with not one but two systems. The first was issued by IDSA (the Interactive Digital Software Association), an organization formed by Acclaim, Electronic Arts, Konami, Nintendo, Sega, Sony, Viacom New Media, and Virgin Interactive. Its Entertainment Software Ratings Board (ESRB) slotted games into a five-category scale corresponding to the film industry's G, PG, PG-13, R, and NC-17 movie ratings.[32] Video game developers generally accepted the ESRB system, but computer game companies formed an alternative body, the Recreational Software Advisory Council (RSAC). Its rating system was based on a self-administered assessment based on variables that included levels of violence, sex, and offensive language. By November of 1994, "both rating systems were in place yet neither was a standard ... retailers didn't care which system was utilized either as long as the games were rated."[33]

These moves averted a government-imposed rating system – or worse, attempts to outlaw violent games. Controversy had been at least temporarily calmed. The industry systems were not mandatory, although to sell in almost any major retail outlet required "some sort of content labelling or rating label."[34] They were also very forgiving. For example, the RSAC rating algorithms have "built-in exemptions" for "strategic aggression" and "sports games" that let wargames or boxing games off relatively lightly.[35]

Whether such systems inhibit violent content or simply legitimize it is arguable. Although ratings may sometimes influence developers to tone down product, their net effect on curbing the "spiralling escalation of competitive violence" is dubious; 1993's *Mortal Kombat* seems positively tame compared to 2000's *Quake III*, *Soldier of Fortune*, or *Kingpin*. "Bad" ratings can simply become a component in the marketing strategy for violent games. *Night Trap*, unfortunately, reminds us of how authoritative disapproval can be transformed into a selling point. Pulled from the market by Sega, it later reappeared as a computer game advertised in *Wired* magazine with a double-page spread of the Stars and Stripes and the caption "Some members of Congress tried to ban *Night Trap* for being sexist and offensive to women (Hey. They ought to know)."

Sega's assault on Nintendo also signalled a new stage in the internationalization of the video game market. In video gaming, as in many other media industries, the high cost of production (developing a game) compared with the low costs of reproducing subsequent copies (making a cartridge or disk) and the high proportion of misses to hits place a premium on maximizing sales, and on expanding into new markets. Moreover, as the industry attracted increasing public attention, fear of governmental regulation (of violent content) or intervention in trade practices ("Japan Panic") gave companies an incentive to diversify their markets.

Nintendo's invasion and revival of the American market had shaped the video game business on a strong Japanese-US axis. The next logical area for expanding video game sales was in the third sector of the developed world – Europe. Nintendo had enjoyed market shares of over ninety percent in Japan and eighty-five percent in the US during the early 1990s. But it only ever took forty to sixty percent of the European market, which it considered "a secondary market opportunity."[36] In the 1990s it started to alter its attitude, partly in response to US congressional inquiries into its licensing and sales practices. But Sega was there ahead of it.

Nintendo had entrusted sale of its products in Europe to third-party distributors such as Mattel and Bandai – toy companies with a limited knowledge of video game markets whose lack of expertise was reflected in limited returns, which in turn diminished their enthusiasm for games, producing "a downward spiral of mounting disinterest."[37] It was not until 1991 that Nintendo set up its first European subsidiary in Germany, and it did not follow in the UK and France until 1993. Sega, by contrast, had already bought back a well-run distribution system from Virgin Communication and was operating as Sega Europe from the middle of 1991. Sega's sales in Europe hit $40 million by 1990, $100 million in 1991, $250 million in 1992, and $600 million in 1993.[38]

Competition in the European market in many ways set the pace for development in other parts of the world. This was particularly the case in the UK, where Sega established its European headquarters, and where many of its most audacious marketing strategies originated.[39] The success of Sega's European campaign was largely attributable to the

relative autonomy given to its managers, compared to the more autocratic style of Nintendo, which "exercised control over the UK from
Kyoto via Seattle."[40] Without success in Europe, it is doubtful whether
Sega would ever have been able to take on Nintendo globally. It was
in Europe that Sega was able to build momentum, both by creating a
big-enough hardware market to attract software developers, and by
finding innovative marketers who would forge the promotional strategies so crucial to its success. Nintendo found itself not only leap-
frogged technologically and outclassed in marketing but also outflanked
on an increasingly transnationalized corporate battlefield.

NURTURING THE MANIFESTATIONS: RENTALS, CHANNELS, PARKS

While Nintendo had energetically developed various forms of cross-
promotion and corporate alliance in selling games, its focus nevertheless remained on the sale of in-home consoles and games. Sega, in
contrast, adopted a much more experimental approach to finding new
locales and new ways of marketing its products. So strong was
Nintendo's grip on standard retail outlets that Sega had to search for
alternatives. Sega's history as an arcade-game maker made it alert to
opportunities other than standard across-the-counter sales. It regarded
its arcade interests not only as a revenue source but also as a research-
and-development laboratory for applications deployed at other sites.
Doug Glen, vice-president of business development and strategic planning for Sega of America, said: "We look at it as a continuum of experience. You can leverage the asset from the highest end down to the
mass-market – the home. We get into the leading edge and nurture the
experience throughout the various manifestations."[41]

Sega sought the "leading edge" on multiple fronts. It allied with
Blockbuster Video, enabling customers to rent games just as they
rented videos: only once the success of this venture had been demonstrated did Nintendo make a similar move. It explored marketing
games by cable TV, partnering with two cable giants, Time Warner and
Telecommunications Inc. (TCI) to create a channel that made a library
of Sega titles available for downloading with the help of a special
"tuner/decoder" cartridge plugged into the Genesis.[42] It was also one
of the first serious pioneers of what was to become "online gaming."
In joint projects with AT&T it investigated the possibility of connecting
Genesis consoles through the phone. AT&T eventually backed out of
the deal, but the project gave birth to one of the first independent
Internet gaming companies, Catapult. Sega sold video game systems
for use in airplane flights, allied with Cineplex Odeon movie chains

to open interactive entertainment centres in cinemas, and partnered with Toshiba to bring games into Japanese hotel rooms.[43]

The most substantial of Sega's out-of-home projects was its investment in video-game-based amusement parks. This was a logical extension of its arcade heritage, which was now to be blended with the all-powerful entertainment-industry example of Disneyland. As Herman points out, "Sega needed to open such a venue in order to get its newest arcade games played."[44] A game such as the driving simulator, *Virtua Racer*, cost about $18,000 – well beyond the means of a typical arcade owner. Only an arcade owned by Sega, and one with a very high through-flow of players, could profit from such a machine. In their design, the Sega theme parks were carefully distanced from the "seedy" image of traditional games arcades. The places were clean and well lit, with "Game Specialists" monitoring the floor. These centres featured simulators, "interactive" rides, and the newest games: visitors could play *Sega Rally* in an imitation racing car that tilted and shook according to what was happening on screen, or wear a "virtual reality" Mega Visor display projecting images that changed according to the direction in which the wearer was looking. Sega later formed an alliance with DreamWorks and MCA Universal to develop one hundred and fifty interactive entertainment centers in North America by the year 2000; one of these is analysed in detail in chapter 10. The Sega "Game-Works" parks were based on the Joyopolis theme park in Tokyo. The largest was to be a 50,000-square-foot centre in Las Vegas. The first opened in 1997 in Seattle, carrying the war into the American hometown of Sega's rival, Nintendo.

Sega's strategy of generating multiple "manifestations" of its games was at once daringly experimental and dangerously scattered. It pioneered new channels and cross-connections that rapidly became crucial to the whole industry. But many believed that such variegation led Sega to a debilitating overextension and loss of focus. For the console wars were now expanding, and it found itself pitted against new combatants.

REACHING FOR THE NEXT GENERATION: ATARI AND 3DO

Console video gaming was now defined as a "perpetual innovation" business, and its pace was accelerating. Sega had overtaken Nintendo by introducing the 16-bit console. Hardly had this established itself on the market before the eyes of the industry turned to the next logical technological revolution – a still more advanced 32-bit game-playing machine, the so-called "next generation" console. With the time lag between new product cycles diminishing, developing a new console, determining the timing of its launch, building attractive game libraries, and creating the requisite marketing buzz were to prove frantically exhausting missions.

Just how exhausting was discovered by two companies that leaped into the video game ring, hoping to repeat against Sega the same "revolution strategy" that Sega had effected against Nintendo. One was an old combatant, Atari, attempting a comeback. Another was a newcomer, Trip Hawkins's 3DO. High hopes attended these companies, but both were to flare up briefly and as quickly burn away in what was now a blazingly competitive market. Hawkins was already a legendary figure in the gaming world. A Harvard MBA, he began his career as a marketing executive at Apple and then in the 1980s started Electronic Arts, which rapidly became the largest of the third-party video game publishers that developed and marketed games for a variety of console systems. Riding high on this success, Hawkins in 1992 started a new company, 3DO, with the ambition to "wrest gaming from the controlling hands of corporate giants, like Nintendo and Sega."[45]

These hopes rested on a new console, the REAL 3DO Multiplayer, a CD-storage 32-bit game-playing machine, released in 1994. Hawkins did not plan to manufacture the device. He licensed production to a consortium of developers and received financial support from firms like Panasonic and Sony, supported by a "strategic alliance" of investors that included AT&T and Time Warner.[46] Backed by these impressive names 3DO's initial public offering generated skyrocketing stock values. The confidence was short-lived. The key to 3DO's strategy was to beat Nintendo and Sega to the punch in marketing a next-generation 32-bit machine. Hawkins insisted that the superior performance of his console would be decisive. Surprisingly, given his experience with Electronic Arts, this emphasis on hardware was maintained at the expense of software. In 1994 Hawkins alienated game developers by doubling royalty charges to offset the ballooning costs of the 3DO. Launched later that year, the console was an impressive system. But it sold at a retail price of nearly seven hundred dollars – about seven times the cost of competing, though less powerful, machines – and was supported by only a sparse number of games. The 3DO hit the market "with a dull thud."[47] Despite desperate price-cutting measures and employee layoffs, by 1997 3DO's console hopes had imploded, and Hawkins reshaped his company as a "boutique publisher" of game software.

A similar fate befell Atari, which had hoped to recreate its glory days by racing to claim the 32-bit market. In 1993 it launched the cartridge-based "Jaguar" claiming it was the first 64-bit console. Competitors, like Hawkins, retorted that it was merely 16-bit. *Aficionados* praised the system's performance. But again, it was supported only by a small number of games. Lagging sales deterred developers from making more games for the machine, setting in motion a vicious spiral that rapidly became meteoric, crashing through the familiar emergency measures of frantic price cuts and the creation of extra peripheral devices. When in 1996 Atari officially announced the discontinuation of the Jaguar, it had in fact been dead for months.[48]

The failure of 3DO and the Jaguar showed that success in the console wars required more than good machines. Both companies had attempted to emulate Sega's digital leapfrogging of Nintendo but had not studied the Nintendo-Sega wars deeply enough. Both missed the crucial lesson that the video game business was not just about machines but also about culture and marketing. Both 3DO and the Jaguar systems were launched without the promotion budgets and campaigns at which Sega and Nintendo had become so adept. Moreover, neither Hawkins nor Atari grasped that hardware systems depended on their association with popular games: good game "tech" had to come backed with good game "text." Both companies based their hopes on systems that had excellent technical performance but were accompanied by very few games. Although players might crave the ultimate game experience, they were clearly not unaware of the costs of their favourite activity, nor willing to leap blind into expensive upgrades without proven advantage, nor confident enough in new companies to abandon those with established track records. Without the blitz of persuasive advertising and promotion that was now the norm of video game culture, neither Hawkins's charisma nor Atari's fading glory could overcome this consumer resistance. But just how truculent the market had become hit home when the next company to enter the next-generation race – Sega itself – experienced an even more dramatic disaster.

WORLD DOMINATION? NOT QUITE

The initial result of Sega's multipronged offensive was stunning success. Over three years it steadily overtook Nintendo in the US market. In 1991 Nintendo lost its position as the top-selling video game company.[49] In characteristically hyperbolic style, *Wired* magazine celebrated the industry's rising star with a 1993 article entitled "Seizing the Next Level: Sega's Plan for World Domination."[50] However, while Sega had triumphantly pushed its way into the market, it did not deliver a knockout blow. Nintendo staged a strong comeback based on lower prices and better software titles. In 1995 one analyst wrote, "In the eyes of many customers Sega and Nintendo had achieved a rough sense of parity ..."[51] This was as good as it got for Sega. Within a year it had snatched defeat from the jaws of victory.

The company fell victim to the very forces of technological and cultural innovation that it had itself unleashed. In the mid-1990s Sega was paranoid about technological lag behind the competition. It knew that Atari Jaguar and 3DO were targeting the "early-adopter" 32-bit market, that an old enemy – Nintendo – would eventually press on its heels, and that a new foe, Sony, was looming on the sidelines. In this tense atmosphere, it thrust desperately for the next-generation market with the development of its own "Saturn" system. Having already upstaged Nintendo in its marketing of the Genesis, it aimed at repeating this move with the Saturn, keeping pace with an aging generation of consumers by targeting eighteen- to thirty-four-year-olds who had grown up with computers and gaming, were more sophisticated in their use of the machines, and expected more from the gaming experience. The validity of this move was later proven by a gigantic competitor, Sony, who was already hovering in the wings of the console market and would later borrow Sega's insights. Sega, however, was unable to benefit from the knowledge; for in its haste to reach the market it fatally botched the launch of the new system.

Afraid of being beaten to the punch, Sega pre-empted itself by releasing a CD attachment for the Genesis, as well as the "32-x," a peripheral that allowed 32-bit games to be played on the 16-bit system. It hoped these devices would keep gamers occupied while the Saturn was in development. Instead they just confused gamers, all the more so since the various releases were not coordinated across global markets, and 32-bit systems were being marketed in Japan while only add-ons, many of which did not work very well, were available in North America.[52] In 1995 the Saturn was finally launched in the US. With Sony breathing down its neck, Sega rushed the system to completion. But the planned pre-emption didn't work as anticipated. So concerned was Sega to make a fast launch that it left many developers in the dark. Consequently, like 3DO and Atari it could deliver only a small library of games of questionable quality. To the degree that the 32-bit machine did catch on, Sega miscalculated how quickly it would erode demand for its now-outdated Genesis machine. At the end of 1996 Sega of America took huge losses worldwide on warehouses full of unsold 16-bit games.

The pace of technological revolution was proving too much not only for producers but also for consumers, who were bewildered by the array of competing and incompatible next-generation systems. In 1994 video game receipts dropped sharply to about five billion dollars, the first downturn in the game industry's history since the big Atari "meltdown" ten years earlier. In many people's eyes, video gaming, which a few months before had appeared to be on the verge of its next generation, once again seemed played out. The erratic marketing of the Saturn was a disaster from which Sega never recovered, a video game company's retreat from Moscow. Just how deeply it traumatized the company can be gauged from the fact that when Sega did attempt to re-enter the console market in the late 1990s, the pre-awareness TV spots in Japan for its 128-bit Dreamcast machine were largely an apology for the company's mishandling of the Saturn launch. Within a few years it had not only lost its ascendancy to Sony but also fallen far behind its traditional opponent, Nintendo, to become a very distant third in the console wars, and it increasingly turned towards its other video-game-based ventures, such as theme parks and cable channels.

THE COMPUTER BYTES BACK

As the console wars raged on, interactive gaming was being quietly transformed by the reappearance of the video game's long-lost relative, the personal computer. As we have seen, home consoles and home computers had emerged in the late 1970s and early 1980s as part of the grassroots innovation that converted computers from the mainframe tool of military and corporate bureaucracy to a household medium: the consoles made by Atari, Nintendo, and Sega were in essence just simplified computers with all their power dedicated to the single function of gaming. But despite this common origin, video gaming and personal computing had followed diverging paths. Personal computers, culturally coded as "serious" machines, were still the domain of the relatively affluent, while lower-income families bought the cheaper, more "childish," consoles.

Nonetheless, a vigorous subculture of computer players, circulating games on floppy disks, persisted.[53] This was in many ways a hard-core gaming culture. In spite of limited graphics and slower chips, computer games had the advantage of being programmable. They often appealed to an older or at least technically more literate group of players. Moreover, these games, unlike their console cousins, were part of the exciting innovations in computer networking. Electronic bulletin boards and online connection provided a semiclandestine arena for those interested in playing, programming, and hacking games. In a community that tended to regard digital networks as electronic commons, "piracy" was frequent. Free-release, or "brown bag," versions of beta versions of game software were a common method of getting the gaming community involved in a new product. Thus, although the world of computer gaming was relatively small in North America, it was experimental, interconnected, and technologically sophisticated.

In the 1990s continuing improvements in design and a decline in the price of personal computers began to widen this circle. By 1994 some twenty-four percent of US households owned a personal computer.[54] Only a handful – fewer than ten percent – had the CD-ROM players that we think of today as vital for gaming.[55] Nevertheless, new microchips, particularly Intel's new Pentium chips, were giving home computers performance levels that rivalled the ability of consoles to handle fast-paced action and doing so cheaply enough to penetrate the households of an important consumer demographic. Another factor adding to the popularity of computer gaming was the take-off in Internet use, which, having grown steadily but slowly from the early 1970s, doubled in size in 1987 alone – and then did so every year thereafter for at least a decade. This process was further accelerated by the creation of the World Wide Web, with its attractive visual interfaces and point-and-click hyperlinks. What had previously been the esoteric domain of researchers and avant-garde technoenthusiasts was by the early 1990s rapidly being opened to a public that, even if it was still largely restricted to affluent North Americans and Europeans, was numbered not in the thousands but millions. All this changed the face of computer gaming. Within a few years a world of hobbyists swapping and stealing programs in what was partially a digital "gift economy" was converted to a target for giant conglomerates eager to dominate the so-called "information highway." As so often happens, the new commercial potential was first demonstrated not by megacorporations but by small and innovative companies. In 1993 two games, both the products of bootstrapped developers but radically different in their ethic, aesthetic, and audience appeal, signalled that something new was happening in computer gaming. They were *Doom* and *Myst*.

DOOMED TO SUCCEED

Doom was the game that blasted hard-core computer gaming to commercial success, propelling interactive entertainment to simultaneous peaks of graphic violence and technological sophistication. It was the creation of id Software, a Texas company created by John Romero, who designed game graphics, and John Carmack, who built the digital "engines" that drove the game-play. Both became legends of the gaming world.[56]

They had established a reputation for groundbreaking verisimilitude in gaming violence with *Wolfenstein 3D*, a game for MS-DOS computers released in 1992. *Wolfenstein* was the first "first-person shooter." The player is an escapee from a Nazi concentration camp (hence morally licensed for unrestrained vengeance) sent to attack Hitler's headquarters: along the way, he shoots enemy soldiers, wolves, and aliens, culminating in a final fatal encounter with the Führer himself. Previous shooting games had, of course, placed the player in the position of, say, a cockpit gunner; but the shooter's viewpoint was in effect static, fixed in relation to the scenes and opponents that scrolled past him. What made *Wolfenstein* revolutionary was that it gave the illusion of occupying the vantage point of a protagonist who moved. Using Carmack's new engine, *Wolfenstein* exploited perspective and parallax-motion effects to create a game where the player's vantage point changed with every move he made on screen. It was a radical game because it gave the experience of seeing through the eyes of its vengeance-bound hero as he raced through a carefully rendered mazelike setting, turning at any angle, moving forward and backwards, dodging the fire of assailants, all the while executing massive carnage with a gun whose effects could be seen in the graphic and bloody wounds opened on target bodies. The result of this 3-D creation was massively to intensify the illusion of actual embodiment and with it the adrenalin rush of the kill-or-be-killed situation.

Throughout 1993 rumours spread in gaming circles that id was working on a game of yet greater intensity, speed, and violence. The release of demos and screenshots had generated wide anticipation. On 10 December the arrival of id's representatives at the University of Wisconsin computer centre to upload the "shareware" version of the game produced one of gaming's mythic moments. So many enthusiasts were waiting to copy the file that their entry was blocked. As the delay built, the fans created an Internet chat channel, devoted entirely to minute-by-minute updates and complaints. When eventually the game was uploaded, the demand on the university's file-transfer site crashed the system immediately, and repeatedly thereafter.

The new game was called *Doom*, and it lived up to its title's apocalyptic overtones. The player is positioned in the game as a Space Marine dispatched to a moon where scientific experiments in "interdimensional transport research" have gone seriously wrong. A gate has opened to Hell, no less, and through it pour legions of infernal creatures – Revenants, Cyberdemons, Arachnotrons – that fuse medieval diabolism and twenty-first-century armaments. Dance-of-death skeletons are clad in SWAT-team body armour and goat-legged mini-Satans fire heat-seeking missiles. Such opponents are, as the instruction booklet pointedly reminds us, "the perfect enemy." They "have no pity, no mercy, take no quarter, and crave none." The only option is violence – extreme violence.

With all the other members of his squad killed immediately on arrival, the protagonist must fight his way out, through a labyrinth of tunnels, pits, and traps exploding barrels of toxic waste, collapsing ceilings, and other environmental hazards. The crucial considerations, detailed on the interface toolbar, are "Arms" and "Ammo" (an arsenal escalating from pistol to shotgun, chain gun, rocket launcher, and plasma rifle and culminating in the coyly abbreviated "BFG 9000"), "Armour" (infrared goggles, radiation suits, Kevlar vests, and titanium breastplates), and "Health" (an index of the amount of punishment the player can continue to absorb, increased by liberal application of the drugs – "stimpacks" and "bezerk packs" – that litter the corridors you careen through).

Although broadly speaking the game interpellates its player in the role of a futuristic soldier – male, muscular, white, Schwarzenegger-esque – it might be more accurate to say that in the *Doom* world the player is the weapon. One's point of view is defined by the alignment of a gun barrel jutting into the screen in front of you. The cross-hair sites orient the player in the *Doom* universe, defining the function of the virtual body as a mobile and pitifully vulnerable support system for the weapon to which it is prosthetically welded. The game experience is fundamentally about mastering speed of manoeuver and precision of aim while cannoning through a disorienting labyrinth of corridors and chambers. The dominant colour is an ominous brown, punctuated by the scarlet explosion of bodies into "gibs," or bloody giblets of flesh. The pace and fluidity of motion are astounding, the aesthetics those of corporeal disintegration, nightmare monstrosity, extraordinary velocity, bewildering disorientation, and extreme fear. Bringing digital brilliance to bear on the most aggressive strands of game culture, the new first-person shooters inevitably intensified the long-simmering debate about gaming violence. Before *Wolfenstein* and *Doom*, developers of violent games could argue they were doing no more than was already done everyday on film and television; Carmack's engines and Romero's graphics changed this. It was impossible now not to suspect that the depth and conviction of computer-generated illusions had entered an unprecedented dimension.

What made *Doom* revolutionary was not only its speed and violence but also its use of the Internet. To market the game id deployed the radical shareware marketing strategy that it had pioneered with *Wolfenstein*.[57] Carmack and Romero aimed at getting the game into the hands of potential customers as quickly as possible. Instead of just releasing a fragmentary demonstration, as other game developers had, they allowed the download of complete *Doom* episodes or "levels," trusting that once gamers experienced the extraordinary graphics, animation, and smoothness of play, they would be hooked and purchase the full off-the-shelf version. This faith was amply repaid. Some twenty million shareware versions of *Doom* were distributed online. But these stimulated sales of over two million copies, making it in 1994 the best-selling game ever.

Even more daringly, id later released the actual source code for the game. This allowed *Doom* players with basic programming skills not only to use various shortcuts and cheats but also to make their own levels of the game. These self-made scenarios could be circulated in turn on the Internet. The result was to generate an extraordinary collective culture of *Doom aficionados* in which, paradoxically, players competed both as destroyers – blowing apart monsters, and opponents – and also as creators, rivalling each other in the intricacy and detail of their sinister architectural constructions. Some of these self-taught designers won worldwide reputations, and a handful were eventually offered paid jobs as game designers with Romero and Carmack.[58]

Nor was this the end of id's ingenious exploitation of the Internet. *Doom* included the possibility of play against both the computer and human opponents on networked computers. In these "death matches," the enemies were not just the demons but also the other gun-toting human avatars. Stories abounded about corporate LANS (Local Area Networks) being clogged by *Doom* players determined to "frag" each other on company time. In fact, *Doom*'s online play option was quite limited, restricted to four players. But its successor, *Quake*, which marked another quantum leap in speed, graphic quality, and gore, allowed up to sixteen players to exchange volleys of virtual fire, taunts, and imprecations: *Quake* "clans," teams that fought together online, became a major Internet phenomenon.

In 1994 *Doom* brought id nearly eight million dollars in revenue; by 1996 this had grown to more than fifteen million dollars. Its impact moved out in ripples far beyond the point of over-the-counter sales. In addition to the immediate profits from these games, id found a lucrative source of revenue in licensing use of Carmack's game engines to other developers at a price of about half a million dollars, depending on royalty agreements. Other companies elaborated on the nuances of the futuristic first-person shooter mode: Apogee Software with *Duke Nukem*, GT Interactive with *Unreal*, Interplay with *Descent*, and Sierra with *Half-Life*. Eventually Romero split from Carmack to found his own company, Ion, developing yet more ambitious and bloody action games. First-person shooters – *Doom* and *Quake* in particular – became inextricably implicated in North American controversies about youth violence that reached boiling point in April 1999 when two teens massacred thirteen people at Columbine High School in Littleton, Colorado.

But id could claim an influence that ran beyond the game world. Its innovations in Internet distribution and marketing anticipated developments that were important to far larger information enterprises. Netscape, for example, picked up the shareware strategy in its "browser wars" with Microsoft, while the distribution of source codes for collective development formed the basis for the success of the Linux operating systems. The id blend of technological sophistication with atavistic bloodlust can thus claim a place in digital capital's hall of fame.

MYST-IFICATION

If *Doom* seemed to amplify the most violent currents of interactive game culture, another computer game released in 1993 pointed in a different direction. *Myst* was also the product of a small development company, Cyan Studios, a tiny enterprise based in rural Washington and run by two brothers, Rand and Robyn Miller. But here the similarities with id's product cease. To many, *Myst* seemed like the good twin of the evil *Doom*. Where one was violent, the other was pacific. While Romero and his disciples at id swaggered with a calculated "badness" that smacked of high-tech Satanism, the Millers were pious but soft-spoken Christians. Where *Doom* was the descendent of arcade shooters, the ancestors of *Myst* were the children's games Cyan had developed for the Macintosh computer, games with titles like *Spelunx and the Caves of Mr Seudo*. Where *Doom* was a celebration of speed and terror, *Myst* was serene and beautiful.[59]

It has been described as "an odd, nongame-like game ... more like a beautiful, moody and slow-moving art film (accompanied by a breathtaking soundtrack) than a traditional computer game."[60] At its outset, players are deposited on an archipelago of islands, uninhabited but covered with the traces of abandoned civilization. They must assist Atrus, a monarchic wizard whose benign inventions are being perverted by one of two sons, Sirrus and Achenar. The player travels from island to island, solving puzzles, performing rituals, manipulating switches, gears, consoles, or sliders in order to gain access to secret books with vital information about the crime. Having finally negotiated these *Mysteries*, players should be able to identify the guilty son. A correct choice frees the king; the wrong one condemns the player to share his imprisonment.

The player is thus positioned not as a warrior-destroyer but as a mythical quester, puzzle solver, observer, and, finally, rescuer. Instead of promising murderous action, the game packaging advertises aesthetic

experience and an intellectual challenge. While *Doom* was pointedly
aimed at adolescent males, *Myst* seemed designed to attract adults,
perhaps (despite the notable lack of female characters) even adults of
both genders. The game's appeal to an older audience stems from two
major factors: "its lack of violence ... and the measured pacing and
narrative."[61] The manual pointedly reassures the novice that "you
don't die every five minutes. In fact you probably won't die at all."[62]
In *Doom*, of course, the only way not to "die" is by perfecting your
virtual killing skills at a frenetic speed. *Myst*, on the other hand,
proceeds at a leisurely and indeed "bookish" pace (the game's plot
centres around books, libraries, and manuscripts) that doesn't assume
that the player possesses reflexes accustomed to video game controls.

Yet *Myst* was in many ways as technologically innovative as *Doom*.
id's nightmare visions were in part a technological sleight of hand –
the product of a pace of play that drove the player's imagination to
fill in the hallucinatory details of attacking monsters from what were
actually quite sketchy graphics. By comparison, *Myst*'s gameplay was
snail-slow. But it had traded speed for visual quality. The textured
spacious scenes of *Myst*'s "gorgeous slide show" were more demanding
creations than id's endless claustrophobic corridors, and its "transpar-
ent and minimal" interface was extraordinarily elegant.[63] Even Romero
recognized *Myst* as a digital masterpiece.

Cyan did not make use of the computer's networking power as id
had. But it accomplished something perhaps as significant by giving
interactive games recognition within mainstream cultural channels,
where they had previously been seen as technological wonders, dangerous pleasures, or tokens of an unbridgeable generational divide but
hardly as serious cultural artifacts. *Myst* changed this. It was celebrated
by *Wired* as the first CD-ROM "that suggested ... a new art form might
very well be plausible," *Newsweek* dubbed it an "instant classic," and
the *New York Times* said it inaugurated "a new art form" and hailed
it as "a landmark" in the game industry: "its reflective, almost cool
aesthetic suggests what is possible: image, sound and narrative."[64] Very
much unlike *Doom*, *Myst* became a legitimate topic of academic study.
It was praised as "a new media form in which the 'reader' becomes
not only coauthor, but also theatre goer, movie goer, museum visitor
and player, all at the same time."[65]

If the real-life analogue of *Doom*'s virtual world is the US war
against demonized Third World enemies, then *Myst* offers a mythically
sublimated version of a yuppy recreational lifestyle. Its lingering attention to the beauties of abandoned industrial machinery in landscapes
stripped of all traces of actual manual labour neatly matches sensibilities that are acculturated to gentrified urban settings where rusted-out

factories become fashionable marketplaces and derelict wharves are converted into expensive residential districts.[66] Remarking that "*Myst* was a grand exercise in virtual tourism," Herz points out how its popularity coincided with that of high-end off-the-beaten-track adventure travel packages and vicarious immersion in high-design environments and architectural stylings promoted by magazines and television shows such as *Metropolitan Life* or *Martha Stewart Living*.[67] All this is then bathed in an atmosphere of diffuse New Age mysticism. In these respects, the game speaks very precisely to a blend of aesthetic and religious sensibilities that are pervasive among the stratum of "symbolic analysts" who have benefited so markedly from the high-technology restructuring of advanced capital. The relationship between *Doom* and *Myst* is not just one of opposition between "nasty" and "nice" but a more complex interdependence, since the idyll of one arguably depends on the violence of the other: the *Myst* archipelago is ruthlessly defended by the Space Marines.

Perhaps surprisingly, *Myst* was more successful than *Doom*, which it toppled as the best-selling computer game. This triumph was repeated when Cyan came out with a sequel bestseller, *Riven*. Yet although *Myst* outsold *Doom* in the stores, the millions of shareware copies of *Doom* distributed on the Net gave it a far greater reach. *Doom* and *Quake* spawned a host of clones, emulations, and rivals and generated, through their network play and design, one of the most powerful, if perturbing, of virtual communities. *Myst* and *Riven*, on the other hand, stand on a rather lonely pillar in the game industry, admired but not widely imitated – a situation lamented by the Miller brothers, who in one interview complain, "Where are the knock-offs?"[68] However one tallies the respective merits and influence of *Myst* and *Doom*, what is certain is that between them they transformed computer gaming and in doing so drew on it the attention of some of the most powerful corporations of information capitalism.

CONCLUSION: CREATIVE DESTRUCTION AND THE PERPETUAL-REVOLUTION DILEMMA

The "mortal kombats" of the interactive game industry between 1990 and 1995 provide a striking example of what Joseph Schumpeter described as capitalism's tendency towards "creative destruction."[69] The fight for competitive advantage opened by Sega's challenge to Nintendo unleashed successive waves of innovation that swept some interactive game companies to the shores of success, threw others onto the rocks of catastrophe, and tossed corporate ships between the two within the space of a few years. These waves of innovation were in part technological, with the successive development in the space of a few years of 8-, 16-, 32-, and 64-bit game systems; as the pace of technological revolution intensified, the console designers were themselves struggling to accommodate to shrinking timelapses between system release dates. As Nintendo found to its cost, innovating too slowly could mean being overtaken by competitors with better technology; but as 3DO, Atari, and eventually Sega learned, going too fast could be even more catastrophic.

But innovation was also the watchword in the cultural and technological circuits of the industry. The other side of "creative destruction" was "destructive creation," as game developers strove to attract the all-important young male market segment by intensifying game violence in productions such as *Mortal Kombat* and *Doom*. Games such as *Myst* suggested other cultural possibilities, appealing outside the testosterone zone, but they were less-widely pursued. Competition also sharply revealed the importance of what in military terminology would be called "psychological operations." From the moment of Sega's brilliant advertising offensive it became apparent that success in the video game market now depended on carefully researched, costly, and inventive promotion campaigns that both responded to and reshaped customer demand within a complex segmented market, engineering transformations in the "soft" field of consumer expectations at least as quickly as in the "hard" field of digital technology.

Successful commodification of digital play was proving a risky and arduous business, demanding not only technological innovation but also a rapidly expanding stock of marketing knowledge and cultural capital. At the same time, under the influence of the new technologies, aggressive advertising, and public attention generated by the Sega-Nintendo wars, the market continued to expand explosively. By 1995 fifty million households worldwide owned video game systems, and computer games were beginning to figure as a serious commercial prospect. Perilous as it might be, this was a growth dynamic that soon drew the attention of some of the most powerful corporations of information capitalism.

7

Age of Empires:
Sony and Microsoft
1995–2001

In the mid-1990s the interactive game industry's position within the circuits of digital capitalism was radically transformed. Since the crash of Atari, the business had made an extraordinary recovery. The console wars had galvanized video gaming, and computer play was on the rise. But interactive entertainment still remained a distinct and rather specialized sector of the post-Fordist economy. It was ruled by a cluster of companies that had made – or lost – their names and fortunes in the field of digital play. Pundits and theorists of the digital business still often overlooked the sector, perhaps because of a lingering suspicion that the obsessions of young male teenagers were not altogether worthy of serious attention.

But this view was to change dramatically. Interactive gaming not only continued to expand, significantly altering the demographics of its market, but it was also transfigured by its connection to an "information highway" whose computer and telecommunication networks promised to transform society and the economy. At the same time, the structure of the industry was reshaped by the entry of two of the largest entities of digital capital. On the video game side the consumer electronics giant Sony strode into the console wars with overwhelming force. In the area of computer gaming, the greatest of all software companies, Microsoft, made an inexorable, if slower, advance.

SONY STEPS UP

The first giant footsteps to be imprinted on interactive play were those of Sony, which in 1995 launched its own next-generation video game console, the PlayStation. Sony was a multinational and multimedia conglomerate in the top fifth of the Fortune 500, with a six-figure workforce and a brand name recognized worldwide. Its reach, resources, and experience dwarfed those of Nintendo and Sega. Sony's reputation had been built on consumer electronics – stereos, TVs, VCRs, cameras, camcorders and the famous Walkman. In the 1980s it entered the entertainment industry, purchasing CBS Records and Columbia Pictures, as well as Loew's Theatres – acquisitions that had contributed to North America's fears of oriental cultural takeover. In fact, Sony's legendary chairman and cofounder, Akio Morita, was no Japanese nationalist but a missionary for "globalization." In his hands the company established large production and research facilities in the US and Europe, partly to reduce the effects of currency fluctuations but also to take advantage of local design and marketing expertise. Sony was in many ways the epitome of a transnational corporation, one that in the search for worldwide profit consciously divested itself of attachment to any local culture.[1] It was also looking for new directions as the ailing Morita relinquished his leadership and a new CEO, Nobuyuki Idei, began to restructure the company as a media giant of the digital age.

Sony's interest in video gaming grew from a failed arrangement with Nintendo to produce a CD-player peripheral for Nintendo's game consoles. Nintendo defaulted on the partnership, but Sony, refusing to be left at the altar, pursued the project on its own under the command of digital engineer Ken Kutaragi, in whose hands it developed into the design for a new 32-bit CD-ROM platform video game system. Sony's company planners initially regarded Kutaragi's project as an "embarrassing" child's toy.[2] It took a fierce struggle within the corporate culture before the company abandoned its traditional model of making profits directly from consumer electronics and instead adopted the "razor and blades" strategy, selling hardware at cost and seeking profits in software and licensing.[3] The Nintendo-Sega console wars had demonstrated that the cyclical competitive world of video gaming was a market with a deeply destructive capacity. Sony, however, had the advantage of deep pockets, vast organizational capacity, and, not least, of learning from its competitors' failures. All this was evident in the detailed and comprehensive planning that went into launching the PlayStation.[4]

The launch is perhaps the most important single moment in the history of a video game system, a small window of opportunity in which to impress gamers, distributors, and developers. High early sales figures foster brand recognition, create "media buzz," and convince retailers to commit scarce shelf space to a game. They also build a system's credibility in gamer culture and can determine whether independent developers and publishers will be interested in making and selling software for the platform.[5] A good launch is therefore considered to be worth multimillion-dollar promotional investments.

The two-hundred-million-dollar launch of the PlayStation stands as a classic demonstration of marketing strategy in the multimedia entertainment industry.[6] Meeting its scheduled launch date, Sony marched into the arena with impeccable timing. Sega gamers were frustrated with the 32-X and Saturn, while Nintendo players were restlessly awaiting the N64, leaving an open audience that might switch to a new brand. Launching in early fall allowed Sony to capture the back-into-the-house season and the money teens might have earned from summer jobs, while allowing time to build up to the Christmas season push.

If Sony learned a lot from Nintendo in terms of console design, it had its eyes on Sega in terms of marketing, reverse engineering the technology of one of Sony's rivals and the promotional strategy of the other. Steve Race, who at the time was head of the PSX project to prepare for the PlayStation launch, went so far as to hire many of the people who had worked on the Sega marketing team.[7] Race is said to have "argued with other Sony executives over how to handle marketing. He wanted a more aggressive, darker ad campaign than the Japanese executives would allow."[8] This was important because Sony wanted to do to Sega what Sega had done to Nintendo, outflanking its rival by aiming at a more "mature" market.

The PlayStation was initially marketed for the twelve-to-twenty-four age range. By going "up age," Sony sidestepped the saturated children's market, while developing a profitable post-baby-boomer segment. Unlike the Nintendo "kiddies," Sony's target audience was likely to have walked to school listening to a Sony Walkman, so the company benefited from its "bankable brand name."[9] This older group was an obvious target because of its relatively high spending power. But the decision also involved psychological analysis. In their consumer research, Sony came to the conclusion that "everyone is 17 when they play [video] games."[10] Younger players "look up to the best gamer who is usually a little older and more practiced and talented. Then there are people who start working and grow up, but when they go into their room and sit down with their video games, they're regressing and becoming 17 again."[11] Targeting twelve- to twenty-four-year-olds directly addressed a group with crucial disposable income, but it also appealed universally to all would-be seventeen-year-olds in the gaming world.[12]

This demographic analysis was reflected directly in the advertising campaigns for the PlayStation, which sought to master a twisting spiral of consumer cynicism and promotional ingenuity. Sony was aiming at a media-savvy audience. These were the techniques used in advertising PlayStation. Sony's campaign blended the "character marketing" approach pioneered by Nintendo for *Mario* with stylish humour and irony that drew on and exceeded Sega's efforts. At the centre of the campaign was the figure of Crash Bandicoot, a Tasmanian Devil-like creature with a propensity for irony, sarcasm, and twisted humour. He is both the star of specific PlayStation games, the *Crash Bandicoot* series, and the figure that gives a consistent identity to Sony's promotion of the console. His adventures in television ads are plentiful: a metal-detector scan holds up the airport security line as he tries to get through with a jet-pack strapped to his back; he taunts nuns in a unique rendition of the Californian surfer accent; he enjoys a coffee at a greasy spoon while telling the locals about his adventures in the waters of the video game *Jet Moto*. This last ad is a smart appeal to gamers as members of an élite youth culture, pitting the allure of virtual experience and sexual vitality against mundane adult routine. The locals are oblivious to Bandicoot's presence as he tells an "unbelievable story" of riding his jet-motorcycle: "So, anyway" – the ad switches to game graphics – "I'm playing *Jet Moto* right, and I'm riding on this jet motorcycle and I'm hurtling through different terrain: snow, concrete, ocean, jungle." The ad switches back to the restaurant as Crash finishes the story. "I mean, when you get that kind of power between your legs you don't know what to expect – anything could happen."

Another ad takes direct aim at Sony's main competitor. Crash drives his dilapidated pick-up truck into the parking lot of Nintendo's headquarters. The contents in the bed of the truck tower in the air, covered with a sheet. He jumps out, grabs his megaphone, and begins to taunt Mario: "Hey plumber boy, mustache man. Your worst nightmare has arrived. Pack up your stuff. I've got a little surprise for you here – check it out." He removes the sheet to reveal a pyramid of TV sets equipped with the Sony PlayStation console. Each screen displays different scenes from the video game. "What do you think about that? We've got real time, 3-D, lush, organic environments. How's that make you feel buddy? Feel a little like your days are numbered?" Bandicoot is escorted off the parking lot by a security guard who seems to question his own loyalty as he gets friendly, asking, "Is that Italian?" Bandicoot responds coolly, "No, Bandicoot, it's an Australian name."

Bandicoot was from his inception both a game hero and a marketer's tool to build brand awareness – a hero brought into being to compete with Mario and Sonic. Not only did he have an arrogance and "attitude" neither of his rivals could match but the ads in which he appears represent a new sophistication in video game marketing. Bandicoot is game content shaped to promotional requirements, advertising preceding play. Game design and marketing strategies were now thoroughly interlocked, with Kutaragi, the inventor of the PlayStation, allegedly "reviewing every marketing decision."[13] But at the very moment that marketing considerations were driving game design more explicitly than ever before, the promotional intention was being artfully effaced as advertisements were made to look ever more like play.

FAST PLAY OR FLEXIBLE PRODUCTION?

In response to the insults of Crash Bandicoot, Nintendo taped a promotional video in which inept villains from Sony and Sega threaten to torture Mario, the Nintendo mascot, unless they are shown the new Nintendo "game secrets." The demonstration leaves the kidnappers forlorn: "Are we in big trouble!" one of them laments.[14] The secrets Nintendo was banking on were those of its own next-generation console, the Nintendo 64. Scheduled for a 1995 release, the "N64" was later pushed back to 1996 to allow more time for the hype and excitement to brew. "Sega, Sony and the PC market had been competing for the market of hardcore, heavy gamers."[15] The linking of the N64 to the newly released *Super Mario* reaffirmed Nintendo's traditional orientation to younger players and more mainstream buyers.[16]

The battle between Nintendo and Sony opposed different strategies not only in marketing but also in the technological circuits of the industry, pitting two production models against one another. The vital factor was the shift from the cartridges that video games had long been based on to the new CD-ROM drives. Sega, 3DO, and Jaguar had all tried introducing the CD option to their next-generation machines, but their inept marketing failed to make its capabilities clear. Nintendo, however, had stuck with cartridges. Cartridges gave a superior speed of play, and Nintendo believed this would give its games a quality edge over CD-based competitors. Sony, however, adopted the CD format and with it a new philosophy of game development. The crucial advantage the PlayStation had over Nintendo's consoles was not how fast games could be played, as it had been with the Sega Genesis, but rather how quickly they could be developed. If cartridges were faster to play, CDs were faster – and cheaper – to make.

While the production costs of a cartridge were in the region of thirty-five dollars, those for a CD were between five and ten dollars. This cost difference was critical for Sony. It was not that lower manufacturing costs made Sony attractive to developers.[17] It permitted Sony to practise what is often considered a hallmark of information age production – "flexible specialization." While Nintendo had controlled game quality through carefully restricted licensing, CD production gave Sony a different, more fluid approach. As Paul Roberts says, cheap CD runs "let the company test different games in small batches and, if a

hit appears, bring huge volumes to market quickly."[18] This flexibility
also gave the company an advantage in another fashionable endeavour,
globalization, because it helped developers "defy cultural barriers":
"Games that do well in Japan often bomb in Europe and North
America. CDs let PlayStation troll for crossover titles by doing short
runs of say, 5,000 copies and monitoring the market response."[19]
Overall, the CD format allowed Sony a supply approach diametrically
opposite to its rival's. Whereas Nintendo carefully rationed the release
of titles, focusing on a handful of games made by so-called "dream
teams" of top-rank designers, Sony made sure there were many high-
quality games available for the launch of the PlayStation, aiming at
immediate satisfaction of a broad cross-section of gamer tastes.

In the PlayStation's first year, Sony released about one hundred
games from dozens of developers, many of which, such as Namco,
Psygnosis, and Ubi Soft, already had an established reputation in the
gaming world.[20] It capitalized on omissions in the Nintendo repertoire
– such as sport games, where the PlayStation rapidly established its
pre-eminence – and challenged Nintendo's areas of strength, such as
role-playing titles, where "Sony put a huge marketing campaign"
behind the *Final Fantasy* series, published by Square Soft, a company
that defected from Nintendo.[21] By contrast, when Nintendo released
the N64 in North America there were only two games available for
the machine – *Super Mario* and *Pilot Wings*. Even by 1998 Sony was
rolling out titles at more than twice the rate of its rival, releasing one
hundred and thirty-one games to Nintendo's fifty. By concentrating on
Mario and a handful of other select titles, Nintendo stood to realize
huge profits through "economies of scale" but risked being over-
whelmed by the greater variety of Sony's "economies of scope."

In the end, however, the fight over post-Fordist production techniques was not fought to the death. What allowed at least a temporary
truce was an equally post-Fordist marketing strategy – demographic
segmentation. While Sony and Sega competed for the dollars of "hard
core, heavy gamers," Nintendo's targeting strategy was based on the
belief that the "bread and butter" of the video game audience is
children.[22] These differing approaches contained the basis for a provisional settlement of the console wars, based on a division of the
market, with Sony catering to the older gamer and Nintendo positioning
itself as a classic firm for younger players.

By the late 1990s the console wars had reached a temporary armi-
stice. The entrance of Sony completely shifted the balance of power in
the industry. After Sega's Saturn catastrophe, there were really only
two major hardware platform providers left standing: in 1998 Sony
PlayStation accounted for some fifty-nine percent of next-generation

sales, Nintendo for thirty-two percent, and Sega for a meagre nine percent.[23] But if Sony transformed video gaming, video gaming also transformed Sony. In 1998 PlayStation, once scorned by company executives as a childish venture, accounted for forty percent of the multinational's profits.[24] By 2001 more than twenty percent of US households had a PlayStation, and some eighty million had been sold worldwide.[25] Although this proportion later declined, interactive gaming had clearly become an integral part of its planned metamorphosis into an integrated-network-era leviathan. The digital game business had moved from a sideshow of digital capitalism into a central theatre of activity.

MAINSTREAMING COMPUTER GAMES: THE *DEER HUNTER* PHENOMENON

Change was also brewing on the computer side of the industry. The massive sales of *Doom* and *Myst* came at the very moment US computer makers faced an unexpected crisis.[26] The market for expensive home computers combining Intel's ever-faster chips and Microsoft's ever more bloated software packages had plateaued. Though 1994 was the first year more than a million home computers were sold, it was also the first year the majority of purchasers were buying a second machine, a clear sign that the market had stalled within a narrow segment. As one analyst put it, "the PC industry, having already stuck a computer in every office and in the easiest forty percent of American homes, needs new customers."[27]

There were two responses to this stagnation. One was the push for machines that cost less than one thousand dollars made by companies such as Dell. The other was the attempt by Microsoft and Intel to find new "high-end" uses for powerful machines. Gaming held enormous promise for both approaches. Hard-core computer gamers were notorious for demanding the fastest, most powerful machines, and sales to these consumers could "lead" business sales, since many corporations, particularly those whose staff worked at home, felt a pressure to match the "state of the art" machines their employees played on. At the same time, while cheap machines might not handle the latest version of *Quake* online, they would be good enough for the kids to play *SimCity* on.

Redefining the PC as a gaming machine was attractive for producers across the entire spectrum of the industry. Intel's Pentium chips were already recognized in the computer gaming subculture as a must. In 1997 the company, facing a problem selling its flagship Pentium II in a saturated business market, launched a campaign with a new slogan: "PC: It's Where the Fun Is."[28] Chairman Andy Grove made it quite clear that gamers were his target market: "No audience is as demanding as a fourteen-year-old boy."[29] As important as faster chips was simpler software. The computer CD-ROM market was plagued by customer difficulties in configuring their computers' sound and video card channels; the 1994 release of Walt Disney's *Lion King* CD, for example, degenerated into a fiasco of bugs, failed installations, and jammed customer complaint lines.[30] But the following year Microsoft issued Windows 95 with features that made PC games easier to install, faster to run, and more enjoyable: "plug and play" was much closer to reality. Microsoft's monopoly power position ensured that its software provided a near-universal standard for game users and developers.

With the technological barriers to easy computer gaming now under vigorous attack by Intel and Microsoft, and with a propitious cultural buzz surrounding the success of *Doom* and *Myst*, the way lay open to market computer gaming beyond the hard-core niche. As members of the "Nintendo generation" reached an age when they might use a personal computer, they realized it could be used not only to work or study but also to play games. Sports gaming, for example, was primarily played by upwardly mobile male professionals in their mid-thirties, precisely the demographic segment likely to own or use a computer. There was a wide expectation that the computer might generalize this trend and "drive the video game market from the typical teenage male to a much broader demographic base."[31] Game developers were already familiar with computer technology and attracted by its rapidly evolving capacities.[32] Publishing titles for the "open architecture" PC, developers did not have to pay the royalties that console companies had levied. Large companies could cross-develop for multiple platforms, such as Eidos making versions of its famous *Tomb Raider* games for the Sony PlayStation console as well as for Macs and PCs. Whereas in 1995 computer games accounted for twenty-six percent of Electronic Arts' titles, in 1997 they amounted to some thirty-five percent.[33]

This drive to expand computer gaming generated a new variety of titles. Some, such as Sid Meier's brilliant *Civilization* and *Gettysburg*, were more appealing versions of computer gaming's traditional preoccupations with war and strategy. Others, like Maxis' *SimCity*, had more civic scenarios, or deployed the improving graphics and speed of personal computers to make genres like flight simulation astoundingly convincing, while yet others pursued the path of ultraviolence pioneered by *Doom* and *Quake*. One sign of an enlarging commercial circuit was the appearance of titles synergistically tied to popular mass entertainment hits, such as the series of *Star Wars* games – *Rebel Assault*, *Force Commander*, *x-Wing Fighter*, *Rogue Squadron* – produced by Lucas Arts, or the *Star Trek* games published by Simon and Schuster.

But perhaps the decisive indication that computer gaming had been propelled beyond the domain of technosavvy hackers was the surprise triumph in 1998 of *Deer Hunter*. A rudimentary game concentrated on the pleasure of slaughtering virtual wildlife, *Deer Hunter* cost $100,000 to make, compared with the average game development costs of $1.5 million, sold for about twenty dollars compared with the fifty- or sixty-dollar price tag on state-of-the-art games, and could be mastered with minimal computer skills. It sold well over one million copies, staying at the top of best-seller lists for months on end. *Deer Hunter*'s success was greeted with scorn by hard-core gamers whose idol, id Software's John Romero, commented, "I wouldn't want to design a game like that, even if it did go on to sell millions of copies."[34] But for marketing managers, "the *Deer Hunter* phenomena" signalled the beginning of a computer-game breakthrough, demonstrating that interactive entertainment had finally reached "American heartland types" and "the Wal-Mart customer."[35]

GO PLAY IN THE STREET:
GAMING ON THE INFORMATION HIGHWAY

As significant in terms of popular excitement, media buzz, and technological romance was another new form of game delivery and promotion that promised to bypass crowded store shelves – Internet gaming. In 1994, at the very moment when *Doom* and *Myst* were grabbing public attention, the US Clinton/Gore administration announced construction of the so-called "information superhighway," connecting the computers, phones, and televisions of North America into a seamless network that would transform everyday life. The immediate result was a burst of merger and acquisition activity by telephone, cable, and entertainment corporations that were jockeying for position in their bid to reap profits from video-on-demand, tele-gambling, and virtual advertising. Computer gaming was immediately caught up in this frenzy, with online play seen as one of the "killer applications" for the new "highway."

Some of the highest hopes for online play focused on "persistent universe" games, or "massively multiplayer gaming." Unlike most peer-to-peer games, which last only as long as the competitors are playing, in "persistent universes" the attributes of characters and environments are stored in databases so that the "world" exists on an ongoing basis, regardless of the entry or exit of any particular participant. Writing of an early persistent world game, *Meridian 59*, Ashley Dunn writes of how such simulations "have a story, a cast of characters and, yet, they have no beginning or end, simply a flow of little plots that intertwine every day ... In this world, everyone is the main character of the book and everyone's life is the main plot ... So much of what happens in this game just emerges from the interaction of many people."[36] Observing the capacity for surprise and unpredicted development contained in such games, in this commentator's view, they may become the true art form of the digital age, "the novels of the future."

The ancestors of commercial "persistent worlds" are the venerable text-based Internet games called MUDs (Multi-User Domains) or MOOs. Players assume virtual identities within a fantasy setting and interact by typed-in commands, such as "Kill dragon." Virtual worlds of this sort flourished in the days before the Internet was commercially colonized, and almost all were free. What made commodification feasible was largely the move from text to graphics interfaces and the creation of worlds in which players see actual depictions of their characters and environment. This change arguably makes such games more accessible and engrossing, "since everyone shares the same vision, sees the same landscape," though die-hard MUDers defend the text-based worlds as richer imaginative exercises.[37] But graphics interfaces also open far greater opportunity for commodification than text, because the support of such elaborate simulations requires software that needs to be either downloaded from a game site or purchased on a CD-ROM. Users can thus be charged for the necessary software, as well as for ongoing entry to their world of choice. However, the same qualities of unpredictability and spontaneous evolution that make "multiplayer" games so attractive can also create unanticipated difficulties for commercial enterprises.

ULTIMA ONLINE: VIRTUAL REBELLIONS

Perhaps the most celebrated example of the profits and problems of pay-for-play persistent worlds is *Ultima Online*. *Ultima* was the creation of one of the most famous of game developers, Richard Garriott, who emerged from the subculture created by the meeting of the *Dungeons and Dragons* role-playing games and early computers. In 1979, at age nineteen, while working as a computer store clerk in Houston, he wrote his first game on Apple II, a solo-play medieval hack-and-slash epic that earned him royalties of some $150,000 and was followed quickly by two similar games in what became the *Ultima* series and the formation of a game development company, Origin Systems.[38]

What distinguished Garriott from other digital *Wunderkinder* was his attempt to transform the game genre that made his fortune. The early *Ultima* games were governed by the straightforward logic of violent conquest and unrestrained pillage. Then Garriott began to tinker with the formula. "I got bored with it," he said. "And while I certainly don't think I was corrupting America's youth by producing this stuff, I had to sit down and consider the ethical implications."[39] *Ultima IV* altered the logic of the play. To win, the player had to conform to the "eight Britannian virtues" – compassion, valour, honour, honesty, spirituality, sacrifice, justice, and humility. Positive game outcomes resulted from truly chivalrous behaviour. As Herz observes, this was an artistic risk, because "ethical consequences were a new frontier in game design."[40] But the gamble paid off. Thousands of players embraced the additional complexities brought to gaming by *Ultima IV*'s karmic logic, making it Origin's first best-selling game. Subsequent *Ultima* games were widely hailed as "a welcome alternative to the mindless violence of many computer games."[41] Garriott, a.k.a. Lord British, eventually sold Origin to Electronic Arts and now lives in a castle in Austin, Texas, regarded not merely as a successful game developer but as a spiritual guru to his virtual subjects in Britannia.

The alliance of chivalric virtues and digital capitalism was seriously tested when *Ultima* entered the world of online gaming. *Ultima Online*, though not the first persistent-world game, was one of the most elaborate. On entering Britannia, or one of the other "shards," or parallel kingdoms, of the *Ultima* universe, the player could not only hunt ogres and dragons but purchase houses, run shops and banks, enter into alliances, create clans, guilds, and other institutions, and build a social environment in which over time entire communities and populations change.

Origin had made a huge investment in *Ultima Online*, which took two years to develop, had a beta-test involving some twenty-five thousand players, and required an array of servers, technicians, and support staff estimated to cost one million dollars annually. It charged some sixty dollars for the CD-ROM and offered an initial one-month period of "free" play. After this, however, online citizenship in Britannia and its associated subkingdoms cost $9.95 month, a cost in addition to the twenty dollars or so a month most people would already be paying to an Internet service provider. *Ultima Online* sold one hundred thousand copies in three months. But from the moment of its launch, it was plagued by problems, some of them technological, others social.

The social problems had to do with the ethics of online play – particularly with the issue of "PKs" – player killings. Although the solo versions of *Ultima* altered the ideology of mercenary plunder that dominated role-playing games set in medieval times, many who flocked to networked play ignored this reform. Exposed to the unrestrained aggression cultivated in other parts of the gaming culture – including the earlier incarnations of *Ultima* – Britannia rapidly threatened to become a dysfunctional society, "overrun with maniacal, brutal, twitchy-fingered *Quake* killers who are ready to murder anyone on site."[42]

In the *Ultima* world, "death" did not eliminate the player from the virtual world, for resurrection and return was possible. But it did involve the loss of laboriously accumulated attributes and possessions, which could be profitably stolen by the killer. Many gamers were appalled when their elaborately constructed avatars were annihilated in seconds. The easiest targets of attack were novice players, or "newbies," who had a life expectancy in Britannia comparable to that of an infantryman on the first day of the Battle of the Somme. PKing became a serious barrier to recruiting new players, who soon grew disenchanted with paying ten dollars a month to be murdered the moment they put a foot across the virtual frontier.[43]

There were digital as well as social engineering issues. *Ultima* suffered from "bugs." Software failures were compounded by the technical difficulties of sustaining online play. The initial popularity of *Ultima Online* swamped Origin's servers, leading to frequent crashes. When the virtual worlds were restored, participants found that hours of online play had simply been forgotten, in a sort of massive social amnesia. The lag between the issuing of an on-screen command and its actual execution also generated unpredictable effects. These problems were intrinsic to the Internet, but many believed they also reflected the inadequacies of Origin's technological support. A mass of exasperated players vented their frustrations in the magazines and online forums of the gaming world.

The complaints spilled into Britannia itself. In his Lord British persona, Garriott had a castle within the online *Ultima* game world. A group of players disgruntled with Origin's lax response to customer complaints staged a virtual revolt and stormed the gamelord's residence. Once inside, they "drank themselves silly, trashed Lord British's throne room, and protested loudly."[44] Lord British watched the proceedings shrouded in a cloak of invisibility. Garriott's response came IRL ("In Real Life") in the form of an "Open Letter From Lord British" addressed to "the present and future citizens of Britannia," published in game magazines, explaining the steps Origin was taking to deal with the various "technical and creative barriers" confronting *Ultima* – "Server capacity," "Too Few Game Masters," "Bug Extermination," and "System Exploiters and Rogue Players" – and declaring that "just as in the US, where we have passed laws to prevent things like monopolies, so too must we slowly improve the 'laws' of Britannia."[45]

However, promises could not stop the rot. *Ultima* became the object of the "first consumer-led class-action suit in computer game history."[46] It was launched by George Schultz, who in the *Ultima* world was known as Bunster, archer-mage, leader of the Silver Dragoons, and who in real life was a forty-seven-year-old lawyer who had worked with Ralph Nader against General Motors and Ford. A fervent *Ultima* player, Schultz shared the general grievances about its inadequate performance. Acting on the encouragement of other dissatisfied customers encountered through the Internet newsgroups, he sought to pursue a legal action on the part of disaffected players. "The software industry is full of shit," he declared. "For some reason software is the only product in America where the public has accepted the product as being like that."[47]

Schultz's suit attracted widespread interest amongst game players and businesses. The *Ultima* newsgroup was fractured by pro- and anti-Schultz postings; online gamesites offered regular bulletins on the progress of the lawsuit, some of them conducting e-mail polls on the topic. Other online game businesses watched the suit with apprehension. *Ultima*'s attorneys attacked Schultz vigorously for his hyperbolic accusations ("He likened the defendants to makers of the exploding Pinto ...") and racist and sexists postings, in essence portraying him as "a psychotic hustler."[48] They also pointed out that Schultz and his coplaintiffs continued to play *Ultima* even while suing the company. Then e-mail became critical to the dispute, deepening the suit's reputation as a test arena for digital-age legal issues. One of Electronic Arts' lawyers produced an e-mail written by one of the plaintiffs to *Ultima* customer support in which they "admitted they loved the game ... [and] said the real problem was not with the game, but with another player who had killed them and stolen their stuff."[49] In late September of 1998, a judge refused Schultz's request for class-action certification, declaring that it would have had a "chilling effect indeed on creativity and multiplayer game playing on the Internet."[50] However, the judge did allow Schultz and several individual plaintiffs who had joined his suit to go forward on a case-by-case basis. By the terms of the 1999 final settlement, Origin and Electronic Arts did not have to pay any damages, make any refunds of cash or free playing time to the plaintiffs, or pay their costs, but they were required to make a fifteen-thousand-dollar donation to the San Jose Tech Museum of Innovation. Nonetheless, Schultz claimed victory, declaring that the suit had served as a wake-up call to the gaming companies, and the computer industry in general, to take issues about "bugs" and customer service seriously. In 1999, shortly before Origin released the latest version of *Ultima Online*, it closed down the message boards on which Britannia's subjects had voiced their grievances.

MICROSOFT: EVANGELISTS AND ARCHANGELS

Though small companies did the trailblazing in PC gaming, it was not long before giants populated the field, too. By far the most imposing presence was that of Microsoft, information capitalism's most aggressive corporation. In one area of software after another, Bill Gates's corporation has established its products as an industry standard and then "leveraged" that advantage in order to colonize another, adjacent area of digital industry, moving from operating systems to software applications to Internet browsers to cable set digital converter boxes.[51] Microsoft's interest in the game industry is a component of this global strategy, albeit one that was rather late in developing. Even in the early 1990s some ninety percent of PCs ran Microsoft software, making it a potentially crucial "gatekeeper" to computer game development. But the corporation did not take advantage of this position, which perhaps reflected Gates's own unplayful disposition. Its initial multimedia ventures focused on serious or educational products were largely unsuccessful: Windows earned an execrable reputation amongst games developers as an unfriendly platform. This situation only changed with the development of Directx, a suite of programs that acts as an intermediary between Windows and games to ensure that the operating system can handle graphics, video, 3-D animation, surround sound, and other multimedia applications.[52]

Its development was the work of a small group of Microsoft "evangelists," a select team blending technological expertise with aggressive marketing skills: "nerds with charisma."[53] The group that created Directx was unaffectionately known at Microsoft's Redmond headquarters as "the Beastie Boys" because of the aggression with which they garnered resources for what was initially seen as an insignificant project. To win over the game-developing community, the Beastie Boys threw extravagant parties with violent, sexist, and macabre game themes, publicly sneered at the conservatism of their colleagues, and disparaged the known inadequacies of Windows technology. These practices did not win them friends within Microsoft's relatively staid corporate culture. But they successfully attracted the interest of game developers who had previously regarded the company with contempt. Directx had such a high level of expectation that the corporation could not renege on the project – which suggests how crucial marketing is in gaming not only in terms of the industry's relation to its customers but also internally.

Directx was an addition to Windows 95 but integrated directly into Windows 98 and 2000, and Internet Explorer. Since its first appearance it has passed through several versions, going from controlling basic graphics and sound capacity into a means of adding 3-D audio, network services, and force-feedback input features into games.[54] The declared goal is to provide developers with a common set of instructions and components, ensuring that their applications run well anywhere

Windows is found. This is obviously a real benefit to both developers and customers. Equally, it enables Microsoft to become an industry standard setter for the PC game industry. A change to Windows now has far-reaching implications for developers: unanticipated incompatibility between a new version of Windows and a game's installation system, for example, can be a huge problem.[55]

Characteristically, Microsoft has "leveraged" the success of Directx. Having defined many of the norms for game development, it is in a strong position to create or acquire programming tools for game developers.[56] Microsoft already produces Visual Basic and a Visual C++ compiler, both used for developing games. In 1994 it paid $130 million for the Canadian company SoftImage – famous for its high-end animation instruments, which were used to create the dinosaurs in *Jurassic Park* – and redesigned the software to run within Directx on Windows NT. It also hired some of the biggest names in the computer graphics industry, including Alan Ray Smith of *Toy Story* fame, to work on creating integrated multimedia applications for game developers. By 1998 over one thousand companies, including Sega, Sony, and Nintendo, were using SoftImage tools for game development.[57] Microsoft is also developing a digital video engine giving immersive experiences – similar to virtual reality – of space and objects.[58] Furthermore, game designers may use management software such as Microsoft Project to organize complex production schedules in their development studios. All this extends into the arena of game development the "lock 'em in and tie 'em down" strategy by which Microsoft has successfully dominated other areas of programming.[59] The Beastie Boys' project has met their "evangelical" goal of controlling "mind-share" – i.e., getting customers "to develop software for Windows or, at the very least, get them to frame their thinking in terms of Microsoft."[60]

Gates's empire has itself become a game maker, "a late but incredibly powerful entrant to the game industry."[61] Initially Microsoft's ventures in this area were fairly timid, focused on puzzles, card games, and flight and golf simulators, staying carefully away from the controversial areas such as violent games. As its Directx initiative began to take hold, however, it began a far more intense and diversified push on the game market. Microsoft bought up SingleTrac as well as Bruce Artwick, a game startup by former designers of military flight simulators, and established publishing relations with developers such as Atomic Games, Rainbow America, and VR-1. It produced or published such games as *Monster Truck Racing*, the *Close Combat* platoon warfare series, the mystery-adventure *Under a Killing Moon*, and a puzzle game by the creator of Tetris, *Pandora's Box*. It is a partner in Sega's game-based theme parks. Microsoft also has a thirty-million-dollar stake in

DreamWorks Interactive, a branch of the studio headed by Steven Spielberg, David Geffen, and Jeffery Katzenberg, which concentrates on making or marketing games tied to high-profile media products, such as games based on Spielberg's *Jurassic Park* and *Lost World*. As Nathan Newman observes, "The Microsoft leaders have made it clear that the company is in a prime position as movie entertainment converges with interactive media."[62]

The next step was networked games. Microsoft's embrace of the game industry in 1995 coincided with Gates's "Internet Tidal Wave" memorandum, which belatedly turned Microsoft's attention to cyberspace.[63] Online gaming was an obvious area in which to realize this strategy. In 1996 Microsoft bought the Internet Gaming Zone, one of the strongest contenders in the area of networked games, from Electronic Gravity and made it into a site in which gamers could play Microsoft games against one another. Its deep pockets allowed it to tolerate heavy losses in a way that competitors could not afford. By offering free play and drawing what revenues it could from advertising rather than charging subscription fees, the Gaming Zone had by March 1997 attracted more than twice the number of registered players that rivals such as Mpath and TEN had won.[64] Observing the success of *Ultima*, Microsoft has constructed its own "massively multiplayer" games, including the mediaeval epic *Asheron's Call* and the space-combat epic *Allegiance*. It also teamed up with game companies, Hasbro among them, to make online versions of popular boardgames such as *Monopoly*, *Risk*, *Scrabble*, and *Battleship*. As Ed Fries, the manager of Microsoft's Entertainment Business Unit, announced, "We want to be the biggest name in online games."[65]

In order to promote its games to consumers, Microsoft used tactics very similar to the "evangelist" techniques that were used to sell Directx to developers. In this case, however, the key constituency was not programmers but taste-setting "super-fans" and the crucial arena was the Internet. Microsoft was taught the importance of online communication to game marketing by one fan Web site. Started in 1997 by Mike McCart, a forty-four year-old Coast Guard veteran living in Alaska with the online moniker of "Archangel," HeavensWeb.com was entirely devoted to *Age of Empires*. It attracted 180,000 visitors a month and delivered 1.2 million "page views." After running the site alone for several months, "Archangel" finally recruited twelve other fan volunteers, or "angels," to join in maintaining the site. Some analysts ascribe the majority of *Age of Empires'* one million sales to "Archangel." The lesson was not lost on Microsoft. HeavensWeb was eventually taken over by GamesStats, a commercial operation that runs a giant conglomeration of fan sites that delivers 6.8 million page views a month. "Archangel" McCart was hired by Ensemble Studios, the Texas-based firm that created *Age of Empires* for Microsoft, to run its site. Meanwhile, Microsoft hired Arbuthnot Communications of San Francisco to do for money what McCart had done for free. Connecting "game makers to taste-setting fans" is how ArbCom's owner Leslie Arbuthnot describes her role: "We make sure we send them very early beta products. You do that and they become your friend for life. We allow them to sign up for press lists, so they get news about their favourite game at the same time the press does. They are all geeks – they want to be first to find out if there is a new patch for their game, or a new level editor. We make them feel very important, because they are."[66] ArbCom cultivates "a network of super fans" that can "drive a game to hit status," including one hundred and twenty fan sites devoted to *Age of Empires*, and other similar sites for games such as *Asheron's Call.*[67]

CONCLUSION: "TEACHING PEOPLE ABOUT PROGRESS"

It is not just the marketing but also the cultural content of *Age of Empires* that is revealing. It is a strategy game in which players command ancient civilizations – Egypt, Greece, Persia, Assyria, Phoenecia, Rome. They build their economic base, explore surrounding territories, and wage war. What is striking is the way the game inscribes the logic of high-tech capital back into the dawn of ages. As Herz observes, "The grand sweep of human events is expressed as a series of technological upgrades."[68] She quotes Bruce Shelley, the game's lead designer: "The people who get the technology fastest often have a decisive advantage. If you've got a Feudal Age army fighting an Imperial Age army, you're probably going to lose. It's a subliminal message, almost, about technology – and technology did advance quite a bit during this time period. On a basic level, we're teaching people about progress."[69]

The point is not just that inventions in weapons, armour, agriculture, and building have a critical influence as players advance through Stone, Tool, Bronze, and Iron "ages." What is more telling is that *all* aspects of civilization are seen as technological in their logic. Thus, religion and class structure are explicitly described as factors on a "technology tree" that can be systematically "upgraded." One can "research" monotheism or the afterlife in a temple (the ancient equivalent of the Microsoft Redmond campus?) in order to convert enemies, or "invent" aristocracy in order to improve the speed with which one's units move. This world is organized according to a gendered logic well understood by Silicon Valley: everyone is male, and "progress" unfolds absent of women; reproduction occurs somewhere else, apparently asexually by "clicking" on Town Centre to generate peasants or soldiers as required. Every aspect of human endeavour appears as an advance in techno-cratic instrumentality whose significance is measured and rated in terms of its ability to overcome rival "empires."

In this way, *Age of Empires* recapitulates (perhaps even with a certain self-conscious irony) the logic of its corporate producer, and, more generally, of the whole interactive game business during the 1990s. With the appearance on the scene of Sony and Microsoft, interactive gaming became inextricably wrapped into the strategies of synergistic multimedia corporations whose expansions were based on development of ever more impressive, perpetually upgraded digital technologies. While Lord Britain's unruly subjects were conducting their peasant revolts in a virtual world, the empires of Gates and Idei were mustering their material forces. Although the game industry remains too variegated for any single company to dominate completely, Sony, on the console side of the business, and Microsoft, on the computer flank, promised to command the economic heights of the business. It was not long before those two empires would, like Rome and Carthage, Greece and Persia in the ancient world, come into direct confrontation. To understand the dimensions of that confrontation, however, we need to have an overview of the digital play industry at the dawn of the twenty-first century.

8

The New Cyber-City:
The Interactive Game Industry
in the New Millennium

INTRODUCTION: THE MATRIX

You can communicate to a new cyber-city. This will be the ideal home server.
Did you see the movie "The Matrix"? Same Interface. Same concept. Starting
from next year, you can jack into "The Matrix"![1]

Ken Kutaragi, designer of the Sony PlayStation

The film *The Matrix* to which Kutaragi refers is a fable about a takeover of the planet by a monstrous techno-entity that sucks dry the
physical and mental energies of human beings while enveloping them
in a world of deceptive simulations. The prospect he offers sounds a
little less appealing, perhaps, than he intends. Nonetheless, the possibility that interactive games might serve as a gateway to a comprehen-
sively networked world of virtual entertainment and services is certainly
a key element today in making them one of digital capital's most avidly
observed, fastest-growing, and hottest new media industries. In this
chapter we conclude our institutional history with an overview of the
structure and dynamics of the interactive game business, summarizing
some of the main themes of our historical narrative while offering a
snapshot of the industry as it stands at the beginning of the third
millennium – a key node in the networked environment of virtual
capitalism. If the medium is indeed the message, then the subtext of
Kutaragi's utopia is the enclosure of cultures of play by the imperatives
of the mediatized marketplace.

Remembering our earlier suggestion that interactive games could be
seen as an "ideal commodity" of post-Fordist capital, we structure the
overview to follow some key discussions about the economic, technological, and cultural implications of the new digital regime of accumulation. We begin by looking at the corporate structure of the industry

and the light it sheds on debates about whether post-Fordism favours small electronic cottage industries or the giant media empires. One element that is ignored in most of the post-Fordist literature is the ongoing but significantly altered nature of interaction between the military and commercial industries in propelling digital development.

Turning to the consumption side of the business, we review the most recent data on the scope and composition of the market for video and computer games. This reveals characteristically paradoxical post-Fordist features of *Fordist* mass media, insofar as gaming is now a widely disseminated, large-scale, and intensively marketed form of commercial entertainment, and of *post*-Fordist emphases of segmentation and niche targeting, as gamers are split up by age, brand, and technological sophistication into a series of interlinked micromarkets. These markets are being expanded, moreover, beyond the limits of in-home privatized domestic entertainment into new cultural spaces as online games and multiplayer worlds are constructed as cyberspatial portals to net-savvy youth markets. All these developments are taking place in the context of an increasingly transnational organization of the post-Fordist market, in a so-called "globalizing" dynamic that, while it reaches widely across the planet, still orbits mainly around the hubs of the advanced capitalist economies. Dramatic as the expansion is, it remains shot through with instability and crisis. Our "state of play" report concludes with an evaluation of the game industry's situation in the wake of the bursting of the Internet bubble and the dot.com crash, events that call into question the viability of the information capitalism of which the game business is now so integral a part.

ELECTRONIC COTTAGE INDUSTRY OR GLOBAL IMAGE EMPIRE?

Central to debates about post-Fordism are issues of corporate size and power. In one popular version, post-Fordism is the era of little high-tech companies. New digital production – so the story goes – gives advantages to small nimble firms that can swiftly adapt to changing technologies and market conditions. In the most optimistic versions of such theory, post-Fordism is portrayed as an era of digital artisanship, based on a multitude of small enterprises, organizing production as high-technology craft-work and forming vibrant community networks of efficient but human-scale business.[2] But attractive as this post-Fordist "small is beautiful" picture may be, it has earned some well-deserved scepticism. Critics point out that in industry after industry the benefits of digitization accrue most richly to large corporations, be it Microsoft in software, AT&T and MCI in telecommunications, or

Time Warner/AOL, Disney/ABC, Viacom or News Corporation in the field of entertainment and news. From this point of view, the passage from Henry Ford to Bill Gates may have involved changes in the technologies and organization of production, but it has not in the least diminished the pull towards concentration of ownership, oligopolies of knowledge, and corporate gigantism of "global image empires."[3] Paradoxically, either side in the debate could cite the game industry to support its version of the post-Fordist digitally mediatized marketplace. Depending on which aspect of the business one focuses on, digital entertainment can be invoked either as a case study in the blossoming of small-scale high-tech industry, or as evidence of the expanding power of media conglomerates.

To understand this complexity, we can look at a map of the industry (diagram 7). At the top are three industrial sectors from whose intersection the digital games business emerged: the computer industry, which produced the technologies, both hardware and software, that are the basis of interactive entertainment; media conglomerates involved in film, television, and music, which provided models for on-screen entertainment content and organization; and the toy industry, which pioneered the commodification of children's play and on whose domain video gaming has increasingly encroached. All three were involved in the genesis of interactive gaming. Atari was a spin-off of the computer industry. Toy companies such as Mattel, Coleco, and Hasbro tried to garner a corner of the early video game market. Warner Communication, a huge entertainment conglomerate involved in movies, music, and publishing and the ancestor of Time Warner, bought Atari in 1976. The volatilities of the new industry, however, particularly the Atari "meltdown" of 1983–84, made many of these established interests withdraw, returning only in the 1990s when interactive entertainment again burgeoned. In the meantime, the game industry evolved its own distinct structures and dynamics.

PLATFORM OLIGOPOLIES

One of these new complexes involved the firms making the platform technologies on which games are played – video game consoles, personal computers, and arcade machines. Pre-eminent amongst them are the console companies – shown on the second level of our diagram – which make the hardware for home video game play, and whose fortunes we have followed in our history of the industry. As noted earlier, they operate on a "razor and blades" model. Since the early 1990s console making has been characterized by tumultuous oligopoly. That is to say, there has always been only a handful of major players,

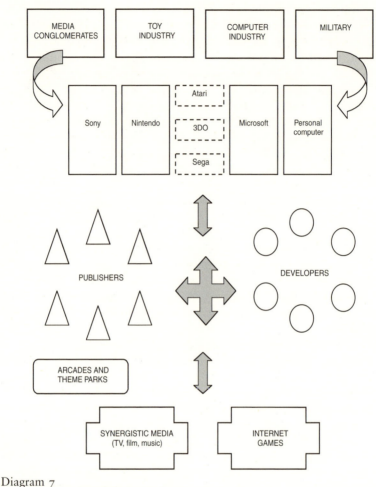

Diagram 7
The Interactive Game Industry

but their identity and number have fluctuated depending on the outcome of fierce internecine competition. In 2000 there were three: Sony,
Nintendo, and Sega.

The other main digital game platform is the personal computer, shown
on the right of our diagram. Because the PC is a multipurpose device,
there is no precise equivalent here to the big console companies. But
gaming is an immensely important aspect of the personal computer
business as a whole. In fact, according to the Interactive Digital Software
Association, entertainment software is the most frequently used application on all computers. At the end of the 1990s games accounted for about
thirty percent of all PC software sold at retail in the US, and some ten to

fifteen percent of total software sales, probably with higher percentages in Japan and Britain.[4] Many commentators suggest that interactive gaming drives the overall trajectory of the computer industry, because the high demands for processing speed, graphics display, and networking capacity made by the hard-core gaming culture set technical standards that later trickle down into the more mundane requirements of the business sector.[5] For these reasons, the game market affects all aspects of the computer industry, from chipmakers to manufacturers to network builders and software producers. Here, Microsoft is pre-eminent. We have seen how over the last decade Gates's empire has moved into the game business in a major way, making or publishing its own games, implanting its Directx technology, authoring tools, and Internet browsers, and running its own online gamesite.

As the interactive game industry entered a new century, a new round of innovation again destabilized its technological circuits. The console seemed on the point of "reconverging" with the computer from which it had separated some thirty years earlier. Sony, Sega, and Nintendo, driven both by internecine competition and by the increasing intrusion into the game market of the PC, were thrusting towards the next "level" in console development – the 128-bit machines that were powerful enough to rival personal computers – while Microsoft broke out of its bastion and entered into the renewed console wars.

The first of the new machines, Sega's Dreamcast, appeared in 1999, complete with Internet connection, a CD-ROM drive, and a Windows operating system.[6] For Sega, it was a last-ditch play. Ever since its catastrophic Saturn venture, the company had been a lagging third in the console field. In 1997 it attempted to acquire Bandai, Japan's largest toy maker, famed for its *Power Rangers* and *Sailor Moon* products. The $1.1 billion deal would have created a corporation more than twice the size of Nintendo but it fell apart because of conflicts between the managerial cultures of the two corporations.[7]

Similar problems bedevilled the Dreamcast launch. Deep differences split Sega's Japanese executives and US managers, the Japanese favouring an Internet-only marketing strategy that would have bypassed the retail networks so painfully rebuilt by US staff after the Saturn débâcle, and awarding Japanese game developers higher royalty payments than US software companies. These difficulties in coordinating the global circuits of the industry resulted in the departure of the head of Sega's US office on the very eve of the Dreamcast launch.[8] Demoralization, loss of support from US game developers, and customer awareness that Sony was going to release its own advanced console doomed the Dreamcast, which sank like a stone. In 2000 Sega, which once seemed set to dominate the interactive game world, announced it would write off $688 million in losses and retire from the console business to specialize in software development and its arcade/theme park interests (shown on the lower left of our diagram), its parables of rise and fall a striking demonstration of the promises and risk of high-technology post-Fordist markets.[9]

This left a three-way fight between Sony with its PlayStation 2, which was on the market in 2000, Nintendo with its GameCube, and Microsoft with the xbox, the last two released in the following year. In terms of technology, the xbox was the most advanced machine, with hard-disk drive, extra memory, and Net ports. Its rivals lacked some of these features, though they could be retrofitted with network connections and more storage capacity at an additional price. However, earlier console wars showed that success does not necessarily go to the firm with the highest performance in the technology circuit but depends crucially on the cultural content of accompanying games, and on shrewd marketing strategy.

Many commentators speculated that the technological powers of the xbox would be of little avail if Microsoft failed to find a smash-hit title with which its new machine would be identified. Here, Sony's extensive portfolio of successful games, such as *Metal Gear Solid* and the *Final Fantasy* series, and the possibility that the PlayStation 2 might eventually be linked to networks accessing multiplayer games such as *Everquest*, give it an advantage – one also enjoyed by Nintendo with its series of established franchises such as *Pokémon*, *Zelda*, and *Mario*.

Although the three companies were predicted to spend a combined $1.5 billion worldwide in marketing campaigns, Sony and Nintendo both had deep experience in hyping games, while Microsoft's essentially conservative brand image seemed ill adapted to the transgressive, twisted ethos of game marketing. Marketing is not just an issue of overall spending, moreover, but of market segmentation: Microsoft was directly challenging Sony in targeting older male players in the eighteen-to-thirty-four age range, while Nintendo was somewhat insulated from this competition by its focus on the younger market.[10]

Predictions about the outcome of the new console wars were therefore mixed. Some analysts favour Microsoft, betting on the fact that the console maker that dominates one generation of game machines has never managed to keep its lead in the next. Others predict that Sony's superior games and its sophisticated marketing knowledge will allow it to come out ahead, with slightly more than forty percent of the world console market, and Nintendo and Microsoft dividing the rest. What is certain is that for all contenders, the contest will be extremely expensive. The new machines, especially Sony's and Microsoft's, are very costly to make. As usual in the console business, they will be sold as loss leaders. Microsoft has to sell eight or nine games for every xbox to break even.[11] Paradoxically, the more xbox units Microsoft sells, the more money it may lose. Some internal sources suggest the company will lose eight hundred million dollars over the xbox's first four years, and two billion over eight years before it can even think of turning a profit.[12]

The stakes that justify this kind of risk are very high, however, and go beyond the control of the game industry. The real promise of the new consoles is that they will be a key node in broadband Internet entertainment networks, accessing games, films, and music, and they will operate as a "Trojan horse" bringing e-mail, Web browsing, and e-commerce to millions of consumers. Although several information empires are jostling for dominance over this crucial cyberterritory, two of the main contenders are Microsoft and Sony. As Sony's Kutaragi said of PlayStation 2, "One of my goals is to take entertainment even further, from games to a fusion of games, music, and movies. Eventually, I'd like to see a world where everyone's PlayStation is connected to a broadband network."[13] In grappling with this tension between niche media and mass audiences, it is worth recalling McLuhan. "If, finally, we ask, 'Are games mass media?' the answer has to be 'Yes.' Games are situations contrived to permit simultaneous participation of many people in some significant pattern of their own corporate lives."[14]

For Sony, the PlayStation 2 is part of a comprehensive strategy to retain and extend its long-established dominance of home electronics into a world of integrated and interoperable domestic networks – DVRs, TVs, cellphones, memory sticks, PlayStations – which would in turn connect to the larger networks of Sony's film, music, and game operations, making the corporation a true Internet-economy giant, what its president, Nobuyuki Idei, likes to term the "first broadband entertainment company."[15] Microsoft is also aiming to control this territory, in which its crucial beachhead is the personal computer, with the Explorer browser firmly welded to the Windows operating system. In this version of the future, Windows-based systems will be at the centre of an expanding array of smart peripherals, as well as acting as the main point of connection to the Internet. Although the xbox is not intended to usurp the central role of the personal computer in this configuration, it is part of Microsoft's overall Windowscentric strategy, an attempt to focus game development within its galaxy and to prevent Sony from seizing control of the game systems that will be a major component of networked entertainment.[16]

The outcome of the Sony-Microsoft battle over a "post-PC" or "PC-plus" future is unpredictable.[17] The protagonists may fight to the finish to become the industry standard setter for digital households, shrink from outright conflict in favour of various collaborative/competitive options, or find themselves outflanked by other contenders seeking to dominate broadband entertainment networks. But while on the surface the conflict appears to be simply one more round in video gaming's recurrent console wars, the terrain over which it is fought has been transfigured, as PCs and video game consoles increasingly converge. At the moment they remain distinct devices and may for some time represent distinct market segments; there is a significant niche too, largely commanded by Nintendo, for small hand-held game devices. But both console and computer gaming platforms are now components in an ensemble of digital communication and entertainment devices that are increasingly integrated and interconnected with one another under the canopy of branded media empires. The game industry is no longer a discrete and distinct sector. All its circuits, technological, cultural, and promotional, have become intertwined with the wider orbits of an e-capitalism betting on digital networks as the critical zone for growth and profit. The advantages clearly lie with giants such as Sony and Microsoft which vie for control of the software standards and network protocols that tie together these virtual collocations of profit.

SOFTWARE CHURN

Where the view of the gaming industry as a showcase of artisanal small-scale post-Fordism finds better support is in the area of software development (on the right of diagram 7). Software – the games themselves – is the lifeblood of the industry. The creation of a digital game can be divided into distinct stages: development, publishing, and distribution. Development is the design and creation of the game. This expensive process may be financed either from the developer's own pocket or by venture capital, or, increasingly commonly, by advances from a game publisher. Publishing involves the overall management of the game commodity – manufacturing, packaging, and promotion. Distribution entails getting the games to retailers and other outlets. In some ways, the game software business resembles the music industry, from which it has often borrowed models for managing and marketing creative talent. The developer is like a band, the publishing house like the label. In the case of games, however, the picture is complicated because development, publishing, and distribution functions can be variously combined or separated. A single company can perform one, two, or all three of these activities. A game-development company may sign a contract with a game publisher, who in turn contracts with a specialist distributor. But a publisher may also develop some or all of its games in-house, or handle its own distribution.

Without sufficient variety of high-quality games, console and computer play alike would die. But "no one company has the resources, in terms of either money or personnel, to monopolize software creation."[18] From the very early days of the industry, hardware companies have found it necessary to encourage developers and publishers of software. For console games, developers must buy rights from console-making companies to make games that will play on their machines. In the case of computer games, the open architecture of the personal computer makes this relatively easy. The result is a very complex system, with a fluid and bewildering set of arrangements and considerable diversity in the scale and organization of companies. The IDSA describes the sector as including "companies that began as poorly financed start-ups and those that sprang full-blown from major media conglomerates; those with a vested interest in feeding their own hardware systems and so-called third-party publishers who develop titles for the dedicated hardware systems; companies who serve a broad range of categories, and those who specialize in individual niche markets."[19]

There is a constant churn of new companies created either by "people beginning their careers or companies formed by developers who leave larger companies to start their own."[20] From the time when Bushnell started Atari with two fellow engineers, each putting up one hundred dollars, the industry has been full of tales of small enterprises fuelled by creative talent and entrepreneurial energy, bootstrapping themselves to fortune: games development is often characterized as "the ultimate cottage industry for the information age."[21] Many of the most important games have been made by what were initially small development enterprises: *Doom* by id, *Myst* by Cyan, and *Ultima* by Origin. But the electronic cottages often serve an important economic – and creative – purpose for the media empires. As Mosco explains, "Although the number of independent [media] production companies grow, these absorb high product risks and labour costs for the giants."[22]

Escalating production and advertising costs drive towards the demise of small software development companies.[23] It is now accepted that the days of the "lone wolf" developer are over. But there are counter-vailing factors: even where large conglomerates have entered the business, they often have difficulty assembling talent and choose instead to arrange alliances with smaller, more creative companies. Star developers who rise to fame working for large production and publishing houses but chafe under corporate deadlines depart, taking whole production teams with them to start their own companies. On the surface, this reinforces the sense of game development as a bubbling field of new homegrown initiatives, although "the result is actually more stagnant (though competitive) than it may seem, as many of these development

teams tend to be ex-employees of the largest companies jumping ship"
and signing deals with their former bosses that leave them freer to
create and deliver at their own pace.[24]

PUBLISHING BOTTLENECK

As in many media industries, the high ground for strategic control of
interactive game revenues lies not in production but in marketing and
distribution. Game publishers (on the left of diagram 7) control the
vital bottleneck through which the game has to flow on its way to the
consumer's dollar. The flood of titles means that, in the words of
Interplay president Brian Fargo, "You've got to scream to be heard."[25]
The result is that access to retail outlets is limited to developers who
have links to large publishers with marketing clout. Since the 1980s,
the battle to master this strategic channel has seen a "cycle of consolidation ... with smaller companies being bought or absorbed by larger
entities."[26] The result of "mergers, acquisitions, and failures [has been]
the rise of a dozen 'super publishers' that control much of the industry."[27] These include Nintendo, Sony, Sega, and Microsoft but also
companies such as Activision, Broderbund, Electronic Arts, Info-
grames, Interplay, and Virgin Interactive Entertainment. As one report
said: "Many are predicting that ... the game publishing industry will
come to resemble the record industry, in which four or five publishers
control 85 percent of the market."[28]

This stranglehold is particularly powerful on the video game side of
the market, where it is consolidated by the console company's proprietary control over which games are licensed for their machines. The
open-platform architecture of the personal computer allows for greater
diversity. Here online publishing offers small developers the possibility
of dodging this barrier to entry into the mainstream market. Selling from
Web pages and using gamesites that have credibility in gaming subculture
to promote and sell games may open important opportunities for niche-
market games – for example, the small but lively sector of "grognard"
war gaming centred on historically accurate military simulations.

As we show in detail in chapter 10, the whole process of game
publishing is increasingly integrated into the synergistic webs of multimedia giants, through film and television spin-offs (all the major
Hollywood film studios have investments in or alliances with digital
gaming enterprises, as do more specialized companies such as Lucas
Films and DreamWorks), the renewed presence of toy companies
(Mattel and Hasbro have major video game divisions), cross-promotions
and merchandising deals, and, of course, as a component of the
projected and possibly proprietarily closed systems of broadband

online entertainment – for example, through the connection of publishers such as Sony and Electronic Arts to AOL/Time Warner. Although the system is at the moment still porous, the dynamics of marketing and distribution push powerfully towards the "mainstream" a relatively restricted repertoire of games promoted by well-connected publishing houses, and towards marginalizing or asphyxiating the projects of developers outside this circle.

The interactive gaming industry thus dramatically illustrates the contradictory forces shaping the structure of the information age corporation. On the one hand, software creation is populated by a varied, changing set of developers whose practices seem, at least at first sight, to support the idea of diverse decentralized digitalized post-Fordist artisanal enterprise. This sector is played upon, however, by powerful forces that drive towards consolidation and concentration of ownership and favour promotionalization and standardization of game content. It is also surrounded by other areas of business controlling crucial routes of access to the marketplace where the benefits of size are much more apparent.

THE MILITARY-ENTERTAINMENT COMPLEX

In April 2000 Japan's trade ministry placed limits on the export of Sony's PlayStation 2 game console on the grounds that the machine's sophisticated software could be adapted as a missile-guidance system, presumably by terrorists of "rogue states" such as North Korea.[29] The incident is a plangent reminder of the military roots of digital play. And it points to a little-discussed aspect of the Fordism/post-Fordism transition: the continuing but changing role of the military in driving technological innovation, spurring economic growth, and setting cultural agendas.

As we have seen, digital games, like the Internet, arose not within the free market but in the Keynesian welfare/warfare state that was a crucial component of Fordism.[30] (See top of diagram 7.) Interactive gaming is a spin-off of the military-industrial complex – indeed a derivative of nuclear war preparations, the two first digital games being a primitive *Pong*-like tennis-playing game made in the Los Alamos nuclear labs, and *Spacewar*, mentioned earlier, a military simulation "hacked" into being by defence-related workers at MIT.[31] The spectre of these military/cold war roots is a chilling confirmation of McLuhan's remark that "the social practices of one generation tend to get codified into the 'game' of the next ... the game is passed on as a joke, like a skeleton stripped of its flesh."[32]

Military derivation is by no means unusual in the history of media. But with digital games we are dealing with more than just originating military influence. The interactive play business is characterized by a far more organic, persistent relation with the weapons complex; not just "spin-offs" from the military to civilian applications but "spin-backs" and "spin-ons." Today, in the flexible, numerically downsized, partially privatized, but very high-tech organization of the post-Fordist military, Pentagon simulation makers constantly transfer technologies to commercial game making, while the military frequently contract services from, adapt the products of, or enter into commercial code-velopment partnerships with civilian industry – making interactive gaming the most persuasive instance of what has been dubbed the "military-entertainment-complex."[33]

US military doctrine now gives digital technology a pivotal role in maintaining superiority over a post-cold war world rife with nuclear, biological, and chemical threat. The ability to move seamlessly between simulation and action, effortlessly reinscribing the lessons from real war into virtually mediated scenarios and vice versa, building up a flexible reservoir of prerehearsed and instantaneously realizable operations, is a critical part of this role. In his *Virtuous War* (2001), James Der Derian maps what he terms the "military-industrial-media-entertainment net-work" – or, as he felicitously dubs it, MIMENET – where the conjunction of advanced military planning, computer simulation, and the simula-tion-designing expertise of film and video game companies is creating "a new configuration of virtual power."[34]

Amongst the sites Der Derian visits on his road trip through this new complex of information-age militarism are the US army's digitized Advanced Warfighting Experiment in the Mojave Desert where US soldiers and their "Krasnovian" opponents fight, wrapped in cyborg systems of portable computers, satellite linkups, and networked sensors that inform them of the lethality of their actions; the "x-files territory" of DARPA where a "Synthetic Theater of War" is being constructed using overhead reconnaissance, orbiting telecommunications plat-forms, and massive parallel computing to integrate virtual and live war simulations; and the "Institute for Creative Technologies" where the Pentagon and Hollywood announced at the University of Southern California a forty-million-dollar collaboration bringing together mili-tary gamers, computer graphics artists, and entertainment executives to explore shared interests in simulation technologies.

Of the many bizarre stories Der Derian tells, one that stands out is the revelation that the war in the Persian Gulf in 1990 had been "gamed" by *both* sides in advance. The Americans had anticipated Iraqi plans on a highly digitized exercise, "Internal Look," whose deploy-ment scenarios were available on the computers of General Norman Shwarzkopf and other US commanders and played a major role in tipping the balance for intervention. The Iraqi's also simulated their attack, using a wargame they had bought from the US firm BDM for use in the Iraq-Iran war, now customized for an invasion of Kuwait.[35]

Der Derian argues that we are entering an era when critical areas of defence and foreign policy are dependent on virtual forms of military planning. This, he suggests, accelerates a logic of *realpolitik* in which capability always outruns morality. Once we "know" – by means of simulation – what the outcome of, say, bombing Baghdad or nuking Beijing or invading Kuwait will be, the impetus to adopt the "winning" strategy becomes almost irresistible. He further suggests that the regime of "defence simulations" – rehearsing military action – and "public dissimulations" – mystifying or denying these plans – creates a "potentially permanent state of interwar" in which the distinctions between war and peace become elided.[36]

As Der Derian points out, a critical impulse behind the new alignment of the military and entertainment industry was the realization by US war planners that "the commercial sector, in particular the film, computer and video-game industries, was outstripping the military in technological innovation." Indeed, "where trickle down from military research on mainframe computers once fueled progress in the field, civilian programmers working on PCs could now design video games and virtual environments that put military simulations to shame."[37] But although Der Derian points to this general dynamic, he does not unpack it in any detail.

A closer look at the interaction between the computer game industry and the Pentagon allows us to grasp the dynamic more clearly. Pearce provides an abbreviated history of this development: "The end of the cold war resulted in 'military downsizing,' which left a whole lot of military contractors in the lurch. So, they looked around and they asked themselves, what can we do to maintain our business? What are we going to do with all these fancy gizmos we've been developing? And the most important question of all, *Who but the military can afford us?* The answer was, of course, the entertainment industry. And thus, the Military Entertainment Complex was born."[38] The innovation exchanges between the military and entertainment sectors has been described as the "war-chest-turned-toy-chest."[39] The military might have had a lead in innovation over other sectors during the cold war, but the end of the war is said to have allowed the entertainment industry "[to enjoy] the use of these toys without having to pay for their development."[40] But in the 1990s commodification synergies behind technological innovation took a twist in the development of a new generation of simulation games.[41] The industry could avoid some of the costs of innovation by persuading the military to put its own research-and-development investment to work not just in war but also in the video game marketplace.

The case of a Massachusetts-based company, MÄK Technologies, provides a good illustration of these new partnerships. MÄK reports that it was awarded the first so-called "dual use" computer game to be "co-funded and co-developed by the (United States) Department of Defense [DOD] and the entertainment industry."[42] Working primarily with defence contractors, MÄK's specialty is a "new way of training, known as SIMinterNET, [using] low-cost PCs, the power of Internet technology and the thrill of interactive video games to teach decision-making, reinforce team-tactics and perform mission rehearsal."[43] In April 1997 MÄK announced a contract awarded by DOD's Small Business Innovation Research Program to develop for the Marine Corps an amphibious assault computer game to be used to train in "warfare capabilities in the areas of Command and Control, Maneuver, Firepower, and Logistics."[44] The contract was valued at seventy thousand dollars, a mere speck in DOD's budget, but it came with an option to increase to eight hundred thousand dollars. It is less the financial amount, however, than the bare fact – and its candid promotion – that makes it noteworthy. MÄK worked with the computer game publishers BMG and Zombie Virtual Reality Entertainment to adapt a version of the game, a tank simulation eventually published as *Spearhead* by Interactive Magic, for the commercial gaming market.[45] "The realistic mode will closely follow military tactics so that it can be used by the Marine Corps for training and education to enhance future amphibious warfare capabilities. The fun mode," on the other hand, "will provide instant gratification to the game player."[46] The company boasts that its games will surpass the quality of previous military-training video games: weapons will be more realistic, as will ammunition supply, the consequences of injury, and the weaponry "power."

Warren Katz, MÄK's chief operating officer, says that the game-development relationship with the Department of Defense is intelligent from the perspective of military preparation because it takes advantage of the highest level of precision design and skills-training technology.[47] From a business perspective it offers a marketable product, a portion of whose production costs have already been absorbed by the military agency. He explains that the contract "represents a major step for the DOD ... in that they are recognizing the benefits of collaborating with a commercial video game publisher from the beginning of the design process."[48] What this kind of partnership allows, he says, is the production of "a video game which is much more realistic than any other game ever produced for this genre, making its commercial success highly likely."[49] Perhaps DOD holds part of the game's license, which would help explain Katz's remark that DOD will earn "the cost benefit of unusually large volume sales for a military training device."[50] It is a strategy to lower the costs of production, and clearly it not only represents new ways of marketizing the military but it also reveals the new mediatized synergies between the market and the military.

DOMESTIC INVASIONS AND DIGITAL DIVIDES

It is in the sphere of consumption, however, that the influence of interactive games has been most dramatic. As we have seen, many theorists suggest that a crucial aspect of the transition from Fordism to post-Fordism has been the capitalist shift in emphasis from "material" to "experiential" goods, expanding markets for incessantly exhausted and renewed entertainment commodities that penetrate into every available space and moment of everyday life.[51] Video games can be seen as an exemplar of this new regime of consumption. Consoles such as the Atari 2600, the Nintendo NES, and the Sega Genesis, relatively inexpensive and piggy-backed on the already pervasive television set, introduced an entire generation to the uses of digital technology and the pleasures of virtual consumption. Later, games were important in creating the mass market for home computers. But gaming is not a global village of universal participation. Thus, we must examine the unequal conditions of the cultural practice of interactive gaming by considering the biases inscribed at the point of access to this medium.

As with other domestic technologies, access to interactive play has been marked by great unevenness and deep digital divides falling along lines of class, gender, race, and age. As late as 1994, according to a study by the research company Alexander and Associates, half of US households had no game-playing system – neither console nor computer. Moreover, there was a sharp market division within interactive play between console and computer gamers. The PC household was older, less likely to have children at home, predominantly white, and possessed of a much higher income. The console-only household was younger, with a significantly lower income (about $20,000 a year less), and was more likely to be black or Asian. Such households were described as representing "middle to lower income" families. Noting the sharp divide between console and computer players, the study observed, "these two groups are radically different ... Infrequently in the analysis of consumer behavior patterns do we see such distinct differences between the owners of products that are often linked together in the same category."[52] The interactive game market was thus not only marked by a split between have and have-not households but also by a clear fracture between high- and low-end consumers.

Since the mid-1990s, however, the picture has changed somewhat. Both consoles and computers are now far more widespread. Aggregate data from a variety of sources suggest that in 2000 some thirty-five to forty percent of a total of one hundred million US households own consoles: if one considers only families with children, the proportion would be much higher.[53] Over a relatively short period access to computers has also grown sharply. In 2000 some fifty-one percent of US households had computers and about forty-two percent had in-home Internet access; government reports predicted that this latter number would rise to more than fifty percent by 2001.[54] Although the personal computer is a multipurpose device, gaming is one of its leading applications. Market research companies report that twenty-three percent of computer households name gaming and entertainment as their most frequent use of the home computer, and twenty-one percent as the second most frequent.[55]

Given these figures, the claim of the Interactive Digital Software Association that in North America some sixty percent of the population, or 145 million people, play interactive games of some sort is not incredible – although such aggregate data say nothing about nuances of practice ranging from casual to obsessive.[56] Alexander and Associates report that by 2001 thirty-six percent of US households owned dedicated video game consoles or hand-held devices.[57] This represents a significant widening out of a market niche that was originally defined almost exclusively in terms of adolescent and pre-adolescent males. The Alexander and Associates study estimates that, as of 2001, 64.9 percent of console-owning households were white, 16.5 percent African American, and 13.5 percent Hispanic.[58]

One of IDSA's more startling claims is that there has been a dramatic shift in the gender composition of the gaming population: its "state of the industry" report for 2000–01 alleges that in 1999 approximately thirty-five percent of frequent console gamers and forty-three percent of frequent PC gamers were female.[59] Again, the failure to identify how these numbers relate to frequency of gaming makes the data rather vapid. In chapter 11 we examine these statistics critically and look at how marketing aimed at boys and young men has biased the culture and experience of interactive play. For the moment it is sufficient to say that while the scale of the IDSA claims warrant scepticism, a variety of evidence suggests that more women and girls are indeed playing interactive games.

Digital play has also spread outside the teen and preteen bracket – a result of the maturing of generations of young players and the successive "up-aging" marketing strategies of game companies. One market research report suggests that while some twenty percent of the gaming population are seventeen or under and eighteen percent between eighteen and twenty-four, nineteen percent are twenty-four to thirty-four, twenty-three percent thirty-five to forty-four, fifteen percent forty-five to fifty-four, and five percent over fifty-five.[60] Details of the occupational and income composition of the gaming population are more difficult to obtain. Alexander and Associates report that, as of 2001, the mean income of an N64 household was US$50,500.[61] So despite the increases in computer and console ownership, digital divides continue to break sharply along lines of class, income level being the single surest indicator of computer and Internet access. Some of the expanding sales of game technologies are occurring through the *intensification* of gaming within affluent multiplatform technologically cocooned homes, complete with "wired" bedrooms and dens for individual members, rather than through *expansion* into lower-income households. Contrary to predictions that the personal computer would take over the game market, the console market seems to have benefited from the competition, as households with computers added a dedicated gaming machine to avoid conflicts over the use of digital devices.[62] Whereas according to a 1994 survey fewer than ten percent of US households owned both consoles and computers, a 1999 study put dual-ownership households at around twenty-five percent of the total and also found that many households owned more than one console. Multiple ownership also appears in the computer market.[63] Nonetheless, the relative cheapness of consoles, the advent of computers that cost less than a thousand dollars, and hand-held game devices are clearly widening the availability of digital play.

All this means that, at least in North American terms, gaming cannot be considered a luxury pastime. Video and computer games are not as common as the radios and televisions that can be found in nearly every North American and European household, but they can be fairly characterized as a "mass" medium. Interactive gaming, however, is far more stratified and segmented than, say, broadcast radio and television were during the 1950s and 1960s. Aggregate counts of the game-playing population include both the hard-core computer gamer with a PC worth more than three thousand dollars, complete with the latest Pentium chips, 3-D accelerator and sound cards, high bandwidth Internet connection, and a library of highly specialized games, to gamers on an "obsolete" 32-bit console with a couple of Nintendo or Sega classics. The interactive game market comprises both these extremes and many intermediate categories. This complexity is compounded by the different targeting strategies of game companies as Sony and Microsoft pursue older video gamers, Nintendo caters to younger players, and computer gaming companies sometimes zero in on fairly specialized gaming cultures. In that sense, digital play, although in some ways a mass medium, is not a standardized and homogeneous "Fordist" market but rather an ensemble of connected yet distinct "post-Fordist" niches within which gaming is socially and culturally structured.

COLONIZING CYBERSPACE

At the same time, interactive play is reaching out from its basis of privatized in-home play to lead the commercial development of cyberspace (shown at the bottom right of diagram 7). As we have seen, networked play – the connection of the machines of two or more players so that they compete or cooperate within a shared game world – has been a component of interactive gaming since its earliest days. But its large-scale adoption represents a quantum leap in the nature of digital play. Instead of games simply connecting human *to* machine, they now also connect humans *through* machines; skills are tested against people rather than artificial intelligence. This interaction occurs within programmed constraints but offers far more unpredictability and surprise than even the best synthetic opponent can supply. As with most other forms of human-to-human contest, network gaming generates social relationships, from abusive taunting to convivial get-togethers, that extend beyond the moment of play. Most of the activities that surround physical sporting events – postgame discussion and argument, formation of teams and leagues, tournaments, game lore and gossip – also surround network gaming, making it one of the most important incubators of so-called "virtual communities."[64]

The promise of online play pioneered by such games as *Quake* and *Ultima* continues to hold a fascination for the industry, even though it is not always substantiated in terms of revenue. Evaluations of the prospects for Internet gaming vary wildly. Johnny L. Wilson, editor-in-chief of *Computer Gaming World*, predicted that "online gaming is going to be the dominant form of gaming around the middle of the next decade."[65] Another researcher observed that "on-line gamers are a bit like Christian fundamentalists; they receive attention that is wildly disproportionate to their numbers."[66] This uncertainty about whether the Internet gaming cup is half full or half empty reflects some basic demographic data. In 2001 about half of North American households had in-home Internet access – or, to put it another way, half didn't. And only a much smaller fraction, about ten percent, had the high-bandwidth connections necessary for reliable online play.[67]

The rhetoric of smooth, instant connectivity touted in advertisements for Internet games masks a reality of considerable frustration, mishap, and expense, including long download times, sudden "crashes," unexpected incompatibilities, and an accretion of extra charges that pile up on top of basic Internet service fees. Players with slow connections are at a notable disadvantage, especially in "twitch" games such as *Quake*. "Latency," or lag, the factor that makes the World Wide Web a "World Wide Wait" and results in "jerky" game play, is a prevalent problem.

A survey by Forrester Research found that only three percent of 1998 game industry revenue could be attributed specifically to Internet play and that only fifteen percent of the owners of *Quake*, usually considered a quintessential online game, actually played in cyber-space.[68] The survey speculated that by 2002 such play would account for close to a quarter of an estimated eight-billion-dollar industry total, but it also observed that there were serious barriers to growth: "There's a learning curve associated with getting online with your game and it's just not that easy to do yet."[69]

Mass audiences for online play, even in North America, depends on major improvements in the universality and quality of digital infra-structures. This places the gaming industry, and digital capital in general, on the horns of a dilemma. Investing substantially in the creation of online content, including games and gaming sites, requires confidence that many households will in fact be "wired" – connected to an extensive, expensive, high-bandwidth infrastructure. But investment in that infrastructure in turn represents a gamble that consumers will pay for carrier services to access virtual content, content that is risky to develop because of uncertainty about the scope of wired markets.

One development that is widely hoped to break this impasse is the most recent generation of video game consoles – the Microsoft xbox, Sony PlayStation 2, and Nintendo GameCube, which are either currently connectible to the Internet or have an add-on capacity promised for the near future. Ever-confident market research companies predict that the number of us households with such consoles will grow from one percent in 2001 to eighteen percent in 2005.[70] While they concede that only a portion of those households will actually play online, they anticipate revenues from such networked gaming in excess of two billion dollars in 2005, about one-sixth of the total video game revenues for both software and hardware forecast for that year.[71] Major game corporations seem to be acting on the basis of such predictions: it is speculated that Microsoft's expensive gamble on the development of the xbox will only pay off if there are major online revenues to be reached through it.

Precisely how to tap these revenues remains something of a conundrum in the gaming world. After the early success of *Quake* and *Ultima*, there was a rush of pay-to-play sites, such as Mpath, Total Entertainment Network, and Catapult, many of which folded quickly. These were succeeded by others that targetted more casual players of simple card and lottery games. In many cases, game developers

initiated online portals not so much for direct revenues as to promote
CD sales and attract advertising, an issue we discuss later in chapter 12.
But between 2000 and 2001 there was a shakeout in which a handful
of larger companies bought up many of the smaller game portals,
concentrating the network game business around a handful of mega-
ventures, such as Electronic Arts, which invested some two hundred
million dollars and signed a five-year deal with Internet giant America
On Line to become its sole game provider; Vivendi Universal's Flipside
site; Sony Online Entertainment's Station; and Microsoft's Gaming
Zone.[72]

Several of these sites are betting on the potential of subscriptions
from massively multiplayer games. Following *Ultima*'s example, major
corporations have leaped into the design of multiplayer worlds,
Microsoft with *Asheron's Call*, Sega with *10six*, Interplay with *Baldur's
Gate*, and Sony with *Everquest*. Each had a development budget of
more than five million dollars and was released with delays and cost
overruns. Some failed miserably. Others have succeeded: Sony's *Everquest*, the leader in the genre, had some 350,000 monthly subscribers
in 2001, each paying almost ten dollars a month in subscriptions –
nearly twice what it takes Sony to administer the game.[73]

In such massively multiplayer games issues of world durability,
player survival, and company profitability converge in a way that
strikingly illustrates the interaction of the cultural, technological, and
marketing circuits of the gaming industry. Because persistent world
games are ongoing and open-ended, they require "constant infusions
of cash and creativity, and buzz."[74] Not only do they have to be kept
going online, a technical challenge, but they need to be regularly
tweaked to add interest or deal with the unforeseen consequences
generated by thousands of unplanned interactions of virtual characters.
Everquest requires twenty-four-hour customer service staffed by 120
people, and between 1999 and 2001 Sony's programmers had to double
the amount of virtual "territory" available for players to explore.[75]
If players become bored or frustrated with the culture of a virtual
world, revenues fall, the infrastructure of technical and creative support has to be reduced, and the quality of the gaming experience
declines, potentially setting in motion a vicious spiral of degradation
and collapse.

If interest and recruitment can be maintained, however, the rewards
are large. Successful online game companies not only sell the software
necessary to enter but they can charge players admission fees *ad
infinitum*. Virtual enthusiasts celebrate the experience of such multiplayer worlds as a radically new form of participant-created entertainment, and there is some justification for this claim. But real-time

multiplayer games also in some ways repeat the mass media model of commercial television in providing to a large audience a common, simultaneous entertainment experience in which, though personal "stories" may vary, the overarching parameters are set according to the technological and marketing logic of profit-seeking corporate sponsors.

GLOBAL GAMES

This process is occurring on an international scale. Most accounts of post-Fordism see as intrinsic to its dynamics some internationalization of markets, breaking out of the Eurocentric or Northern confines of Fordism for wider consumerist domains. The phenomenon is hyperbolically represented in popular accounts of globalization, where digital games are often held up as an example. For instance, Kenichi Ohmae, theorist of a "borderless world," speaks of "the Nintendo kids" as a cosmopolitan echelon of youth who are "forging links to the global economy, turning their backs on older generations and traditional values, and using new technologies, such as the Internet, to circumvent government restrictions."[76] Such accounts can point for substantiation to the hybrid genealogy of industry. Video gaming began as a US industry, annihilated itself when Atari crashed in a welter of substandard software in 1984, and was revived in North America by a triad of Japanese companies, Nintendo, Sega, and Sony, in a way that ignited Western "techno-orientalist" paranoia about "silicon samurai."[77] Of course, the actual outcome of this traffic has not been a Japanese takeover but a complex "global enterprise web" made up of Japanese-owned but US-oriented multinationals; even more transnationally oriented US media companies; and various smaller locally based software developers, all tied together in an elaborate system of alliances and partnerships where the major corporations compete simultaneously in North American, European, and Japanese theatres of operation.[78] Video game companies contend in an international arena. The global market is divided into three segments – North America (primarily the US), Europe, and Asia (principally Japan, but with eyes on the possibilities of Southeast Asia and China). In 1992 the US accounted for some fifty percent of the video game market, Japan thirty percent, and Europe fifteen percent.[79] Over the course of the 1990s, the proportions altered, eroding the pre-eminence of the US market; IDSA estimated global entertainment software sales at about ten billion dollars, roughly shared between the North American, European, and Asian markets.[80] Many US entertainment software firms take forty to sixty percent of total revenue from overseas and regard foreign markets as among the most promising for future growth.[81]

These economic forces underpin a cultural hybridization in which video and computer games meld Japanese *anime* cartoon styles, American superheroes, and British neocolonial nostalgia. Such hybridization demonstrates what cultural theorist David Harvey terms "time-space compression," caused by the intensifying scope and speed of post-Fordist production processes, which splice cultural products from around the world in an eclectic postmodern mix. Companies such as Sony shuttle titles across the Pacific, turning Japanese misses into US hits and vice versa, and in the process "the cultural streams of East and West swirl into the Tastee-Freez of global entertainment."[82] Russian developers make the strategy game *Cossacks: European Wars,* a simulation advertising itself as being made simply by "Germans" and selling in the US. The language of "placeless profit" flows freely from the lips of industry leaders. The head of Nintendo's game development says, "We don't find any difference in kids' feelings nationwide or worldwide. Our R&D is thinking about the world as a target for each of their products."[83] Nintendo's president, Hiroshi Yamauchi, declares, "We do not see borders in this business. Some countries may be too poor or have heavy tariffs on imports, but with those exceptions we will go anywhere in the world. There are no borders."[84]

Nearly all the game industry's sales are within advanced capital's triadic core of North America, Europe, and Japan, but there are attempts to penetrate beyond that core. Fearing that cultural industries as a whole face an "entertainment glut" as film, television, music, and games compete for the attention and spending power of consumers, the industry has also looked beyond the "developed" capitalist world to Asia, the Middle East, and Latin America as potential markets.[85] In 1993, for example, Nintendo licensee American Softworks partnered with Pepsi Cola to launch the first Spanish-language video game, *Chavez,* endorsed by lightweight Mexican boxing champion Julio Caesar Chavez and aimed at Latin American markets and Latino consumers in the US.[86] Electronic Arts recently made its first marketing ventures in Thailand.[87]

In South Korea, one of the most "wired" of Asian countries, digital play has become an important part of youth culture; games such as *Starcraft* are enormously popular; ace players are celebrities; there are organized leagues; and some games are broadcast on cable television. In December 2001 the first "World Cyber-Games," dubbed "the Olympics of Computer Games," were held in Seoul, where four hundred players from thirty-seven countries contended for a piece of the $300,000 purse.[88]

But sales in areas outside the hyperdeveloped world remain negligible. Taking a truly planetary perspective, only a fraction of the world's population participates in the digital game culture. Video games are a global but not a universal commodity. It may be possible to find them in almost any zone of the planet, but in most places only a minority can afford to play. The statistics are basic but inexorable. For one-sixth of the world's population, a Sony PlayStation 2, which costs more than three hundred dollars, represents a year's total income; even a hand-held Nintendo Game Boy Advance represents three months' livelihood. The world's 250 million child labourers have no time for gaming; only some six percent of the global population has Internet access.[89] Even if transnational marketing means that many of the world's children dream of the adventures of Mario or Lara Croft, life for a majority centres on a precarious struggle to fulfill the basic need for food, water, and shelter. The $8.8 billion annual revenues of the US video and computer game industry alone is worth slightly less than the estimated annual additional expenditures needed to provide clean water and safe sewers for the world's population – slightly more than would be needed to give basic primary education to everyone on the planet.[90]

CONCLUSION: ICEBERG.COM?

For nearly a decade, interactive gaming has been humming with the "new economy" euphoria that has made delirious investment in information technology the driving force of the US, hence the global, economy, "with more and more precious investment capital being thrown into this tiny sector at the expense of all the rest."[91] But at the turn of the new millennium the fate of the whole project suddenly became deeply uncertain. Volatilities in the values of high-technology companies revealed that e-fever was beset with anxiety. Nagging fears began to surface: that the Net might actually be intractable to the commodity form; that people might not adapt to "e-commerce" in a networked environment where they were used to free experiences; that profits might be insidiously sapped by piracy; that markets constrained by digital divides might not be large enough; and that transnational expansion might be stalled by a "backlash" against globalization. Given the gigantic expense involved – often wildly in excess of confirmed demand – in the creation of digital content and in the laying of telecommunications infrastructures, these risks had a serious effect on e-investor confidence.

As dot.com companies with massively inflated share values showed an alarming lack of actual profit, uncertainties were magnified as a perverse result of the very capacities for online activity promoted by e-capital. Online investors and day traders sold frantically. The collective euphoria and irrational exuberance that had buoyed up high-tech markets flipped into panic capitalism, jittery, nervous, and suicidal. Dot.coms became "dot.bombs," the Nasdaq lost fifty percent of its value in a year, and the new economy rapidly assumed the very old fashioned contours of a collapsing bubble. Three trillion dollars of high-tech investment circulated through the marketplace, only to disappear into the icy North Atlantic storms of the cybereconomy.

Evaluations of the Internet bust vary, the diagnoses ranging from terminal doom to standard industry shakeout.[92] In comparison to the telecommunication and computer sectors, the interactive game industry has probably charted a steadier course than most. Revenues in 2000 were down from the previous bumper year; many small game developers closed; sales of the new generation of consoles were slower than anticipated; Sega, as we have seen, withdrew from the market, writing off the millions it had invested in the Dreamcast; Vivendi shut down its new online gamesite, Flipside.com; and many major media corporations cut back on game-related Internet operations. But there were no immediate signs of large-scale rescue missions. Nonetheless, the bursting of the Internet bubble revealed that the course of digital capitalism was far from smooth. If we look more closely at the circuits of today's interactive entertainment business, we find in them tensions and contradictions, some specific to the video and computer game industry, others symptomatic of more general problems in virtual capital. In our next section we set out to open some critical windows on these issues.

PART THREE

Critical Perspectives

Our analysis reflects critically on digital play as a cultural industry and cultural practice wrought within the contradictory dynamics of today's mediated markets. The interactive game business, like all of digital capitalism's cultural industries, is characterized by paradox, tension, and uncertainty. We reiterate a point made earlier: to define the interactive game as an "ideal" post-Fordist commodity is not to say that it presents a problem-free profit opportunity for business but, on the contrary, to emphasize how it crystallizes within itself both the dreams and nightmares of information capitalism. It is precisely because video and computer gaming show us how an information age industry confronts conflict and controversy and attempts to negotiate them – succeeding brilliantly at times, failing catastrophically at others, often improvising partial and provisional compromises – that they make such a worthy case study.

In this section we recast some specific issues that emerged from our historical narrative, making use of the tools available to critical media analysts that we outlined in part one of the book. Specifically, we organize our investigations into the interactive game industry's unresolved problems by again invoking our model of technological, cultural, and marketing circuits.

Thus, in chapter 9 we look at unanticipated problems arising within the technological circuit of the industry. Such difficulties spring from the human subjects who make and use the digital devices on which the interactive game business depends, but who at times act in ways that oppose and subvert the discipline of the market. At the point of production, in the making of interactive games, this independence manifests itself in new forms of labour unrest that can threaten the smooth flow of software and hardware from studios and factories. At the point of consumption, where

game users acquire computers, consoles, and CD burners, it appears
in the piracy and hacking that hemorrhage value from the gaming
companies' coffers. We argue there is a link between work and
"warez" that has to be grasped in terms of the global scope of the
industry's circuits for producing and consuming gaming technologies.

Chapter 10 addresses the ambivalent success of the game
industry's marketing circuit. Video game companies have been
pioneers of the branding and synergistic marketing practices that are
now considered vital for commercially managing the highly fluid
trends of contemporary youth culture. The industry's legions of
marketers and cultural intermediaries have spun around its virtual
worlds a vortex of high-intensity television advertising, media spin-
offs, merchandising and licensing agreements, product placements
and urban location-based entertainment venues, all constantly
updated by vigilant tracking and assimilation of the latest, coolest
cultural currents. But this very marketing triumph has negative long-
term implications for creativity and experimentation in digital play.
Escalating advertising fosters cynical audience resistance, creating a
vicious spiral of "symbolic exhaustion," in which styles in commercial media culture grow stale quickly so that advertising designers
must struggle to keep up with their increasingly savvy audiences.
Furthermore, massive marketing budgets drive towards the consolidation of the industry around a handful of giant publishers.
Promotional dynamics tend to favour the development of sure-fire
franchises, repetitive clone lines, and metagenre games calculated
for appeal to the lowest common denominator across important
market segments. We use the example of Nintendo's blockbuster
Pokémon franchise to show how the new digital cultural industry
has failed to overcome tendencies towards massification, and how,
moreover, the imperatives of synergistic marketing work their way
back into the very "digital design practices" of the industry. Such
hypercommodification contains within it a paradoxically self-
destructive dynamic that threatens the basic play experiences of
spontaneity, wonder, and exploration on which interactive gaming
must draw.

In chapter 11 we turn to the cultural circuit and examine the
content of video game media, specifically controversies about
violence and gender that have troubled the industry from its earliest
days. A variety of forces converge within gaming's digital design
practices to bias interactive entertainment towards what we term
"militarized masculinity." This bias privileges themes and representations of warfare, fighting, combat, and conquest along with the
subject-positions of aggressive, active male characters. Though most

extreme in the notorious "first-person shooters" and martial arts games, militarized masculinity has structured virtual experience across most of gaming's major genres, constructed by design and marketing practices aimed at the industry's most reliable customers – adolescent boys and young men. We argue that the feedback loops through which interactive game companies "coevolve" games with the participation of their most loyal hardcore players can amplify and deepen the predisposition towards scenarios of violence. Militarized masculinity has been key to interactive gaming's testosterone-built success. The rise of militarized masculinity presents a lucid example of what media theorists Innis and McLuhan referred to as the cultural "disturbances" that are produced by the diffusion of a new medium of communication. These disturbances are amplified, we suggest, by the marketing practices that are used to negotiate with youthful consumers in the cultural marketplace. Yet they are also sites of contestation. For example, violence in video games has attracted continuing criticism from those who are distressed by the uncertain social and psychological effects of virtual violence, and from others who are outraged by the long-standing exclusion and marginalization of girls and women in digital play. In some cases, these concerns have given rise to alternative gaming practices. Nonetheless, interactive gaming's persistent experiential bias towards militarized masculinity may prove to be a barrier to the industry's continuing growth; we conclude the chapter by looking at the forces that are now both impelling and inhibiting new departures in the digital design of violence and gender.

In these three chapters, we assign each theme to one of the circuits in our model of the mediatized market – *pirates and strikers* to the technology circuit; *militarized masculinity* to the culture circuit; and *synergized commodification* to the marketing circuit. We hope this provides a useful way of focusing the issues at stake in each discussion. But it is clearly schematic. Useful as the distinctions between technological, cultural, and marketing circuits are for discerning and naming the forces shaping the interactive game industry, in concrete cases they implode in on one another. All of these cases ultimately involve all of the circuits. For the managers of the industry, the complexion of the "interplay" of the three circuits is a main concern in their struggle to stabilize profitability in a highly volatile marketplace.

Piracy, for example, entails not only the hacker's use of the copying capacities of digital technology but also the transgressive subcultural ethos of game culture, and its violations of a marketing system that maximizes profit on the basis of intellectual property.

Militarized masculinity is a matter not only of designing cultural
narratives of violence and gender but of computer technology's legacy
of military applications and – of special importance – marketing
practices aimed at commercially valuable hardcore male players.
Likewise, synergistic commodification is an issue of how marketing
practices are linked across "multi" media environments, unified by
digital technology, and of how their promotional logic comes to
inform the very design processes that shape the content of game
culture. Thus, it is the *connections* and *contradictions* between
circuits that our analysis seeks to explain in order to do justice to the
multifoliate paradoxes unfurling through today's mediatized market.

In chapter 12, we draw the strands of our critique together
through a discussion of the hit computer game *The Sims*. In this
interactive saga of domestic life in advanced capitalism playfulness is
framed by the virtual consumption of virtual goods constructed to
meet virtualized needs in a virtual environment – all hermetically
sealed by the institutional forces that organize around digital technology. *The Sims* reveals a foundational prognosis in our wired
globe: the continued expansion of commodified culture. But we
believe our study of the interactive game industry helps expose
important contradictions that are inherent in the process, which in
our view is beset by frictions and tensions, the outcomes of which
are uncertain and contested.

Our study of the making of the interactive game reveals three
paradoxes within the circuitry of the digital play industry that is
now so central to what we term "Sim Capital": in its technological
circuit, between "knowledge enclosures" and "democratic access"; in
its cultural circuit, between "militarized masculinity" and "digital
diversity"; and in its marketing circuit, between "synergistic
commodification" and "creative playfulness." Against the conventional triumphalist information age prognosis for a radiant digital
future – one that attempts to deny contradictions and silence the
paradox – we suggest that the industry, now trying to steer a path
amidst economic crisis and global war, has no easy resolution for
these dilemmas. In the coda, "Paradox Regained," we give a closing
overview of the book's perspective on digital play.

Workers and Warez:
Labour and Piracy
in the Global Game Market

INTRODUCTION: TECHNOLOGICAL CIRCUITS
AND HUMAN DISRUPTIONS

When we purchase a video or computer game we probably do not think about how it arrived on the store shelf. But our copy of *Zelda* or *Starcraft* did not spring into being ready-made. It is the outcome of a production process, of the combined labour of hundreds of people. That we forget about this is a telling example of what Marx called "the fetishism of commodities" – the tendency for the market to present us with goods as if they arrived by magic, hiding the mental and manual toil that goes into their making.

This process touches all commodities, be they bananas in the supermarket or jeans in the fashion boutique. But in the case of games it is peculiarly intense. Play, after all, is the opposite of work. Games are "fun" experiences. Every bit of game marketing and promotion actively *discourages* us from associating them with such mundane and boring realities as jobs, management, and labour relations. On game boxes or in the pages of gaming magazines, we find snippets of publicity about "star" game developers, or articles about how "cool" it is to work in the game industry. But these are exceptions that prove the rule. In such depictions making games is itself shown as play – work as fun. The blurring of boundaries between labour and leisure so that not only consuming games but also producing them is represented as a continuum of endless fun is a part of the interactive game industry's hip self-image.

Some of us, of course, didn't buy our games in a store, or online. We didn't buy them at all, or at least we got them for next to nothing, paying far less than regular price. We probably do know something, if not about how games are produced, certainly about how they are *reproduced*. This is because we downloaded our games for free from a "warez" network, or swapped illicit copies through peer-to-peer file sharing, or illicitly burnt our own onto CDs, or got cutrate bootlegged copies under the table. In short, we are pirates. And though we probably know there are legal penalties for what we do, and that game companies are loudly threatening us with dire consequences, we probably reckon that the chances of getting caught are pretty small. We may even get a charge out of cracking the various technological systems while Microsoft or Sony try to keep us out: hell, it's just another level to the game. If work-as-fun is the interactive game industry's wet dream, then piracy-as-play is its worst nightmare.

In this chapter we look at turbulence in the technology circuit of the mediatized global market. We have described this circuit as the process that connects those who make digital devices and those who use them in cycles of innovation, diffusion, and adaptation. The turbulence arises not so much from the failures of machines as from the disobedience of human subjects. Such disobedience can occur at either end of the circuit. It can arise in the development of technologies if the workers who create software and hardware are dissatisfied, slow down, sabotage their product, or go on strike. It can arise in technology use when subjects adapt machines in ways that are unanticipated and unwanted by developers, for example by using games technology to hack and pirate.

So in the first part of this chapter we look at the new post-Fordist work organization the games industry has adopted, and at some examples of labour unrest it has generated. In the second part, we examine the game industry's piracy problems. The two sorts of conflict, both of which arise from the disruptive activity of human agents in the technology circuit, may seem quite distinct. But there are connections, we argue, between striking and hacking, workers and warez, and these connections create a dialectic of discontent that plays across both the making and use of game technologies.

WORK AS PLAY?

The Interactive Digital Software Association, the main lobbying arm of the digital play industry, claims that "the entertainment software industry directly employs 90,000 workers, many in highly skilled positions, with a growth rate of 26 percent (compared to a 2 percent decrease in employment among Fortune 500 companies)."[1] It does not give a detailed breakdown of these figures. They can be assumed to include positions such as full- and part-time game testers; marketers and public relations personnel; shipping and distribution staff; writers and editors of video- and computer-game magazines; box designers; and the phoneline services that give tips and advice to frustrated players. Clearly, this is a labour force with a wide variety of skills, security, and rewards. But the most highly publicized aspect of video game production is game development. It is around this labour that the industry has developed an alluring aura as a business in which "work is play."

Game development, the dynamic core of the business, requires a synthesis of narrative, aesthetic, and technological skills to conceive, plot, and program virtual worlds, deploying the combined expertise of digital coders, graphics designers, software testers, scriptwriters, animators, sound technicians, and musicians.[2] Production is done in studio conditions by teams of six to fifty, the projects taking one to two years to complete. The mobilization of this "new élite workforce" of digital artists and technicians has made the games industry a central arena for experimentation in teamwork, charismatic leadership, ultraflexible schedules, open-space work areas, flattened hierarchies, stock options, and participative management.[3]

For an example we can look at Nintendo's game development process. Software is "something of a religion," according to its president Hiroshi Yamauchi, and under his command the company has "pushed its software development almost to the extreme."[4] To do so, it uses techniques Katayama calls "me-too management … emphasizing individuals more than organizations, oddballs more than cooperators, and relaxation more than intensity."[5] Game development is done by R&D teams, working on the "big room" principle, without spatial barriers between separate sections, relying on the "power of place" to promote discussions between hardware and software specialists. Bureaucracy is minimized. The basic philosophy is that "software developers have expertise and experience that you can't put into writing" and that giving them autonomy will "bring with it enough responsibility to keep Nintendo's development periods within reasonable bounds."[6]

Despite this apparent relaxation, however, the process is relentless. Nintendo's teams compete with one another. Final product is subject to a ruthless assessment process. As in other software companies – Gates's at Microsoft is a prime example – winning the approval of a corporate boss of near-mythic status is a compelling force. Yamauchi's personal attention to software means that "months of work can be disposed of with a scowl."[7] But within the boundaries of what Sheff concedes is an "autocratic, often brutal system," Nintendo offers employees great latitude to explore game design.[8] One R&D leader says, "You can build what you want to, and if it's a hit, you can bask in the applause. That kind of satisfaction is everything."[9] Nintendo's virtuoso

developer Shigeru Miyamoto, creator of *Mario* and *Zelda*, speaks with
deep satisfaction of the moments when "you, as a developer, are aware
of being involved in something that has never been seen before," and
of the pleasure of being able "to lose yourself in that sense of new-
ness."[10] Such pleasures, and the hope of attaining Miyamoto's kind of
celebrity, are undoubtedly integral to game development work.

Of course, game development labour involves repetitive unglamor-
ous coding tasks. But as J.C. Herz concludes from her encounters with
young "virtual construction workers," it is possible to take a "craftsman's
pride" in, say, recreating the furnishings of the Titanic, or generating
"a high quality virtual environment" with "artful dimensions, good
doorway design, well-placed obstacles, easy-to-reach ammunition."[11] If
there is any truth to such reports, successful multimedia companies
depend on harnessing a bona fide enthusiasm for game creation – a
rather maniacal and macho (not to say masochistic) enthusiasm, perhaps,
but nonetheless a digital labour of love.

This is a youth industry that recruits from the culture it has created:
a culture of male adolescents, fascinated with technology, familiar with
game design not just by constant play but by the editing capacities
that allow players to design and share their own levels of games such
as *Quake*. Employment seems like a chance to get paid – quite well –
for fun.[12] Many young employees see it as a first step towards founding
their own company. The industry is full of tales of enterprises fuelled
by little more than talent and energy bootstrapping themselves to
fortune. There are any number of stories such as that of the British
software company Rare, creator of the hit *Donkey Kong Country*,
which was produced by "a few working class youngsters in a terraced
house next to a newsagent." Games development, as noted earlier, is
often characterized as "the ultimate cottage industry for the information age."[13] Growing costs and corporate concentration press against
such hopes, but they cannot be entirely eclipsed as long as the industry
remains fluid.

Yet as recent analyses of "net slaves" show, there is a dark side to
this scene.[14] Management harnesses youthful technophilia to a com-
pulsive-obsessive work ethic, one-dimensional character formation,
and a high rate of burnout.[15] Examining New York companies engaged
in Web design and game development, Clive Thompson notes that their
work ethic is one of studied nonchalance; "playing *Quake* on the
computer LAN, hanging out in a funky office with your dog" is "*de
rigeur*."[16] They embody "the master narrative of the New Work,"
which "has to do with making work seem a lot more fun and thus a
lot less like a job."[17] However, this "ultra-cool" appearance "covers
up a seldom-discussed truth": that the jobs themselves are often deeply

exploitative. Staff work up to eighty hours a week, staying up all night to meet deadlines. Job security is "near zero"; wages lower than the average in other media; stories of stock-option success are largely fabulous; the programmer shortage that supposedly gives high bargaining power exists only for those – usually the young – who are willing to endure driving and unstable conditions.[18]

Thompson's conclusion is that "the studied hipness of new media is a fascinating and rather devious cultural illusion." By making work more like play, "employers neatly erase the division between the two, which ensures that their young employees will almost never leave the office."[19] Instead of becoming enraged, or unionizing, multimedia workers "smile and thank their lucky stars for being part of the digital revolution." For employers "it's a sweet deal: you can't *buy* flexibility like that." Paradoxically, these young multimedia workers, "touted as the most renegade – the most entrepreneurial – generation in years," are actually "amazingly subservient: the ideal post-industrial employees."[20]

All this casts doubts on the myth that game making is "fun." Such labour does not live up to rose-coloured post-industrial visions of knowledge work.[21] But nor does it match the straightforward picture of deskilling and degradation painted by the neo-Luddite left. What emerges is more contradictory. The creation of a new creative high-technology industry has required management to recruit a post-Fordist workforce whose control requires the use of techniques that are very different from the rigid routinization and top-down discipline of Fordism. They involve a high degree of soft coercion, cool cooption, and mystified exploitation. But although adulatory accounts of digital development should be heavily discounted, it probably does offer some young men (and a few women) more rewarding and interesting work than the assembly lines to which they might have been consigned a generation ago. Where there is discontent, it is more likely to express itself through the "churn" of mobile employees leaving to join other companies or found their own, or perhaps through the planting of the occasional malign "Easter egg" in a game, than through organized protest.

PLAY AS WORK?

The "work as play" ethic of game development has another dimension. Many games, especially the good ones, are the product of communities that extend beyond the workplace. The paid workteams of corporate developers – "the A Web" – are only the core of a much wider circle of creativity – "the B Web" – that includes a diffuse swirl of unpaid creators, test subjects, expert informants, and volunteer labour.[22] One of the most striking aspects of the industry is the way it incorporates the activity of consumers into the development of games – creating what Toffler terms "prosumers."[23] If one aspect of the interactive entertainment industry is its representation of work as play, another is its conversion of play into work. We can identify five points in this process – the marriage of gaming with market research; the "laboratory" model of interactive entertainment centres; the use of game testers and expert gamers by major manufacturers; the use of shareware and player editing to add value to games; and the role of gaming culture as a training-and-recruitment arena for the industry.[24]

There is nothing new in the use of market research to guide production decisions. But in the game business it is exceptionally important. Selling gamers a continuous flow of new software puts a premium on knowledge about their changing expectations and preferences. It has become an industry commonplace that success depends on market research and on quickly incorporating trends of youth culture into game content. In-house research is supplemented by the polls, focus groups, and surveys of specialist companies such as Alexander and Associates, DFC Intelligence, NPD, Jupiter Communications, Forrester Research, and Yankelovich. The claim by market researchers that their investigations ensure that consumers get the product of their choice – "We are reflecting what they are telling us as opposed to leading them" – represses the role of cultural industries in building and shaping the tastes they supply.[25] But it is fair to say that in the interactive game business research leads a strong positive feedback loop between players and developers.

One way to gather such information is to attract gamers to places where their play can be observed and their responses evaluated. Some twelve hundred "hard-core" or "expert" gamers visit Nintendo's headquarters just outside of Seattle on some weekends. Electronic Arts invites "local schoolchildren to come by and play games and then they interview them."[26] The logical extension of this approach is to create special location-based entertainment research projects. Digital-games companies have been deeply involved in projects that bring together business, scientific, military, and academic researchers with games players on research sites equipped with "real-time feedback loops" designed to elicit dialogue and measurable response to prototype designs.[27] These include the IST, Pasadena Art Center College of Design, and the Edge – a system that "allowed the exhibits to change every day based on input and feedback from the previous day's interaction with the audience ... A cross between MIT's Media Lab and the local playground."[28] Such ventures are celebrated in the industry as experiments in "unrestricted imagination," as a democratic "open laboratory" and "collaborative learning projects."[29]

A further step towards incorporating consumer knowledge into game development comes in the testing process. There is a notorious saying in the computer business that "the customer is the beta-tester." This is usually understood as meaning that it is the unfortunate purchaser of software who first finds what is wrong with it – a meaning that certainly applies often enough in regard to games. But in the interactive entertainment industry there is a second meaning in that many paid testers constitute a part-time workforce that floats back and forth across the line between employee and player.

Testing is crucial to game development. Failure to "debug" a game can send it "straight to the remainder pile."[30] The testers' role in "balancing" the game – adjusting its rhythm and texture – can be vital.[31] Most big game developers have some permanent testers but supplement them from a temporary pool called in as occasion demands. According to Herz, the Sega Testing Lab in San Francisco "trawls the Bay Area ... for eighteen- to twenty-eight-year-olds by posting flyers and taking out ads ..."[32] Recruits are subjected to an arduous screening process that includes English and basic math exams, because they not only have to play the game but communicate the experience in a "detailed debriefing document."[33]

There are two reasons why game companies rely on casual testers. Firstly, it is cheaper to use part-time flexible labour to whom benefits don't have to be paid. Secondly, games are actually improved by exposure to players outside the core development process. Brandon James, designer at id Software, says the company uses outside testers to catch bugs its inside testers miss: "Everybody out there has a different and distinct playing style. You get enough (testers) who have a knack for going to the most unlikely places in a level and doing the most obscure things and eventually you'll have covered all the paths and areas that an average player will follow."[34] Andrew Goldman, president and CEO of Pandemic Studios, says: "After spending two years on a project you're no longer able to see how new users will react to the product. The only way to maintain an appropriate per-spective is to learn to look at it through the eyes of your first-time users."[35] Indeed, problems sometimes arise when games become so familiar to professional testers that they underestimate its difficulty and demand changes that frustrate genuine first-time players. Paradox-ically, then, it is precisely the "non-professional" know-how of a contingent workforce that developers require to introduce diversity and freshness to development.

A further step in the injection of consumer know-how is to give players the technological means to create their own contributions and distribute them across the Internet. The landmark here is *Doom*. By releasing chunks of the game's source code, id turned every player into a potential programmer who could create his or her own levels of the game. Similar editing capacities have subsequently been included for many other games. Because *Doom* and its successor, *Quake*, can be played online, entire virtual communities have arisen around them as players share the levels they create. One *Quake* "clan" can invite another for a "death match" on digital terrain of its own terrible invention. Participatory design thus becomes an added-value component, opening an ever-expanding vista of worlds created by other players.

These tactics for harnessing the creative energy of virtual communities have implications for the training and recruitment of professional game developers. In the case of *Doom*, online recognition of the talent of some of the amateur "Doom Babies" led to their being hired by id or its rival, Ion, as paid employees, abruptly catapulting from, say, a janitorial day job in Norway to membership among the digital élite in Dallas.[36] More generally, the proliferation of shareware games offers games companies a huge reservoir of play-trained design workers, a pool of labour in which "your only résumé is your level, or your 3-D model, or your new evisceration animation."[37]

A further advance in this logic is Sony's Net Yaroze, a seven-hundred-and-fifty-dollar version of the PlayStation that allows gamers to write their own code. It comes with a kit that includes a software library, and access to the Yaroze Web site, which is used by as many as ten thousand programmers to upload and download their productions. Phil Harrison, a Sony vice-president, depicts the technology as creating a "virtual community" of collaborative digital production, marking a return to the "golden age of video game development, which was at home, on your own or with a couple of friends, designing a game yourself."[38] But as Herz observes, "it's a canny bit of strategy" from Sony's perspective: "A good number of college-age Yaroze programmers will go on to jobs in the industry (some already have), and they are cutting their teeth on Sony hardware. Thousands of bright bulbs have essentially become Sony's junior development community. They have also become a vast unpaid division of Sony's R&D department."[39] Although freedom from commercial constraints may, as Harrison claims, generate "some radically new forms of creativity (that) break the conventions that are holding the business today," Herz points out that "these radically new forms of creativity will be Sony products." Games developed in this way will play only on the ten thousand or so Yaroze units, not on the millions of PlayStations available around the world. To make a game commercially viable, it has to pass through the normal publishing and marketing strategies, including the crucial licensing fees charged by Sony. Sony's perspective, in Herz's words, is that "radical creativity is good, as long as it can be contained. Rogue ideas are necessary, but they must be incorporated into a carefully orchestrated product release schedule."[40] As we shall see, problems arise when the sort of generalized digital aptitude represented by the interactive game industry's unpaid workforce refuses to be contained within these boundaries.

NIMBLE FINGERS

There is one area of game industry work that is far from fun. As we have seen, while the industry freely uses the rhetoric of globalization, nearly all its sales occur within advanced capital's triadic core of North America, Europe, and Japan. Where interactive gaming really does participate in a world market is not in consumption but in production. As a product of an international division of labour, video gaming consoles and cartridges crystallize in their tiny circuits two dramatically contrasting types of work. Both involve digital labour. But we are talking of different digits: in one case, the binary codes of zeros and ones manipulated by mostly male programmers in the developed world, in the other the "nimble fingers" of a global pool of primarily female cheap labour. This division is characteristic of information capital as a whole, and of the computer industry in particular. All games-playing systems, consoles and computers, share a vital component with other parts of the so-called digital economy – microchips. The products of a worldwide semiconductor industry, these are often manufactured in *maquiladoras* and enterprise zones in Mexico, Malaysia, the Philippines, Taiwan, or Korea by a predominantly female workforce recruited specifically for its supposed docility and disposability and subjected to ferocious work discipline under conditions that destroy health within a matter of years.

The game industry also has a more specific involvement in such settings. Games are not just digits. Crash Bandicoot, Lara Croft, and the mutant hordes of *Half-Life* see the light of day courtesy of a material apparatus of consoles, joysticks, and keyboards. This hardware base is produced at quite different sites from the software, often by subcontractors, in Mexico or southern China's Shenzen Special Economic Zone, Taiwan, Thailand, Indonesia, and Malaysia. It is also in these places that the cartridges and consoles into which the microchips are incorporated are manufactured.

The tendency to seek low-wage areas goes back to the industry's early days. When Atari was facing collapse in 1984, one its first moves was to lay off production workers in Silicon Valley and shift manufacturing overseas to Hong Kong and Taiwan, where assembly workers were making about $1.20 an hour, compared to $9.00 an hour for US counterparts.[41] The company was subsequently forced to distribute $600,000 to laid-off employees who in an unprecedented case sued successfully for violations of job security guarantees.[42] This was the first in a series of offshore migrations that gutted Silicon Valley as an electronic manufacturing centre while leaving its advanced research capacities intact.

The pattern was continued by the Japanese-based multinationals. In 1995 Sega transferred nearly all production of video game consoles for the Japanese market to subcontractors in Taiwan, China, Thailand, Indonesia, and Malaysia.[43] In 1994 Nintendo of America laid off 136 US workers involved in assembling games and machines at its Redmond headquarters and relocated operations to Mexico. Although the company denied that the move was related to the recent North American Free Trade Agreement (NAFTA), the fired employees were deemed eligible for compensation benefits under the NAFTA worker adjustment program, a decision a Nintendo spokesperson termed "frustrating."[44] In fact, Nintendo has divested itself of most manufacturing workers. The *Economist* describes it as not only a "fabulous" company from the point of view of profitability but also a "fab-less" company in that it contracts out its fabricating operations.[45] President Yamauchi explains: "Entertainment products are not necessities. You're heading for trouble if you start building your own factories and engaging in your own manufacturing just because there is demand."[46] Noting that Nintendo's annual capital outlays amount to only one-fiftieth of its total value, Katayama suggests that it is "because it does not have any production capacity, [that] Nintendo is able to devote its resources to software development."[47]

Nintendo's consoles and cartridges become visible to the eyes of North American journalists only when they appear at its highly automated just-in-time distribution facility in North Bend, Washington. This is intended to "centralize all inventory, process product immediately after it arrives from factories in the Far East, and quickly deliver product to retail sites."[48] It operates with a computer system that communicates its orders by radio frequency, automated guided vehicles, pick-by-light racking systems, bar code scanners, and a panoptic surveillance system to monitor orders and worker performance. Here, Herz says, "a stream of cheaply assembled product pumps in through the pulmonary artery of the Third World manufacturing sector," while "orders flow back to the warehouse through the venous channels of digital inventory databases and thence to the factories of Central America and China."[49]

ANGRY FISTS

A glimpse of what can go on in these factories is offered by the unusually publicized case of a Nintendo subcontractor. Maxi-Switch is a
manufacturer of computer keyboards (it boasts being one of the top
global companies in that area), control panels for exercise equipment,
and cartridges and circuit boards for the video game industry. Originally a us firm, Maxi-Switch in 1990 became a wholly owned subsidiary of the On-Lite Group, a Taiwanese enterprise with manufacturing sites in Malaysia, Taiwan, and at both Shanghai and Shenzen in China.
It has headquarters in the us, "language configuration and distribution
centres" in Ireland and France, and manufacturing operations in Mexico.
Maxi-Switch's motto is "Your World Partner."

The company started making Sega Genesis game cartridges in 1992
and game-related components for Nintendo in 1993. In 1995 a labour
dispute broke out at a Maxi-Switch plant in Cananea, Mexico, which
produced Game Boys. Game Boys are one of Nintendo's most profitable
products: miniature, portable, hand-held consoles that are popular with
both children and adults. First introduced in 1989, they were "frequently seen in first-class compartments on cross-country flights, in
corporate lunchrooms, and in desk drawers and briefcases."[50] President
Bush, in hospital in 1991 for minor surgery, was pictured in newspapers
commander-in-chiefing a Game Boy. As with all console products, hardware sales of Game Boys are only the beginning. Software cartridges
– "the size of a saltine cracker" – bring in the real profits.[51]

The Cananea factory was one of three Maxi-Switch plants in Mexico.
In 1996 they had a combined workforce of about three thousand,
depending on seasonal demands for game products.[52] The plants are
in the *maquiladora* zones to which North American firms, especially
in the consumer electronics industry, have flocked over the last two
decades, attracted by cheap labour, lax environmental regulation, special taxation concessions, and proximity to the North American market.[53] Maxi-Switch's low-wage manufacturing facilities in Cananea
were twinned with higher-paid "bag and ship" distribution centres
across the border in Tucson.

At the Game Boy factory the workers, many of them young girls
in their teens, worked ten-hour days for us$3.50 a day in poorly
ventilated conditions. Local health officials made three to four ambu-
lance trips a day during the summer months to rescue those who had
collapsed on the production lines.[54] In 1995 a young woman, Alicia
Perez, led an effort to unionize. More than three-quarters of the
workers in the plant signed up. Perez was fired after reportedly being

punched and knocked to the floor by company thugs. Three other union leaders were also fired. In 1996 the Mexican labour boards denied the union recognition on the grounds that a phantom local of the notoriously company-friendly pro-government Mexican Workers Confederation had already signed a collective contract with Maxi-Switch. When workers asked for details of their new representative, they were denied copies of its statutes.

A group of Mexican and US unions turned the situation into a test case for the labour provisions of the North American Free Trade Agreement. They filed a complaint under NAFTA charging Mexico with failure to enforce its own labour laws. The case proceeded slowly. Two days before a hearing before a NAFTA panel, Maxi-Switch recognized an independent union. But it refused to hire back Perez and the two other leaders. In the eyes of the Mexican government, anxious to avoid any precedent-setting rulings that could jeopardize the attractions of *maquiladoras* to foreign investors, the issue was resolved.

This incident was exceptional because it tore the cloak of invisibility that companies such as Nintendo and Sega draw over the manufacture of their hardware. But the conditions it exposed are probably not unique. Sony, the leading producer of games and consoles, also operates *maquiladora* factories in Mexico. One of them, a plant producing tapes and disks at Nuevo Laredo, has been the site of a major labour struggle similar to that at Maxi-Switch. Sony also has electronics factories in China and Southeast Asia.[55] Recently, Indonesian workers (again, mostly women) at a Sony electronics plant struck, seeking the right to sit rather than stand all day: the company responded by threatening to relocate to Vietnam. Labour advocacy groups have cited Nintendo and Sega, along with toy companies involved in video games such as Mattel, Hasbro, and Bandai, as being "actively engaged in subcontracting production in Asia."[56] Manufacture of Microsoft's xbox console is outsourced to Flextronics, a company incorporated in Singapore with management offices in San Jose, factories worldwide, and industrial parks that house inventory in Brazil, China, Hungary, and Mexico – where the xbox is made.[57]

As many analysts have pointed out, post-Fordism results in a highly polarized pattern of employment. While the top end does sometimes correspond to what Lee terms the "ideal post-Fordist model" of skilled knowledge workers, the bottom end – of labour power cheapened by automation and global mobility – is far closer to the experiences of workers in the early capitalist period of "primitive accumulation." This dualized occupational pattern often follows lines of gender and ethnicity and is of course transnationally organized. The dependence of digital industry on low-wage labour in the planetary South for hardware assembly and chip production refutes its self-promoted image as a business where work has been transformed into fun. It may also prove a long-term source of risk and instability. *Maquiladora* electronics factories, like the sweatshops of the clothing industry, are today becoming the target of new planetary movements outraged by disparities between the global polarization of wealth and poverty. It is, moreover, the low purchasing power of the low-wage zones where "nimble fingers" live that ultimately limits the global market for digital products of all sorts, a striking demonstration that even in the information age, capital has not escaped its own destructive paradoxes.

SKULL AND CROSS-BONES

Such paradoxes are not limited to the making of technology. They also characterize its use. We have already seen how the interactive game industry has been at the centre of corporate "softwars" over copyrighting and patenting technologies. But there is another front to these wars – the battle waged by the industry against software "piracy." Piracy is regarded as a serious problem throughout the digital economy, particularly in the areas of office software, music, and video and computer games. Antipiracy organizations like the Software and Information Industry Association and the Business Software Alliance claim that more than thirty-seven percent of programs loaded on computers worldwide are pirated.[58] It has been estimated that one-third of piracy is "garden variety unlicensed copying"; one-third "Far Eastern-style counterfeiting" by large and small-scale for-profit criminal enterprises; and one-third "warez" networks, driven by bragging rights and barter economies.[59]

Digital gaming, with its origin in the unauthorized play of MIT programmers, is a child of hacking. And while information does not want to be free anymore than it wants to be paid, there are plenty of people who want free information, and free games. According to the Interactive Digital Software Association, game pirates released approximately $3.2 billion worth of packaged goods in 1998:[60] this figure is only for packaged software and excludes Internet traffic in games, for which, according to IDSA president Douglas Lowenstein, "there are no hard figures."[61] Since worldwide sales of legal games are approximately estimated at seventeen billion dollars, this means that pirated games are equivalent to just under twenty percent of total business. IDSA also releases lists purporting to compare financial losses from computer and video game piracy with the much-lower total losses from crimes against property such as shoplifting and bank robbery.[62] The figures rest, however, on the improbable assumption that the games would all have been bought at the normal market price.[63] Game makers' associations have an interest in overstating the problem in order to persuade government to take action against pirates.[64] But even allowing for hyperbole, illicit free software is clearly having a major impact. According to Lowenstein, "Piracy in all its forms represents the biggest threat to the continued growth of the industry."[65]

Recently, two pirating technologies have attracted special attention – "emus" and "modding." Emus, short for emulators, are software programs that enable software for one platform to be played on another, so that a Nintendo game could be played on a PlayStation or personal computer. This attacks the proprietorial basis of licensing fees that is basic to the industry. In particular, emulators threaten the division between consoles and computers, rendering "dedicated gaming boxes technically superfluous."[66] Sony and Nintendo have both launched suits against emulator makers. Yet despite legal threats, "today, almost every piece of computer hardware – from obscure products like the Nintendo Virtual Boy, which flopped on the market, to the Palm platform – has been emulated."[67]

"Modding," short for modification, enables people to copy and play game CDs. Many consumers burn their own pirated games using common CD drives. With the right CD-ROM burner, anybody can make a copy of a PlayStation title in about thirty minutes. To make the copies look authentic, one can even print original CD covers found on several Web sites. A standard PlayStation will not read ordinary CD-ROM. But that can be changed with the purchase and insertion of special "MOD" chips. The popularity of modding is a direct reflection of the international nature of the game market, since it first took hold among American players who were reluctant to wait for the official import of Japanese games. But modding opened a large vulnerability in the CD game market. Indeed, one of the reasons why Nintendo decided to stay with cartridges for the Nintendo 64 rather than move with Sony to CDs was that cartridges are harder to duplicate. While the decision placed Nintendo at a disadvantage in relation to their main competitor, it was apparently deemed worthwhile to reduce the risk of piracy. In the long term, it may prove a shrewd decision: a recent article in the Canadian *National Post* tells how it took a Toronto reporter only a few hours to track down and buy "MOD" chips for both Sony PlayStations and Sega Dreamcast consoles, as well as several pirated games at between fifty percent and twenty-five percent of the normal retail price.[68]

But emus and modding are only the latest manifestations of a piracy threat that is endemic to the industry. The threat arises from what Peter Lunenfeld refers to as the "commerce of tools."[69] As he observes, the "word processors, nonlinear digital video editing systems, database managers, Web server softwares, interactive multimedia systems, and even esoterica like virtual reality world-building kits" that are among the most attractive offerings of digital capitalism are not only consumer goods but also "the tool commodities of technoculture," which enable "new commodities and new work." So the relationship between producers and consumers is no longer simply "a case of sellers and buyers" but of "a relationship between hyphenates: between manufacturer-producers and consumer-producers." This process, says Lunenfeld, pushes what Marx termed "the social character of private labor" to an unprecedented intensity, so that "although the commodity still retains its awesome power, the 'made' character of the technocultural commodity is consistently foregrounded for the consumer-producer."[70]

This "defetishization" results, however, in real practical problems for commercial industry. As Ken Wasch, the president of the Software Publishers Association, acknowledges, "Computer software is the only industry in the world that empowers every customer to become a marketing subsidiary."[71] The informal takeover of the means not just of production but also of near-instantaneous and costless reproduction by immaterial labour constitutes a major dilemma on the world market in the era of digital technology.

WAREZ NETWORKS

This problem is inseparable from the intensified speed and scope of cybercapitalism's preferred means of circulation – the Net. A highly sophisticated, competitively organized system of online game piracy has flourished for years, using private File Transfer Protocol (FTP) servers, Internet Relay Chat, and, to a lesser extent, short-lived Web sites to distribute "cracked" titles.[72] In the late 1990s the most prestigious pirate groups – Razor 1911, DOD, Pirates with Attitude, the Inner Circle, TDT/TRSI (The Dream Team/Tri Star Red Sector Inc.), The Humble Guys – "were tightly knit clubs whose members have known each other for years."[73] Pirate bulletin boards can be elaborate operations: famous warez boards such as the Pits and Elusive Dream operated with as many as twenty-three incoming phonelines. According to a source who in the early 1990s was active on major warez networks, as much as ninety percent of their activities involved games. These were usually supplied by employees – perhaps those who had a test copy – within the development or publishing company.

On the boards used by pirate groups could be found "zero hour" software – games available at the same time, or before they became available in stores, complete with manuals and full downloading instructions. The boards also included "crack fixes," patches to remedy problems in the pirated games. Many games were available on warez boards in complete form, including such big games as *Wing Commander 2* and *Ultima 6*, which could take as many as twenty-three disks to download. Weeks before the release of *Doom II* the entire game was available from a pirate BBS, reputedly duplicated from a review copy sent to a British PC magazine. Its producer, id Software, then adopted a maximum-security policy, abolishing prelaunch reviews and external beta reviews. Nonetheless, *Quake*, the eagerly anticipated successor to *Doom*, turned up in final beta test version on an FTP server in Finland three days before the release of the official version.

More recently, the increasing size of CD-ROM games with elaborate video, speech, and music components has made the online pirating of games more complex. One common practice is "ripping," in which the game is digitally lifted from the CD, stripped of much of its video and sound capacity, and then uploaded in a "lite" version. The stripped-off elements may subsequently be offered as add-ons. The skeletal versions of games may be treated as previews, an opportunity to test the game before committing to purchase. Indeed, some warez pirates justify their activities as a form of consumer service necessary in an industry that allows dissatisfied buyers no returns on purchases.[74]

While some pirate BBS operate on a commercial basis, the true warez culture is a non-profit venture: "Warez crackers, traders and collectors don't pirate software to make a living: they pirate software because they can. The more the manufacturers harden a product, with tricky serial numbers and anticopy systems, the more fun it becomes to break. Theft? No, it's a game ..."[75] Warez may be offered as gifts – "testimony to the power and stature of the giver" – or as part of an intricate barter economy operating through select groups where membership is dependent on a demonstrated ability to contribute to the collective store.[76] Cracking is seen as a game of wits played against authorities, or as an anarchic or libertarian political gesture, releasing for general use the potential digital superabundance of information that the computer industry itself has created. The peer-to-peer (P2P) explosion will multiply this problem. Although the music giants have been the first in the firing line, interactive games companies will be next, as video-capable P2P networks such as Swapoo emerge.[77] "I think Napster and Gnutella are pretty serious threats to the games industry," Lowenstein says. "As you get to more broadband, I think they become even more dangerous."[78]

The attempt to control this process is throwing the interactive play business into contortions. Game capital is waging war on itself (Sony, for example, has sued emulator companies that make its console games compatible with PCs); on other parts of e-business (Electronic Arts, Sega, and Nintendo filed a lawsuit against Yahoo accusing the search-engine portal site of ignoring sales of counterfeit video games on its auction and mall areas); and on its own fan base (IDSA lawyers have threatened Web sites that offer free versions of classic games like *Space Invaders*, even though most such games are no longer commercially sold).[79] The irony, of course, is that game companies, like other digital businesses, have encouraged and created the preconditions for the "gift economy" but they are now trying to stamp it out, partly by promoting digital technologies with inherently "piratical" powers of copying and distribution, and partly by their own marketing strategies. When id Software distributed its freeware copies of *Doom* in the early 1990s, it inaugurated a marketing practice – promotion by give-away – that took inspiration from pirate practices and institutionalized it as a Net economy business model subsequently or simultaneously copied by other digital capitalists, from Netscape to Red Hat. The idea was that distributing digital products for free would create the basis for an expanded market for full or recent versions, services, manuals, and spin-off products.[80] In this way, the Net "gift economy" could be contained within – and indeed propel – e-commerce; "dot.communism" could exist but only as a supplement or margin to Net capitalism. The danger is that the genie of "free goods" may refuse to stay in the bottle.

BLACK MARKETS

Significant as the gift economy and warez networks may be in North America and Europe, the major breeding-grounds for contraband games are probably in the black markets of the world, China, the Russian Federation, Southeast Asia, and other emerging, or declining, markets. For some time, the most notorious of these black-market zones was in the Far East. Bob Johnstone summarizes the situation by saying that for Nintendo and Sega, Asia outside of Japan was "more of a nuisance than a market."[81] Within weeks of their introduction, new video games had been copied by bootleggers and "sold region-wide for less than half the price of the real thing."[82] In Taiwan, where the read-only memory chips for the cartridges are mass produced, the semiconductor industry was reputedly nurtured by the copying of game chips, which were then smuggled to Hong Kong or Singapore for assembly. In 1993 Nintendo of America and seventy other US companies called formally for American retaliation against Taiwan, which they identified as the "center for video game piracy throughout the world."[83]

If Taiwan was the manufacturing centre of game piracy, Hong Kong was the distribution hub. An *Economist* report describes the numerous shopping arcades that blatantly sell pirated software. The oldest and most notorious is the Golden Shopping Centre in Shamshuipo, "a grimy backwater in Kowloon" that contains "several hundred small shops selling all sorts of computer paraphernalia."[84] While some of the stores sell original packaged software, neighbours with illegal copies undercut them for a tenth of the price or less. The pirates offer thousands of titles, including CD-ROMs containing the latest fifty-dollar computer games that cost just forty dollars in Hong Kong currency (equivalent to about $5.40 Canadian). The pirated products are up to date: a lawyer hired by the Business Software Alliance to halt the trade reports that "one games maker gave out five evaluation copies of a new motor-racing game – to his distributors – but it still turned up on sale at the Golden Shopping Centre."[85] The Customs and Excise department of the Hong Kong government carries out raids, confiscates stock, and prosecutes offenders. But such action does not squash the trade, much of which is in the hands of the infamous "triad" criminal gangs, so that "detectives hired by software firms often receive death threats."[86] Despite this, some games makers have launched civil actions against pirate shops.

At one time it was believed that the corporations' best chance of squashing the Hong Kong pirates lay with Hong Kong's integration with Mainland China and the imposition of authoritarian state social-ist discipline. That hope has faded, however, as the liberalization of China's economy spawns its own thriving bootleg businesses, many of which were allegedly operated by state agencies such as the People's Army. Beijing had its own "Thieves' Alley" where software pirates congregated. In 1995 China and America teetered on the edge of a trade war over the widespread counterfeiting of software, music, and video. Eventually, the Beijing government agreed to act and closed many of the plants. However, many believed that the only effect of the crackdown was to push production deeper underground. In a 1996 interview, one of the Chinese "computer insects" who merge piratical entrepreneurship with politically heterodox ideas offered this opinion: "I rip you off, then you rip me off. Popular software products – regardless of who developed them – all contain some fishy things. So much intermarriage has gone on over the years that nowadays every-one's related. It's ridiculous for these stinking foreigners to pick on China like they do. We're just following the general trend by pirating some of their stuff ... To be honest, they've been ripping off the Chinese for ages. What's all this stuff about intellectual property?"[87]

More recently, the former Soviet Union has emerged as another piracy hotspot. Towns such as Vilnius in Lithuania and St Petersberg in Russia have been identified as prime sites where you can "snatch up fresh-mint copies of *Half-Life: Game of the Year Edition* in unas-suming jewel cases from hawkeyed hagglers for a buck."[88] RASPA, a

Russian anti-piracy group, declares that as many as ninety-eight percent of PlayStation titles in that country are counterfeit.[89] The Business
Software Alliance says that piracy rates in Russia, the Ukraine, and
other areas in Eastern Europe are at about eighty-eight percent, compared with twenty-five percent in the US, and higher than anywhere
else in the world, although the Asia/Pacific region still accounts for the
highest dollar loss.[90]

But the contraband game problem defies location. IDSA accuses more
than fifty countries of either aiding counterfeiters or failing to establish
or seriously enforce adequate protections against theft of intellectual
property.[91] Much of the product shipped through Hong Kong, Para-
guay, or Lithuania goes to countries all around the world. If the
interactive game industry is now globalized, so too is the shadow world
of pirate enterprise that haunts it.

CONCLUSION: MUTINY ON THE TITANIC?

Piracy losses and labour relations seem completely separate issues –
one a problem of consumption, the other of production, one of illicit
users, the other of unhappy developers. But it is not necessarily so.
Hackers and workers do not come from different worlds. They may
be connected; they may even be one and the same person. In some
ways piracy is the shadow aspect of the interactive play industry's own
labour practices. It is the flip side of the structure of skills, ideologies,
habits, and rewards (or lack of rewards) that the game corporations
systematically cultivate in their various workforces. Paradoxically and
circuitously, the very information age strategies that have been so
successful in enabling game capital to repress production-side labour
problems give rise to consumption-side piracy problems. But piracy in
many ways arises from conditions that the games industry has itself
contributed to – including both the "hacking" culture of computer
programming and the international division of labour that underlies
multimedia production.

If we look first at piracy of the warez type, it is obvious that in a
sense it is merely a logical, though unanticipated, extension of the
industry's work-as-play ethic. The point is not just that hackers and
warez networks use the very technological skills that the games business and digital industry in general promote. They also manifest, albeit
in "perverse" form, the same attitudes it fosters. As we have seen,
game development depends on cadres of digital knowledge workers
who are encouraged to blur the lines between labour and leisure. This
blur legitimizes the long hours, crazed schedules, and obsessive preoccupation with programming that are so productive for an innovation

economy. It is hardly surprising that some members of this labour force, inculcated in the work-is-play ethic, come to treat property itself as a game. Gift-economy pirates make software commodities into the counters in a contest of technological wits, where score is kept in terms of "bragging rights" in much the same way as arcade-game players might seek to top their rivals' scores. A business that seeks technologically to commodify play can hardly be too surprised by the emergence of countercultures that technologically play with commodities.

Tiziana Terranova has recently analysed the unpaid Net-workforce, "simultaneously voluntarily given and unwaged, enjoyed and exploited," involved in building Web sites, modifying software packages, reading and participating in mailing lists, and building virtual spaces in MUDS and MOOS.[92] She observes that the transition from Fordism to post-Fordism, while liquidating or displacing the old industrial working class, has also produced generations of workers who have been socialized as "active consumers" of cultural commodities. Capitalist managers need to recycle these sensibilities back into production to provide the look, style, and sounds that sell music, games, film, video, and home software. But this can only be partially accomplished by the recruitment of paid workers. Media capital is obliged to harvest a field of collective cultural and affective endeavours, which it "nurtures, exploits and exhausts," selectively hypercompensating some and ripping off others.[93] Free Web work, such as that performed by AOL chat-room hosts, or *Quake* online architects, represents the moment where the "knowledgeable consumption of culture is translated into productive activities that are pleasurably embraced and at the same time often shamelessly exploited."[94] Terranova's conclusion – that "free labor is structural to the late capitalist cultural economy" – at first blush seems to affirm another element of cybercapitalist success.[95] But the socialization of production on which it rests has a clandestine implication: erosion of ownership. The obverse of free labour is that great bugbear of digital industries – free goods, or piracy.

The connection between piracy and the game industry's labour structure is even clearer in the case of so-called "Far Eastern counterfeiting." As we have seen, the computer industry as a whole is a participant in an international division of labour that allows for the low-cost production of chips and hardware. It is the beneficiary of a global pool of cheap labour that necessarily has low, sometimes subsistence-level, consuming power. The global marketing campaigns of information capital saturate rich and poor alike and stimulate demand for games (and every other sort of consumer good) even in those areas where the majority of people cannot afford them. Yet as one industry observer notes, "Asian kids love to play video games just as much as their counterparts elsewhere."[96] How surprised should we be when the global mobile networked multimedia empires of interactive gaming encounter in China, Central America, and Eastern Europe not just the occasional angry fist of striking workers but also the silently subversive hoisting of the digital Jolly Roger?

Pocket Monsters:
Marketing in the
Perpetual Upgrade Marketplace

INTRODUCTION: GOTTA CATCH THEM ALL!

"Gotta catch them all!" declares the campaign slogan that exhorts the child-consumers of *Pokémon*, the most successful of all current video game products, not only to acquire all two hundred and fifty variants of the game's mutant monsters but to pursue them across a range of media, running from wireless-connected mobile game devices to home consoles to collectible cards, television shows, films, books, comics, and toys. Whether or not players are triumphant in their hunt, it is a quest that will certainly be worthwhile for the owners and licensees of the *Pokémon* commodity. By 2001 the cumulative amount generated by the *Pokémon* franchise for Nintendo and its partners over some five years was estimated at about fourteen billion dollars, a tribute to the extraordinary scope of the interactive game industry's marketing circuit.[1]

We have already noted the innovative promotional ethos that stands out as the unwavering mainspring of the digital gaming sector. The hyperreality of video and computer gaming is also a "hyped" reality where marketing managers and advertising agencies practise their best moves on youthful consumers they aim to enlist for a lifetime of purchasing experience. From the mid-1990s on, the interactive game industry began cultivating an expanded dialogue with its most regular consumers by working on the symbolic synergies that exist in cultural domains, with a concerted effort to penetrate cultural space. From playground-swappable card collections to MTV advertising spots to *Tomb Raider* movies to Sega theme parks, game marketers were filling every nook and cranny of youth culture that they could find.

The widening role of saturation marketing and creative advertising management has made the interactive game industry a model of synergistic marketing in the wired mediascape. Indeed, the industry provides a perfect case study of what Andrew Wernick terms the "promotional vortex" that "multiply interconnects" various "spheres" of cultural and commercial practice.[2] Promotional practices govern both the transformation of signs and activities into commodities (e.g., as cultural industries make businesses out of entertainment and play) and the transformation of commodities into signs (e.g., in advertisements and in designing appealing commodities) in the contemporary electronic marketplace.[3]

In this chapter we chart the process by which interactive gaming is becoming dominated by its marketing circuit. We said earlier that this circuit involves the activities of and interactions among marketers, commodities, and consumers. Marketing is useful as an entry point for examining the interactive game for a number of reasons. We have already discussed the importance in the "circuit of capital" of delicately balancing the spheres of "production" and "consumption." It is in the ongoing struggle to maintain this equilibrium that marketing plays a pivotal role. Indeed, at a very basic level one of the chief goals of marketing is to actually *make markets.*[4] Here we can recall Vincent Mosco's point: in the media industries, the term "audience" actually refers to a "market" of consumers. As our history has shown, the audience for this new medium didn't spring into being ready made but was carefully cultivated in a dialogue between gamers and workers in the marketing and cultural circuits.

Marketing practices have both facilitated and intensified the commercial organization of communication in market societies. In a mediatized marketplace, political economist Graham Murdock observes that "[m]arkets are always systems of economic transaction as well as of symbolic exchange. The trick" for communication analysts, he goes on to say, "is to find ways of exploring their interplay."[5] Devising these "tricks," we note, is precisely the job of marketers. In the workaday business of culture, marketers set out to maintain a line of communication between the interests of video game companies to make profits, the volatility of technological innovation, and the fast-changing cultural preferences of the youthful video game audience. In the course of negotiating these various demands, we can see how marketing plays an especially important role in managing the interplay of the three circuits. Marketing has become so important that some commentators consider that marketers and other "cultural intermediaries" are in the "driving seat of economic institutions."[6] Not so much in the driving seat, we suggest, as giving directions, helping to steer a route to profitability. Striving to widen the cultural scope of gaming and amplify the frequency of game-related purchases, marketing in the game industry makes use of a range of practices to "negotiate" with its target audience at the level of both game design and marketing communication design.

But this story is beset with paradox. The ever-expanding marketing apparatus is necessary, at least in part, to handle problems that it has itself created – to "manage the cultural tensions provoked by that same extension of the commodity-form which produced the one-dimensional world of consumerism itself."[7] In the high-intensity promotional market there is a dynamic of semiotic escalation. Competition within an advertising-saturated environment pushes marketing agencies to extremes of audacity, sensationalism, and "coolness" to seize the attention of jaded audiences.[8] The very intensity of these strategies, however, not only adds to sign-clutter but often provokes new levels of cynicism and indifference in the audiences. This is especially true in a youth culture whose advertising-savvy participants see through marketing ploys even as they succumb to them. The dynamic is redoubled in the perpetual innovation economy of the game industry, where the relatively short-lived play value of software and the successive waves of hardware innovation in the technology circuit create an incessant upgrade dynamic of new commodity releases. The launch of each new game and console burns up financial, symbolic, and creative capital by the gallon as promotional innovation and consumer resistance pursue each other in a rocketing upward spiral.

The lavish funding and innovative techniques and practices of promotional communication seem, moreover, to have done little to enhance the creativity and diversity of interactive games. Indeed, the opposite may be true. As one game marketer argued, it's not even necessary to manufacture a high-quality product because "more often than not, marketing is the necessary driving force that can turn a mediocre video game into a successful one."[9] Gamers, as we have seen, do make demands on the quality of their gaming experiences. But in the oligopolistic market of the third millennium, branding platforms and establishing synergistic connections with other branches of youth culture become essential for cultivating, consolidating, and expanding a loyal user base. As expensive marketing campaigns become a *sine qua non* of commercial success, promotional expenses eat more and more of game budgets, companies that cannot afford such costs are driven to the wall, and the industry consolidates around giant publishers. Moreover, considerations of branding and synergistic marketing enter into the very conception of game design, as the production of blockbuster hits becomes a priority. As the "digital design" of games becomes increasingly governed by strategic marketing, the diversity of audiences and the creativity of game content seem to diminish.

In this chapter we describe recent activities in the marketing circuit and how they intersect with the cultural circuit. We begin to look more closely at "digital design practices," showing how marketing considerations work their way back into the game development process, leading to the creation of games that are from their inception conceived as franchises whose marketing potential can be extended into multiple cultural spaces and constantly renewed over time. In terms of our model of the game industry's circuits, we can say that the hypertrophy and acceleration of the marketing circuit begins to exert a powerful force on the cultural circuit. Meeting the demands of marketing eventually intersects with a "design" process, on the practice of designing the games themselves. Our history of video games confirms the finding of Paul du Gay and his co-authors in their study of the Sony Walkman that "designers occupy a place as important cultural intermediaries at the interface between production and consumption."[10] This "interface" involves a complex set of negotiations between processes of creative game design and the pressures of selling games in the marketplace. We describe some of the more palpable instances of the "encoding" of marketing imperatives on game texts – an important aspect of "digital design practice." *Pokémon*, whose digitally-translated, biotechnologically-mutated pocket monsters have so effectively picked the pockets of parent-and-child consumers around the world, provides a telling illustration of this dynamic.

MANAGING BRANDS

The US youth market in 2000 spent a record $164 billion, a large portion of it on entertainment.[11] In maximizing their share, youth marketers say the main challenge is to "strategically position our products to acquire cachet" by building "brands that will attract and entice one of the toughest, most fickle markets in the world."[12] One marketing guru, Geoffrey Moore, stresses the role of branding in "selling high-technology products to mainstream consumers" – especially in moving from niched to mass markets. Marketing branded products, he claims, is a terrain where corporations "develop and shape something that is real, and not, as people sometimes want to believe, to create illusions."[13] In the video game market there are only a few console makers to choose from. This means that the efforts of marketers "to establish and fix a set of meanings" around a console and its users is a vital cultural practice for building a loyal consumer base for a console.[14] According to Moore, "the efficiency of the marketing process ... is a function of the 'boundedness' of the market segment being addressed. The more tightly bound it is, the easier it is to create and introduce messages into it, and the faster these messages travel by word of mouth."[15]

This marketing effort to create a "bounded" promotional environment is an ambition taken to new levels in the interactive game industry. As the brand-war tactics initiated by the competition between Nintendo and Sega developed, game marketers became more and more audacious in attempting to combat rivals and break through the cluttered commercial media environment with a unique voice. In developing brand strategies Nintendo and Sega drew on the example of companies such as Disney, while Sony capitalized on an already well developed corporate image when it entered the video game market. But there were distinct features of interactive gaming that drove branding to an exceptional intensity.

One was the dynamism of the technology circuit of the business. About every three years the video game market is "revolutionized" by a new generation of technology: first 8-, then 16-, then 32-, then 64-, then 128-bit consoles. Launches such as those of Nintendo's NES, Sega's Genesis, Saturn, and Dreamcast, Sony's PlayStations 1 and 2, and Microsoft's xbox are moments that can make or break corporate fortunes. At each launch, the promotional intensity of the industry escalates. Striving to maintain profitability when a console can be "post-dated" within three years, in a market where what's cool and entertaining is almost immediately exhausted among its most devoted fans, has galvanized the marketing circuit of this industry.

To this logic of perpetual technological upgrades is added the dynamic of changing consumer demographics. In the earlier years of the industry the market was firmly centred on adolescent and pre-adolescent males. Today, many "twentysomethings" have grown up with video games, and this older audience has become a prime target for the brand-intensive advertising campaigns dedicated to the industry's hit titles. "Up-aging" the market was a trend Sega started in the early 1990s, only to be leapfrogged by Sony later in the decade. By 2001 IDSA reported that fifty-eight percent of all console players, and seventy-two per cent of all computer gamers, were more than eighteen years old.[16] Paradoxically, however, this dynamic also favours the targeting of much younger consumers. Since what is at stake is now a potential lifetime of software and hardware game purchases, there is a premium on winning brand loyalty early. What this means is that at the same time that console and computer technology is transforming at a relentless speed, target audiences are changing too, and promotional codes are correspondingly being revolutionized, requiring yet new levels of symbolic investment to multiply market segments and create numerous entry points into the world of interactive games.

Video game marketers have, moreover, to sell their product not only in competition with rival game systems but against a whole range of other entertainment experiences.[17] As a new medium, video games had to claw their way into a market where television, film, and music already jockey fiercely for the time and attention of youth audiences. Dan Stevens, manager of corporate public relations at Sega, explains, "We're not just competing with another game platform – we're competing against TV. You could either be watching *Seinfeld* or ER, or playing a game. So our marketing has to be done fully aware of that."[18]

Survival demands that companies excel at symbolically and thematically differentiating both systems and games. It is this process of product positioning that marketers believe exerts "the single largest influence on the buying decision."[19] Even within the oligopoly of video game makers, symbolic production in advertising design, and the symbolic differentiation between similar products and media experiences, can be a crucial variable in determining market share. Behind the escalating marketing expenditures of the interactive games business lies the attempt to discover new strategies for riding rapid technological change, coping with depleting experiential values, and prying open time and space in saturated media environments. If unsuccessfully managed, these conditions spell total disaster, as Atari, 3DO and Sega learned. If game companies successfully "ride the chaos," however, fast product-replacement cycles, rapidly changing consumer demographics, and synergistic connections across dense, complex media markets can be excellent for profitability. For carefully managing the interaction of technology, culture, and marketing can accelerate the exchange of game-related commodities.

TELEVISION AND GAMES

For the interactive entertainment industry, it is television advertising that provides direct contact with core consumers, absorbing the lion's share of the promotional budgets.[20] As soon as one looks closely at the advertising of digital play, one of the many paradoxes of interactive games leaps out – the dependence of this new so-called "demassified" computer-age medium on the old "mass" medium of television. This is of course embodied in the connection between the video game console and the television set that has provided its display unit. But it is also highlighted by the consistent reliance of video game marketers on television as the main way to reach potential consumers.

Despite the popularity of interactive entertainment, kids continue to watch TV for two and a half hours per day and go to the movies in much the same way they always did. Since the mid-1980s, therefore, the video game industry has turned to TV advertising to fan the flames of excitement about a new product, and to saturate youth culture with its branded properties. Advertisers use TV as the primary communication channel because it reaches the broadest swath of potential gamers.

TV is also cost effective because youth media audiences are concentrated in demographic niches. As a result, TV ads become a central part of marketing programs, used simultaneously to develop a brand identity for the game maker and to communicate the attractions of the game play experience.

The lively thirty-second spots must not only catch the attention of these fickle consumers but also convey in some engaging way a constellation of meanings that deepen the viewer's involvement with the product concept by making interactive entertainment seem cool. Game marketers have adopted increasingly stylish innovations to establish points of identification for their chosen audiences. At Nintendo of Canada, marketing manager Ron Bertram says that because kids largely determine the purchase of game systems, Nintendo doesn't have to appeal to parents as buyers. "We don't market to parents," Bertram flatly claims. "We market to our target group which is teens and tweens. Parents may be highly involved in the purchase decision, but ... it's the kids that are driving it."[21] As one youth marketing specialist explains: "the challenge of marketing to people twelve to twenty-four goes way beyond trying to figure out what 'cool' is ... 'today's youth (are) the media savvy products of *Much Music* and a chaotic, ever-changing world.' And this makes them more difficult than ever for marketers to reach."[22] The marketing vice-president at Sony Computer Entertainment says ads have to demonstrate game visuals or appeal to humour not only to catch the broadest possible audience in a cluttered promotional landscape but also because the target audience "doesn't like to be advertised to, they can smell advertising a mile away. They don't want to be told how they should feel. Ultimately, we realize there is an attitude in the elements of game play and we have to recognize and understand that this attitude is the real motivator."[23]

Given their media-savvy audience, one of the game companies' most consistent strategies in the late 1990s was to downplay the fact that the messages were ads – that is, to make the sales pitch as entertaining as possible. Bertram says that Nintendo's ads are "fun, irreverent, different, enjoyable, exciting and cool – without ever saying it."[24] The increasingly "active" youth audiences therefore contribute to a "game" played between ad designers and game consumers: stylistic innovations in advertising occur at an increasing velocity in an effort to intrigue audiences and to sustain the "cool" currency of a brand's identity. To target their segment and build brand image, video game advertisers have for a long time been making commercials that look less and less like commercials and more and more like entertainment. According to video game advertisers, "Short of actually playing, the best way to get a feel for a game is watching it being played – something that can only be done on a TV. Thus advertising video games on TV seems like the perfect solution."[25] Now the strategies are to infuse the ads with music, irony, and twisted humour, or fill a commercial with game graphics, to the point where it looks as though you're watching a skilled friend play a video game or watching a music video rather than a thirty-second interruption of a TV show.

The purpose of this style of ad is symbolically to represent, in McLuhan's terms, the sense-ratio and psychic qualities of the interactive media experience. The intention of the TV spots, say video game marketers, is to encourage the audience to connect with the game on a "mental and emotional level." A researcher with Sony explains that, in focus groups, "when we play the ads we watch kids stop for a minute and think about what they've seen. They're going, 'I've gotta deal with this.' It makes them participate mentally and emotionally with our product, that's what we set out to do."[26] The desire for involvement and challenge inscribed in the ads is precisely the reference back to the feelings in game play that marketers seek. As Sony marketer Haven Dubrul says, "One of the best comments we've gotten was that when they watch the TV ads they felt like they were going to feel when they play the games."[27]

This seamless dissolve between game promotion and game experience speaks volumes to the hidden continuities between old and new media. Because the television screen displays both video game play and advertisements about such play, the game and the image of the game fuse. Advertisers "play" their audiences, attempting to push them to the next level of expenditure on games products, in the same technological medium in which the consumer "plays" the product – attempting to reach the next level of virtuosity as a virtual airplane pilot, deadly kickboxer, or acrobatic hedgehog. A post-Fordist individualized interactive entertainment form has come into being on the back of, and in continuing relation to, the Fordist one-way broadcast technologies – and by a route that both maintains and sophisticates the very "mass" marketing techniques digital futurists like to pretend have now been surpassed.

SYNERGISTIC MARKETING: LICENSING AND MERCHANDISING

Perhaps less obvious from the high-energy advertising campaigns is the way the industry explored the synergistic cultural relationships that are the invisible subtext of the interactive entertainment environment. Synergy – a word that flies quickly from the tongues of corporate executives and business pundits – is about the intensification of growth and expansion through integration across the many spheres of production, technology, taste culture, and promotion. It is a strategy to manage growth in a fluid cultural marketplace through interplay between different sites of culture. The synergistic marketing of interactive games involves the coordination of promotional messages to saturate diverse cultural niches, a heightened emphasis on the binding of game design and advertising, and the weaving of a branded network of cultural products, practices, and signs to create multiple entry points for consumer-players, hence multiple revenue streams for game corporations. It is also about moving out of video game play in a traditional sense into other media properties that represent new appeals, new audiences, and, again, new revenue streams. If a gamer can only buy so many games, then game culture can be channelled to other commodified forms of popular culture. The basic logic in synergies is that, just like a high-quality interactive gaming experience, one path always spawns ten more.

Since the early 1990s there has been a concerted effort on the part of video game console makers to saturate youth media culture with their branded products. They have capitalized on the recognition of video game characters within their audience. The branding of platform games and characters is integral to the burgeoning subeconomy of licensing because these marketing practices cultivate the very symbolic value upon which licensing agreements depend. Nintendo has featured Mario's image not only in dozens of its games but also on T-shirts, snack-food packaging, and in movies, spawning an entire franchise of licensed products that span *Mario Bros.* movies to fishing poles sporting Mario's image. Each agreement awakens a new chain of promotional associations with Mario, and a new flow of income for Nintendo.

Console makers and game developers also seek out licensed properties to bring into video games. For example, Electronic Arts, a publisher that specializes in the sports genre, has exclusive licensing agreements with the NHL and NFL, giving it permission to use team logos and player names in its games. Game developers have also obtained licenses that allow them to use storylines from blockbuster movies, such as *E.T.*, *Star Wars*, and *Titanic*, and from hit TV shows, such as *The Simpsons*, *South Park,* and *Beavis and Butthead.* By capitalizing on narratives drawn from popular culture, marketing costs are drastically reduced because the most expensive marketing work – building awareness – is already done. Games are developed around characters and narratives that have already been tested in the entertainment marketplace. The pursuit of licensing agreements is therefore a marketing-driven approach to game design that aims to minimize investment risk.

Again, Electronic Arts has recently been pre-eminent in such arrangements. In 2000, it signed an agreement with America On Line to manage online games for the giant Internet service provider. When AOL shortly thereafter acquired the media giant Time Warner, Electronic Arts was uniquely positioned to become part of the mammoth synergistic web created by the fusion of the two companies. The first major test of the marketing power of this monster was the film *Harry Potter and the Sorcerer's Stone* in 2001. Electronic Arts acquired the exclusive rights to *Harry Potter* and produced four games featuring the boy hero's adventures, for play on personal computers, Sony PlayStations, and two versions for Nintendo's Game Boy. Electronic Arts then followed up on this licensing triumph by acquiring the rights to AOL/Time Warner's second movie blockbuster, *The Lord of the Rings*, and producing a series of games slated to appear concurrently with the release of the second film in the trilogy in 2002. These deals connecting Electronic Arts to AOL/Time Warner make its interactive games a component of the largest synergistic media empire in the world.

The pressure to strike a good license, whether to a Hollywood movie or a professional sports league, is bound up with the rising pressure on console makers and developers to achieve a "mass market" hit game. The children's culture critic Marsha Kinder describes this growing emphasis on "synergies" wherein corporations partner "to reach out to a larger audience by positioning the world of video games within other, more familiar contexts."[28] Rather than "free-floating" icons in popular culture, then, the video game character is a "proprietary symbol" whose appearance is managed through strategic licensing agreements. The video game companies attempt to capitalize on the recognition and symbolic value of their characters by "repurposing" them into licensed products, which open new revenue streams.[29] As the marketing critic Naomi Klein has noted, such "brand extensions are no longer adjuncts to the core product or main attraction; rather, these extensions form the foundation upon which entire corporate structures are being built."[30]

SPINNING MEDIA

The gaming industry is not just a big spender in TV advertising, and a major licenser of characters *from* film and television. It also spins games *into* film and television. Game companies have licensed and co-produced television cartoons, such as *Sonic the Hedgehog*, *Mario Bros.*, and *Digimon*, based on platform game themes and characters. They also license themes to Hollywood films – like *Mortal Kombat*, *Super Mario*, and *Street Fighter*. The success rate of these ventures was not generally impressive, the transfer of video games to film in particular clocking up a remarkable record of box office duds. In 2001, however, two productions altered the picture: *Tomb Raider* and *Final Fantasy*.

We have already remarked on the extraordinary success of Lara Croft and her centrality in attempts to recruit both male and female gamers. In 1997 Paramount Pictures obtained a license from game publisher Eidos to produce a feature-length *Tomb Raider* movie.[31] Eidos was to pocket ten percent of the film's gross box-office receipts and gain by the increased interest in its game. Guided by the synergistic logic of cultural marketing, the movie's director interpreted it as "a unique opportunity to develop Lara's personality and bring her to a wider audience. The prospect of bringing Lara to life while putting a fresh spin on this genre is very exciting."[32] The film was stuck in development for several years, partly because of disagreements between Paramount and Eidos over scripts. Eventually it went into production, with the actress Angelina Jolie recruited to star as Lara only months before she won an Academy Award for her role in *Girl Interrupted*. The shoot, at locations in England, Iceland, and Cambodia, cost nearly one hundred million dollars.

When the film was released, reviews were generally negative. Critics saw the story of Lara's battle against the evil secret society of the Illuminati as little more than an episodic characterless action-adventure flick. But the box-office verdict was far more favourable, and *Tomb Raider* became one of the top films of the summer of 2001. The soundtrack, featuring such bands as Nine Inch Nails and the Chemical Brothers, was broken out into a successful CD. Meanwhile, Eidos, Sony, and many other companies were using the image of Lara Croft to cultivate recognition for other products. Eidos was already using the character marketing of Lara Croft to promote all the company's games: "Don't cherry-pick our lineup, give us full line support. We've got some excellent branding ads banking on the awareness of Lara Croft ... So people know, 'Hey, the company that brought you *Tomb Raider* is bringing you *Fighting Force* and *Deathtrap Dungeon*.'"[33] Croft's image was licensed for use in ads for other youth-oriented products, from fruit juice to Nike running shoes.[34] The DVD release of the movie includes a special trailer for the next *Tomb Raider* game, which Eidos marketers promised would "completely relaunch the franchise."[35] The jinx on video game Hollywood crossovers appears to have been broken.

Games like *Tomb Raider* are licensed by their creators to established movie studios for translation to the big screen. But the film of the enormously successful role-playing game *Final Fantasy* represents a departure from this logic in that it was made through an independent film studio established by a game developer. Square is one of the largest independent game software developers and dominates the console role-playing sector: its *Final Fantasy* series, created first for Nintendo platforms and then for Sony's PlayStation, is largely responsible for popularizing the genre in the US. To make the film of *Final Fantasy*, Square spun off its own film company, complete with a fifty-million-dollar studio in Honolulu, which then entered into a multipicture deal with Sony's Columbia Tri-Star Pictures. The *Final Fantasy* film transferred to the cinema all the advanced animation features of game design – including "actors" who were completely computer generated. Thinking synergistically, Square plans to leverage the success of the movie with licensing and merchandising deals, and to establish a brand reputation across games, films, Internet, and perhaps TV animation. In particular, the film serves as a promotion for the forthcoming *Final Fantasy XI* massively multiplayer online game in which thousands of players will participate.[36] With the success of *Tomb Raider*, and Square's direct entry into film production with *Final Fantasy*, the synergistic integration of interactive games and film appears to have taken a major step forward.

SEGA CITY @ PLAYDIUM: RE-BRANDING THE ARCADE

The game industry's synergistic ventures are not limited to film and television. They also include the creation of new game-based urban amusement sites that radically update the attractions of one of the industry's original incubators, the arcade, and cross it with the appeal of the Disneyesque theme park. As we have seen, Sega has for a long time been at the forefront of innovations in the video game industry. The trend in "location-based entertainment" centres has been no exception. Sega's foray into this new concept was the result of finding itself in a losing corner in a competitive home-console market. Sega turned to the theme park sector in the mid-1990s around the time of its "symbolic exhaustion" and product overload crisis. "The weakness of the game market" was "forcing Sega to pursue other business opportunities" and "to channel existing revenues from its game cash cow into these new ventures to lessen its exposure."[37] One of these ventures eventually involved Sega in the Playdium franchise, a chain of video-game-based entertainment complexes building their way across Canada.

Two former real estate developers, Jon Hussman and Steven Warsh, started Playdium Entertainment Corporation. The vision, Hussman says, consisted in "marrying Hollywood with technology. The power of the computer has enabled the creation of the most incredible experiences imaginable."[38] Hussman and Warsh came up with the Playdium idea by noticing a gap between traditional urban video game arcades and the monstrous theme parks like Canada's Wonderland. Gaps won't be tolerated. This is where commodification finds a comfortable breeding ground. The partners considered that distance and expense were the two main barriers to regular visitor participation in this kind of entertainment. Their strategy was to get repeat customers by designing a location-based entertainment centre that was affordable.

But the vision needed money, a lot more than Hussman and Warsh had. So they sought investment partners. Nina Kung Wang, reportedly Asia's richest woman, invested a huge chunk. "In one transaction," Hussman says, "we got the capital we needed for expansion in Canada and struck a strategic alliance for international expansion. Nina had the ability to finance our growth."[39] In 1993 Hussman and Warsh were put in contact with Sega's chairman, Hayao Nakayama. Sega was already building entertainment centres in Japan and saw a partnership in Playdium as another wing in North American expansion, and a strategic opportunity to spread resources and investment risk beyond the domestic environment. "The deal," explains Hussman, "was we were fifty-fifty partners. Our job was to North Americanize the concept."[40] The initial contract called for fifteen high-tech centres to be built in major cities, from Montreal to Edmonton. Sega would put up fifty percent of the required capital – some $82 million. Playdium would design and run the facilities, which were to be built around Sega's hottest new games.

Playdium quickly became an emblem of the "strategic cooperation" that is commonplace in the synergistic media sector today. In 1996 the deal expanded with the involvement of Spielberg's DreamWorks. With access to this new pool of capital, technology, and creativity, Playdium expanded its plan to build forty centres across Canada and later struck a deal with Famous Players Inc. to open hi-tech game centres in thirty-two movie theatres.

In September 1996 the first Playdium opened on the outskirts of Toronto on the grounds of a huge shopping mall. In their first year of operations, Playdium @ Sega City had over one million visitors and six million dollars in revenue. It is a 33,000-square-foot entertainment park designed, according to the promotional material, as an "indoor games sensorium." The complex includes hundreds of arcade-style video games and simulation games. It also has outdoor activities like beach volleyball, miniature golf, rock climbing, and go-karting on a track designed by racer Mario Andretti. There are indoor baseball batting cages, the "Blue Jays Clubhouse," endorsed by the then Blue Jay's coach, Cito Gaston.

The decision to put Playdium @ Sega City on the grounds of a major shopping mall in a largely middle-class region of Toronto was not difficult to make. The location would mean access to an audience that was already in consumption mode. Playdium was about taking advantage of this consumption site, transplanting Hollywood's conception of the "entertainment experience," and designing an enclosed environment around video games. Hussman and Warsh's experience as former real estate and commercial developers helped in the planning. "The idea," says Hussman, "was to develop a unique environment that was very friendly, very secure and highly themed, and offer the latest and the greatest in technology-based attractions." Interactive entertainment was in turn layered in other consumption situations: "And then marry that with physical games, food, a licensed beverage area and throw in some retail."[41]

In designing these environments, the governing principle is to create an intense immersive experience. The enclosed settings makes them, quite literally, a world within a world. Long gone are the days of the single arcade staff member who handed out quarters. In these environments, it's all about "show business" where "the employees are actors who reinforce the theme and the mood."[42] When we toured Playdium in Toronto, our host, the marketing manager, turned at the entrance and said: "When you walk through that door you leave reality – and enter virtual reality." It is the designer's job to "lead the guest's eye through an experience, and you do that by carefully positioning everything – the guests, the displays, the vehicle they're riding in, and the things they perceive, whether it be visual media, software, dimensional stage sets, or the tableaus of a dark ride."[43]

The PlayCard is the debit-card system used to pay for the attractions. According to the promotional materials, the "convenient pay as you play system lets you pay for only the attractions you choose. The Playdium PlayCard system is hassle free – no flimsy tickets or tokens. It's just like using your bankcard. Refill your cards at our convenient self-serve terminals. The PlayCard eliminates time: no waiting." Parents can "program the cards to restrict access to games." Visitors can use the PlayCard to buy food and merchandise. But there is more to the PlayCard than customer convenience. The magnetic strip records your balance and the length of your stay. These in-built surveillance mechanisms are great for market research; the company then knows exactly which games are being played and which ones to phase out. The PlayCard, like a credit card, gives visitors a sense that they aren't spending real money. It's virtual money.

The Playdium demographic is mainly between nineteen and thirty-four years of age. Visitors spend an average of twenty dollars and stay about ninety minutes. Each attraction is priced between $1.60 and $4.95.[44] Playdium had seven hundred thousand visitors in its first ten months of operation; families plan entire day trips around video games. But the most recent target demographic is the corporate audience. Business luncheons and corporate events make up thirty percent of Playdium's revenues. This newly discovered demographic is made possible in part because of the work-as-fun ethos we discussed in chapter 9. Leisure, especially among the upper echelons of the corporate world, is becoming acceptable, and in the spirit of the new Playdium work ethic happy workers are productive workers; the corporate world reveals its ultimate openness in making time for fun.

As video games are integrated into leisure practices outside the domestic sphere of gaming, the Playdium ventures show us how this medium is disturbing, or reverberating in, a range of cultural practices in consumer culture. The popularity of video games is changing the way investors and developers think about traditional sites of consumption. The megatheatre, themed restaurant, and video game theme park now form an entertainment ensemble that is used to draw young people into shopping malls in cities and suburbs alike. Game theme parks are by now almost standard in mega-shopping complexes. Mall developers say that they "are recognizing the need to inject new value into the traditional shopping experience."[45] Shopping-centre developers increasingly demand that entertainment centres be included in their investment sites. As the Famous Players vice-president says: "Developers see us in a totally different way. It's a more philosophical change. They recognize that their environments must be made more exciting."[46] Interactive games are thus becoming a key component in the new complexes of consumer- and entertainment-oriented space that urban critics such as David Harvey describe as characteristic of the emergent post-Fordist cityscape.[47]

SPORTS AND MUSIC: MONITORING TRENDS IN YOUTH CULTURE

The interactive games industry expands its appeal by weaving into games cultural practices that its young target audience already enjoys. As in the fashion and music businesses, the video game industry has had to contend with the style-conscious "youth communities" that cluster around preferences for brands and genres. These communities are by nature volatile: their constituents are always looking for the "next new thing." Research on youth culture has influenced the kinds of video games that are produced and the way that video gaming is promoted. Trends and fashions move at increasing speeds, as shortening

cycles of production decrease the delay between new cultural themes
– say, skateboarding, or antiglobalization protests – appearing on the
radars of marketing "cool hunters" and then being translated into the
design and development process and launched as a game commodity.[48]
Detailed monitoring of changing tastes and market patterns allows
potential emerging "hits" to be identified and integrated into synergetic
marketing campaigns. As one youth marketing specialist explains: "We
need to make a concerted effort to listen to the underground culture.
If we want to successfully package our products for teens and tweens,
we have to scrounge around in their world to find what's important
to them. Historically, marketers have borrowed from counter cultures
and woven those ideas into the mainstream. Consistently, those loves
have been sports and music."[49]

To make video gaming more attractive to people in their late teens
and early twenties, for example, marketers have integrated snowboarding and electronic music. Industry commentators praised Electronic
Arts for looking to subcultures as a source of video game concepts:
"When skateboarding became a very popular sport, EA did its home-
work, spotted the trend early, and delivered *Skate or Die*, one of its
biggest hits."[50] Managing the interaction of the marketing and cultural
circuits in this way illustrates what cultural critic Richard Johnson
describes as the moment in cultural commodification when "reservoirs
of discourses and meanings are in turn raw material for fresh cultural
production."[51] The process by which game marketers and game designers attempt to harness popular youth cultural practices is even more
marked in the industry's relation to emergent forms of youth music.
For some time, music has been an important dimension of the virtual
game experience. As we have seen, Electronic Art's Trip Hawkins found
in the music industry a model not only for the corporate structure and
conceptions of work in the video game industry but also for marketing
strategy, packaging its games like albums and promoting its developers
like rock stars. The ever-inventive Sega even started a spin-off label,
Twitch Records, which specialized exclusively in video game music
"with its own A&R reps, its own roster of bands, its own recording
complex, and its own executive producer"; Twitch positioned themselves "to be part of the entertainment industry, not necessarily just the
game industry."[52] As Herz explains, the synergistic marketing formula
was that "if a hot Sega game sells a couple of million copies, that's a
couple of million teasers for some band that also has an album out on
Twitch Records."[53] Twitch eventually flopped, but it signalled big
changes in the promotional role that music would play in video games.

Today, music is a carefully planned aspect of game marketing.
Gamers have come to expect soundtracks in their video games that are

up to date with popular music trends. The incorporation of current musical styles potentially accelerates the speed at which a game's value is exhausted – if the music is outdated, the gamer gets bored – and this increases the turnover rate of game purchases. Game music provides another way to identify and target potential audiences with precision; if a game has a punk rock soundtrack, a gamer likely knows whether or not they are the intended consumer. For example, in Sony's attempt to reach a slightly older audience, its successful flight simulator game, *Wipeout XL*, capitalized on the growing popularity of electronic music by including a full soundtrack in the game. The game developer Psygnosis partnered with the Astralwerks record label to release a compilation music CD that gamers could purchase at retail music stores featuring artists such as Chemical Brothers and Underworld, who were then celebrated by video game reviewers as the latest "underground" artists. Gaming magazines advised gamers that the *Wipeout XL* video game and CD would both be "a great way to introduce yourself to … electronic music."[54]

The promotional campaign for *Wipeout XL* offers a general sense of how video game media are engendering change in youth culture. Sony distributed copies of the game to DJs at major clubs in London and New York. DJs were asked to play tracks from the CD. Sony also made PlayStation consoles available at the clubs. This helped spread product awareness, but more significantly, it presented both gaming and the PlayStation brand to an older audience. Sony has since increased its presence in club culture by acting as a sponsor of raves. From these attempts to expand brand recognition, new cultural practices have been generated: graphics from video games are often projected onto walls at raves and console playrooms are often set up. Thus, electronic music has changed video gaming, but the marketing of video games has also changed the culture of electronic music.

These shifts in marketing and cultural practice are in turn reshaping how some entrepreneurs understand the video game industry.[55] Leaving their jobs as club DJs and record label executives, for example, some folks decided to reinvest their subcultural capital in Rockstar Games, a pre-eminent video game development company, responsible for such titles as *Grand Theft Auto III*. Marketing synergies are at the centre of Rockstar's business strategy. As the CEO explains: "I feel that a lot of games are marketed like toys or like technology, and they don't have the marketing edge that goes with an album or goes with a movie. But they fill the same place in people's leisure time and people's minds. So they should be pitched at that level."[56] This means replicating the Hollywood marketing model: soundtracks, product tie-ins, and media events. Rockstar concentrates on forming alliances with subcultural practices that are part of very specific taste cultures. It obtains licensed material from hip-hop artists and DJs as they strive to cultivate a symbolic field around their brand's games that, in the graceful words of the CEO, is "cool as fuck."[57] These fusions allow them to position video gaming before a niche urban audience in its twenties. But they also leverage their licenses as cultural vessels to engage new audiences and open new revenue streams. Rockstar sponsors nights at leading clubs in New York and London, promotes a line of skateboarder clothing, and commissions graffiti artists to design packaging for its games. As two music writers recently remarked of the new linkages between popular music and video games, "It's difficult to decipher the cart from the horse."[58]

PRODUCT PLACEMENT

As traditional media outlets and cultural spaces become more cluttered with advertising, merchants in several sectors have turned to "product placement" – the integration of a product logo into media programming. This practice has become rampant in Hollywood. It is increasingly common in games, where youth-oriented merchants now see an opportunity for more "reach" for their branding dollar in comparison to a thirty-second TV spot. Through game-based advertising, companies attempt to transfer symbolic recognition from the real world of consumer goods into the virtual world of video game play. In Sega's game *Fighting Vipers*, the Pepsi logo was "texture-mapped" onto a skateboard and Pepsi billboards fill the background.[59] In the sports genre, the hoardings at baseball stadiums, hockey rinks, and football fields feature billboards for youth-oriented brands. These corporate logos are moving closer and closer to the centre of the game. For example, Electronic Arts entered a licensing agreement with Volkswagen when the car manufacturer relaunched the Beetle. Electronic Arts designed an entire racing game around the car, *Beetle Adventure Racing*. Volkswagen marketers decided the game would be an effective medium through which to reach their target market.

The synergistic licensing deal that was struck between Nintendo and the colossal clothing merchandiser Tommy Hilfiger shows how youth-oriented industries cooperate to saturate the youth market with their brands. Hilfiger and Nintendo designed an elaborate licensing deal because research indicated that they target a similar demographic and that their audiences share a preference for both companies' products. The president of Nintendo put it simply: "Nintendo game players wear Tommy Hilfiger."[60] The licensing agreement was a means for each company to gain access to one another's brand-loyal audience. As part of the deal, Tommy Hilfiger retail stores added arcade stations where their shoppers could play Nintendo games. Tommy Hilfiger also designed the Nintendo logo into a clothing line. Nintendo saw an opportunity to capitalize on Tommy Hilfiger's symbolic value and brand recognition within its own video game audience, while Tommy Hilfiger saw an opportunity in Nintendo to amplify the exposure of its brand.

While the deal was being negotiated, the game *1080° Snowboarding* was in development at Nintendo. Tommy Hilfiger's audience also enjoys snowboarding. Taking advantage of the flexibility of digital game design, Nintendo made some adjustments that expanded the licensing deal. The director of marketing at Nintendo of Canada explains: "The game was far enough along in its development, but we were still able to say: 'Yeah, if we add the Tommy Hilfiger brand to the *1080°* game, it will add value and credibility to the game.' So we decided to do it."[61] "Tommy Hilfiger" was integrated into the game's subtitle; Tommy Hilfiger logos were added to the banners that lined the snowboarding course; and designs from Tommy Hilfiger's latest clothing line were texture-mapped onto the body of the snowboarder whose back the gamers watch as they play the game.

This example illustrates how marketing decision-making and game design collapse upon one another under the guidance of a licensing agreement. The industry has come full circle: conditions for its spectacular growth were set in existing youth-oriented media niches; now, gaming is itself poised to create marketing opportunities for other corporations that are seeking to target the youth audience. The transformation of game space into ad space is yet another moment in a gradient of commercialization, in which marketers have continually adjusted their methods of influence to take full advantage of the characteristics of each new medium of communication. Whether it be a hockey rink or the clothing on a snowboarder, many video games are now regarded as incomplete without advertising appearing in the right place. For game designers, designing products and logos into games puts them one step closer to the Holy Grail of "realism." In one informal survey of video gamers, most respondents said that they prefer games that feature familiar brands. As one gamer answered, even if a game was more expensive, "I would most likely buy the video game with the license in it. The one without the license? No. It's just not right."[62]

THE METAGENRE: DIGITAL DESIGN IN THE ERA OF HYPERMARKETING

The synergistic connections we have described are becoming vital for success, or even survival, in the game market. As Edward Herman and Robert McChesney note in their study of the entertainment industry: "Firms without cross-selling and cross-promotional potential are at a serious disadvantage in competing in the global marketplace."[63] A game companies' licensing arrangement with the National Hockey League, for example, surely is not a low-cost venture. Coupled with multimillion-dollar marketing budgets, the emphasis on licensing has made it difficult for smaller game developers without licenses and small publishers without deep pockets to survive. The mushrooming marketing costs are therefore related to consolidation in the interactive entertainment industry, which finds "larger publishers buying smaller ones and also buying or creating partnerships with top independent developers."[64]

But although we have emphasized how assiduously the interactive game industry has tried to "close the loop" between producer and consumer through marketing synergies and research, the continuing high rate of failure in game production shows that they have no record of sure-fire success. Indeed, the management of demand for symbolic goods in the postmodern marketplace has been far from flawless, consistent, or uniformly effective, especially as media channels have expanded and cultural markets have become saturated with commercial signs. Rather, the marketers' efforts in some ways increase the very uncertainty they seek to control by creating ever-more sophisticated and jaded young customers. Fads pass, styles change, values realign, markets mature, and boredom and overload increase with saturation and repetition, making management of symbolic value a very risky venture.

In a further attempt to stabilize these volatile dynamics, developers are increasingly found in a situation in which synergistic marketing possibilities are guiding the very design of the game. In terms of our model of the games industry's "circuits," we can say that the hypertrophy and acceleration of the marketing circuit begins to exert a powerful force on the cultural circuit. This goes well beyond simply making a spot for strategic product placements, extending to game themes and subject-positions. Before they support a new title as a platform game, developers and console makers want to be confident that it has a storyline and character that can be morphed into a "serialized" franchise, media spin-offs, and product-linked merchandise.

Paul Baldwin of Eidos explains that just to "stay afloat" in today's competitive cultural environment, game developers have become focused on building a franchise around a single game or releasing a series of sequels.[65] Action, combat, and fighting themes, once proven, repeat and proliferate. Heroes and celebrities, once popular, can be used to maintain audiences. It is true that thousands of games are available. But the logic of economies of scale and the fear of failure favour the serialization of success. As industry commentators have noted: "Publishers are relying more heavily than ever on classic mass-market techniques such as using well-known brand names, sports and character licenses and sequels to proven hits," so that on the console side of the business "80 percent of the top 20 titles in 1997 fell into one of those three categories."[66]

The possibilities for greater diversity in games are therefore constrained by the pressure both to cast a wide promotional net over a mass audience with a single game and to consolidate brand identity through licensing agreements. One manifestation of this dynamic is the tendency of game design practices to coalesce around the notion of a "supergame" – that is, a hit with all or most of the desired attributes of sports, racing, strategy, role-playing, and combat games. Synergistic marketing encourages the emergence of such hybrid "metagenres." As one Nintendo marketer commented: "We predict that the combination action/adventure/racing theme of *Diddy Kong Racing* will have a much wider appeal than straightforward racing simulations."[67] This metagenre trait already runs through a range of video game categories. Sports and racing games increasingly include role playing, adventure, and fighting elements (*Shaq Attack*, NHL *Rock the Rink*), while fighting games include dialogue, cinematic cut scenes, strategy, and scoreboards (*Soldier of Fortune, Resident Evil*).

Such metagenre games often culturally consolidate the niche of "militarized masculinity" that we will discuss in the next chapter; games that combine role playing with strategy and weapons or sports with fighting share a thematic nexus revolving around issues of domination, mastery, and conquest.[68] But there are other marketing advantages to the emerging metagenre. If gamers are always looking for novel experiences, then by adding elements from other genres, the designer enhances the appeal of the game concept and format, while retaining the successes of the old. Furthermore, though there are *aficionados* with a marked preference for strict genres such as fighting games or shooters, by adding elements from a diversity of game genres the designer extends the appeal across a wider contingent of the gaming audience. The metagenre therefore allows a games company to appeal to the broadest possible swath of gamers.

The metagenre games often feature an already popular video game brand avatar, such as *Mario* in a range of Nintendo games and Crash Bandicoot in a string of PlayStation games. This strategy reduces the costs of developing a new character and all the research and design work that goes with it. The benefits from the increased brand recognition and bandwagon effects accrued in the marketing for blockbusters and platform games, such as *Mario Kart, Final Fantasy,* or *Tomb Raider.* The lead video game character can take the starring role in advertisements and spin-off merchandising, gaining efficiencies for the synergistic marketing campaigns. Thus, the metagenre might be seen as part of the diminishing diversity underlying the development of a blockbuster orientation in the industry. Synergistic marketing strategies – targeting, branding, media saturation, licensing – have not only become the point of fusion for the digital design process but have increasingly meshed it with trends in other cultural sectors because their promotional campaigns have had to be executed in a mediated cultural environment prefigured by its own structure of audience segments. For this reason the marketing of video games, far from overthrowing the patterns of mass-mediated culture, has often served to strengthen the new symbiotic relation between various media as allied carriers of commodified youth culture.

POCKET MONSTERS

In 1980, referring to the rise of the "demassified media" and their consequences for culture, the techno-utopian Alvin Toffler optimistically declared that "the day of the all-powerful central network that controls image production is waning."[69] There is, however, no better example of how gaming has both extended and been integrated into the mass-mediated cultural marketplace than the most popular video game of the new millennium, Nintendo's *Pokémon*. As children's culture critic Ellen Seiter suggests: "*Pokémon* points to the shifting patterns of a globalized children's mass culture with its migration from Japan to the US, its movement between digital and analogue forms and its potentially new strategies for licensed characters."[70] Celia Pearce simply calls *Pokémon* "the home run of transmedia."[71]

Implemented on the Nintendo Game Boy, *Pokémon* was introduced in Japan in 1996 to provide a low-end entry-level portal to the Nintendo brand. While Sony and Microsoft vie to position their recent platforms around a core market of teens and young adults who can afford expensive state-of-the-art systems, Nintendo marketers have sagely countered by focusing on kids while they are young. Since the hand-held game systems and software are cheaper, the Game Boy not only has a much wider user base but also a younger and less-gendered gaming audience. With *Pokémon*, Nintendo targets kids from six to ten years old.[72] Nintendo reported in 1999 that *Pokémon* had become its "fastest-selling game ever," accounting for forty-five percent of the games sold for the Game Boy that year.[73] In the same year the cumulative global sales of *Pokémon* games and licensed merchandise reached a staggering six billion dollars.[74]

The *Pokémon* franchise got its start in the electronic cottages of the global interactive gaming industry. Now in his mid-thirties, the

Japanese designer Satoshi Tajiri was *Pokémon*'s chief architect. Tajiri acquired his fascination with games as a hacker. Speaking of his mid-teens, he said, "I just started by taking apart the Nintendo system to see how it worked."[75] The Tokyo-based company he heads, Game Freak, is a "tiny firm" that, in its early days at least, worked on a "shoestring budget" and with "only four programmers."[76] The time spent working is still extreme: "I sleep twelve hours and then work twenty-four hours. I've worked those irregular hours for the past three years," Tajiri says.[77]

Tajiri's *Pokémon* is a cross between a role-playing "electronic pet" and a fighting game, with a collecting element. The game is focused on the adventures of the main character, Pikachu, and his friends, as they attempt to capture imaginary creatures called Pokémon – or "pocket monsters" – and then "train" them for battles, in which violence is minimal. The child-gamer is called a "trainer." The captive Pokémon are added to the trainer's "team" and help the trainer to capture more Pokémon, each of which has unique powers. The aim is to collect as many of the 150-plus Pokémon as you can and care for and train them. When they are trained, the pocket monsters can enter contests that test their strength. The trainer tries to increase his rank by battling other trainers. So the game has multiplayer capabilities, using a game-link cable so two Game Boys can communicate with one another. Trainers must deploy strategy to win battles, which involves remembering which tactics work best for certain Pokémon. Beaten Pokémon are knocked unconscious, or appear to faint, but are not killed in these tournaments.

One media executive behind the *Pokémon* phenomenon calls it "the largest child-driven phenomenon of the decade."[78] Yes and no. Working just as feverishly as the child-gamer-trainers who are glued to their Game Boy screens preparing their pocket monsters for virtual battle are Nintendo's own "consumer-trainers," who have put the mediatized marketing circuit into high velocity in the corporate battle for the children's market. A massive promotional campaign surrounded *Pokémon*, integrating mass-mediated marketing techniques like television and print advertising but also extending far beyond to produce an entire *Pokémon* "franchise," including licensed media spin-offs such as movies and TV cartoons, Internet promotions, and product tie-ins that range from collector trading cards to clothes, snacks, and toys. By 2000 there were more than fifty *Pokémon* licensees, selling hundreds of different products.[79] *Pokémon* was embedded in the webwork of a synergistic, multimedia, globe-spanning distribution network. As one toy industry analyst remarked, "It used to take a while for trends to make their way to Kansas. Now, wherever the kids are, *Pokémon* is there, too."[80] In making this possible, the marketers have strategically synchronized "mass media" and "new media," so that they spiral around one another, referring to and reinforcing one another, deepening the presence of the marketers and the reach of *Pokémon* symbolics in children's culture.

Nintendo Japan promoted the *Pokémon* Game Boy line with a feature-length animated film called *Pokémon: The First Movie*. It was the fourth-highest-grossing film in Japan for 1998.[81] Helping to spark the enthusiasm for *anime* in North America, the launch of the *Pokémon* game here in 1999 was also supported with a movie spin-off. But the Japanese version of the film was adapted for American audiences. To do this, Nintendo worked with 4Kids Entertainment. 4Kids, it is important to note, is the exclusive *Pokémon* licensing agent: it is the kind of company that one finds at the cutting edge of postmodern information capitalism. As one commentator said, 4Kids "doesn't make or sell anything except the right to make toys with certain images."[82] 4Kids' president, Norman Grossfeld, also happened to be the producer of the American version of the film, in charge of "'Americanizing' the movie."[83] Recollecting the initial talks at Nintendo about importing the *Pokémon* "property" to America, one Nintendo executive said, "Nintendo felt that American children could appreciate the same qualities that made *Pokémon* such a tidal wave experience in Japan – it literally saturated their cultural landscape." From background to street signs to the soundtrack, Grossfeld says the "Americanized" version "combines the visual sense of the best Japanese animation with the musical sensibility of Western pop culture."[84] This "sensibility" extended further to include a spin-off soundtrack featuring the biggest so-called boy- and girl-bands in American popular music, from NSYNC to Britney Spears. The film made US$87 million at the box office in Canada and the United States.[85]

Pokémon exemplifies the game industry's saturation marketing strategies and reliance on mass media to reach its audience. A *Pokémon* cartoon, for example, was launched in 1999 on North American television. According to 4Kids Entertainment, "The TV show was created to help you play your Game Boy game."[86] *Pokémon* runs in the US on Kids' WB! network; it's their "number one kids show."[87] The network recently saw their ratings go up two hundred percent, a rise they credit largely to *Pokémon*.[88] In Canada the show is broadcast on the YTV network, attracting two million viewers a week. "*Pokémon* is our friend," said one of YTV's lead programmers.[89] Reported to be the top-ranked series on broadcast television among kids aged two to eleven,[90] the *Pokémon* phenomenon precipitated changes throughout the children's television market, as networks went on a quest for child-oriented *anime* to attract kids to mass media.[91]

This electronic game reverberates still further through the mass media market thanks to the home-video release of *Pokémon: The First Movie*, by Warner Bros. They described the marketing launch as "the biggest of its kind in the history of Warner Home Video and will generate over nine billion consumer impressions."[92] Weaving together mass and new media marketing, the campaign involved the customary barrage of ads, including TV, print, in-store, and Internet advertising, as well as cross-promotions with a plethora of child-oriented merchants and media companies. Leading toy retailer FAO Schwarz even launched a *Pokémon* boutique at its prestigious Fifth Avenue store in New York, a distinction reserved for the most popular and longest-lasting toy brands such as Barbie and Star Wars. On the Internet, the *Pokémon* Web site receives fifty million page views per month, and fan-based Web sites surge.

Whether a complete play experience or not, *Pokémon* has clearly been a synergistic marketing success, changing the landscape of youthful digital culture. It didn't hurt the Nintendo empire, either. Sales related to the *Pokémon* franchise and Game Boys compensated for the Japanese firm's otherwise slow sales of the N64 console in 1999.[93] Due to this craze, according to one Nintendo executive, "We're expecting our best year ever."[94] Thus, while sales of the Sega Dreamcast floundered, *Pokémon* games exploded. In 1999 *Pokémon* titles accounted for four of the top five best-selling games.[95] But Nintendo wasn't the only one to profit from *Pokémon*. The affluent class of information capitalism also profited from the new rules of information property on cultural signs. In a telling example, one investment executive said that, after researching 4Kids Entertainment, the *Pokémon* licensing agent, he purchased four hundred shares for his eleven-year-old child. The share price nearly doubled. Referring to the lead character in the *Pokémon* game, he said, "This little yellow Pikachu made my daughter $8,000 in two weeks."[96]

There is much that *Pokémon* shows us about the dynamics and paradoxes of the circuits of the interactive game in the global cultural marketplace. *Pokémon* is a cultural narrative, originally written as a piece of entertainment software, that has swept, however unevenly, across the global children's culture. Citing *Pokémon* as an example of a cultural commodity that has been "launched both from and into the US market," Yuezhi Zhao and Dan Schiller note that, in some respects, the "character of the global audiovisual industry grew more multicultural" in the 1990s.[97] But they are careful to point out that "far from heralding a new era of democratic choice ... the transnational audiovisual industry is willing to 'parasitize,' rather than flatten, cultural differences."[98] In the process, by blending the immersive qualities of electronic gaming with the sociality of gamer culture, the marketing circuit and the cultural circuit have corroborated one another, mobilizing the social aspects of gaming for the pocket-picking interests of the media monsters.

It is important to note that the fabulous success of *Pokémon* was built on the faddish success of a game property designed for the low-cost hand-held systems. While Microsoft and Sega scrambled for market share in 1998 at the end of the 64-bit technology cycle, the lower price-point of hand-helds provided an alternative gateway to Generation Y. As Seiter explains, "even more unusual is the fact that girls as well as boys have become avid fans, and *Pokémon* has promoted cross-gender play."[99] The fascination with *Pokémon*, which rests perhaps on the fact that these Japanime figures were cute and trainable, came from both boys and girls, which in turn helped bolster the sales of licensed goods from playing cards to furniture. *Pokémon*, it seems, illustrated a corollary to the "perpetual-upgrade marketing" of Sony and Sega, which we might call the perpetual "down-age" dynamic of game marketing: younger and younger users of gaming technology must be anticipated by the marketer of lifelong branded entertainment.

CONCLUSION:
COMMODIFICATION OF CHILDREN'S PLAY

Throughout history, play has been a cultural form valued as an energy release and a way of teaching skills, and for its role in physical, cognitive, emotional, and social development.[100] In the era of industrialization, the emphasis was on productive leisure, making play the "work" of childhood. Although early industrial capitalists had been more than happy to exploit children as a pool of cheap and vulnerable labour (a process that continues today in many parts of the global market), their successors were equally willing to profit from the newly emerging definition of childhood. Parents became concerned about their offsprings' leisure time, establishing the concept of "childhood" as a time of carefree play. Play as a cultural form was valued as an organic, imaginative, free-ranging, democratically rule abiding social interaction. Though play is always active and self-producing, different forms of play permit varying degrees of creativity and experimentation, as well as some questioning of social roles. Free play – in which children construct meaning for themselves in the natural world – became the idealized form of play during the twentieth century and a much-celebrated aspect even of formal schooling.

Throughout the last century the idealization of play became increasingly commodified. Psychologists thought that if play was the work of

childhood, then industrial-age children needed tools for their socialization – "playthings." The entertainment industries formed and prospered
in the twentieth century as toys, games, and bicycles, as well as sports
and play equipment, were proffered to facilitate the socialization and
development of children. Through playthings, the "invention of childhood" was linked to the growth of enterprises that provided the consumer goods vital to emerging visions of domesticity and family life.
Later, films, radio, and television joined the rush to entertain and educate
kids. Among these cultural intermediaries was a cadre that specialized
in marketing leisure products directly to children. The child-subject, once
defined as a mini-worker, was now redefined as an audience and mini-
consumer. Advertisements, branding, and promotional communication
were devised to "acquaint children with the basic framework of consumer behavior – with the need for money to purchase things they want
in the stores."[101] In the age of marketing, play comes to serve a new
function as "the templates through which children are being introduced
into the attitudes and social relations of consumerism."[102]

Paradoxically, commentator Jeremy Rifkin recently argued that in
postindustrial hypercapitalism, "play is becoming as important in the
cultural economy as work was in the industrial economy."[103] Yet the
historical process of commodifying play is shot through with tensions,
because "the assumptions and rules governing play are quite distinct
from those traditionally governing work."[104] What Rifkin calls "pure
play" is socially spontaneous, participatory, and intimate, "not instrumental to an end but an end in itself," while the commodified play of
the market is managed, purchased, and rationalized.[105] So, "the kind
of play produced there ... is only a shadow of the kind of play
produced in the cultural sphere."[106] The expropriation of play by the
forces of the market threatens a devaluation of the cultural meaning
of play, and with it a loss of the cultural legacy that derives from and
was nurtured by noncommodified play activity.

One of the central themes of this book is that the interactive media
create a historical intersection between the commodification of child's
play and the technification of culture and social communication. Young
people have been offered expanded zones for their leisure, yet at the
same time their free play has been subject to the enclosures of digital
commodity forms offered by the interactive entertainment industry.
Looking at this nexus of digital play, we acknowledge that there are
some reasons for hope about interactive play culture. Networked
gaming, in particular, affords young people a novel social experience
of the mediated flow of meaning from the screen – one that is very
different from their experience with TV. Online delivery, moreover, does
grant to digital audiences a certain flexibility in their selection of

entertainment. Like books and video rentals, gamers are not dependent on broadcasting schedulers who determine what they see, and when. With faster modems and better networks, downloading time will soon become irrelevant in digital cultural markets, making entertainment-on-demand a proximate reality. Yet the growing valorization of child-play has in turn meant that it is anything but "free."

The paradox of information capitalism is that even as it encourages an expanded enclave of freedom and self-development of "pure play," it begins to undermine that enclave by commodifying it. For the more play became distributed in the marketplace, the more its forms and boundaries were set by a commercialized media system. As playful leisure absorbed the lion's share of children's energies, television aimed at children became a leading sector in the entertainment economy. As the marketers of playthings acquired special importance in the 1980s, "free play" has been encroached upon by a very un-free regime of invitations, pressures, surveillance, and solicitation from marketers. So too over the last two decades the positions of "player" and "consumer" have become tightly bound up with the technological points of access to play culture. To be a game player, after all, demands that one has the ability to buy the expensive hardware and software, creating a digital divide between families of differing class positions that have access to various levels of gaming technology.

Games from *Pong* to *Barbie* are built on deep cultural wellsprings of children's imaginations. *Pokémon*'s creator, Satoshi Tajiri, describes how the idea for his innovative game came from childhood memories of his own creative play. Tajiri says he loved walking in the woods and collecting insects in jars.[107] He remembers watching those strange lifeforms and finding the insects inspirational and moving. Viewed through the glass, he imagined his pets as characters in science fiction films in which they battled to save the world in a grand stadium. Contemporary children, immersed in virtual playgrounds, do not have those same experiences of free play. Their social encounters revolve increasingly around avatars. And now thanks to *Pokémon*, the bugs they encounter are virtual lifeforms, preprogrammed replicants of mutant ideas that exist only within a screen culture. The ultimate paradox of this technification of play, however, may be that in the intensifying identification of "gamer" with "consumer," the digital marketplace may undermine the imaginative wellspring that makes play "fun" and "free."

Designing Militarized Masculinity: Violence, Gender, and the Bias of Game Experience

In July 2000 the Canadian province of British Columbia slapped an x-rating on the computer game *Soldier of Fortune*, making it in theory purchasable only by adults.[1] *Soldier of Fortune* was one of a new crop of first-person shooters that included *Rainbow Six*, *Rogue Spear*, SWAT, and *Counter-Strike*, games broadly similar to *Doom* and *Quake* but much more realistic. Premised on scenarios of anti-terrorist or mercenary operations, these games emulate the tactics of small-group urban warfare. Opponents are not demonic monsters but plausible-looking humans, on whom the effects of high velocity and automatic weaponry are demonstrated with extraordinary verisimilitude. In *Soldier of Fortune*, one shot kills, or maims. Bodies have twenty-six "hit locations," and they writhe, bleed, and die differently depending on where you direct your fire. You can kill quickly with a shot to the head or blow off limbs one by one. "Show blood" and "show dismemberment" options activate cascades of gore and protracted death agonies.

Although *Soldier of Fortune* was not much more extreme than other games of its genre, the Canadian x-rating gave it a certain notoriety. Its developer, Raven Software, was undeterred. Previews of *Soldier of Fortune II* in game magazines promised "exploding heads, dismemberment, arterial spray, and convulsing body parts a-flyin," enabled by a custom-built technology that modified the *Quake II* engine "to deliver the most realistic carnage possible."[2] The leader of the game's development team, John Zuk, said, "While we are aware of the climate that exists about game violence, we simply forge ahead."[3]

No aspect of interactive entertainment has received more attention than such ultraviolent games. From its very beginnings, the industry has been dogged by accusations about the effects of "video nasties" on young minds. As early as 1982, the US Surgeon General declared that video games were dangerous and aggressive generators of "aberrations in childhood behavior."[4] Many parents and teachers are disturbed by the violence of the virtual games their children and students play. These charges reached a recent peak in the aftermath of the 1999 Columbine school shooting with the revelation that Eric Harris and Dylan Kelbold, the trenchcoated killers, were avid *Doom aficionados* who rehearsed their plans on specially adapted versions of the game. For some, this merely confirmed the claims of David Grossman, a professional soldier and psychologist who, on the basis of his ongoing studies of military-training simulations, had been tracing a direct link between youth violence and first-person shooters and condemning the interactive games industry for providing a "training to kill."[5]

Facing such criticism, both industry apologists and hardcore gamers strenuously deny any connection between real and virtual violence, declaring that "*Doom, Quake,* and their progeny aren't murder simulators, they're paintball without the welts."[6] They point their fingers at other causes, minimize the significance of ultraviolent games in overall industry production, or plead that such games provide a surrogate for – rather than a stimulus to – actual aggression.[7] They too have academic supporters, such as the French researchers Alain and Frédéric Le Diberder who argue that the protests not only rest on an unproven case but also ignore the familiar media mayhem on film and television and deflect attention from more plausible culprits such as the American gun lobby.[8]

In this controversy, both sides deploy empirical studies that vary enormously in scale, criteria, and methodology, and produce contradictory results, enlarging the vast but inconclusive literature on the effects of violent representations on television and other media.[9] We do not pretend here to resolve this debate. Instead, we bear in mind Williams's trenchant criticism that studies of "media effects" tend to reduce complex social processes to "a displaced and abstracted cause" – "the media."[10] "Effects," he maintained, "can only be studied in relation to real intentions."[11] Our analysis of the technological, cultural, and marketing circuits of a new media cannot unravel the conundrum of individual psychological responses and consequences. It can, however, illuminate some of the reasons why and how interactive gaming has cut itself a cultural channel or groove emphasizing what we term "militarized masculinity." We set out to show what forces have worked to generate interactive game design practices that are focused around strongly gender coded scenarios of war, conquest, and combat; how this bias has been amplified by the industry's ongoing negotiations with a base of young male hardcore fans; how recently this cyclical self-amplifying pattern has been challenged and disrupted, from a variety of directions but with uncertain consequences; and why the outcome of the tension between "violence" and "variety" in game design is critical to the future of interactive entertainment.

THE VIOLENCE GAME

Video games are not the first medium to capitalize on the riveting power of violence. Celebrants of digital play reach well beyond prime-time television or Hollywood action films to invoke the *Iliad* or *Beowulf* as evidence that "as long as we've been consuming narrative entertainment, we've thrilled to the exploits of blood-streaked warriors who hack limbs off their opponents."[12] These are the same people who in other contexts confidently declare that interactive technologies make everything new. To develop a new medium in directions that departed from violent traditions would certainly have required an effort of cultural creativity by game developers and marketers. Equally certainly, it never happened. Digital games did not merely continue inherited traditions of violent entertainment but drove them to new levels of technologically enhanced intensity.

Many commentators find the roots of this direction in the institutional origins of the industry. As we have seen, video games sprang from the high-technology military-industrial complex where simulations of mass destruction were routine. *Spacewar* was the product of a culture dedicated to the everyday contemplation of nuclear mega-death. Depicting violence, moreover, was an easy programming task for the simple computers on which early interactive gaming depended, partly because the machines were conceived and designed with precisely such military purposes in mind.[13] Both cultural and technical forces thus ensured that when game pioneers entered the commercial market, it was "natural" for them to create games like *Tank*, *Periscope* or *Space Invaders* based on scenarios of war and shooting or strategizing skills. The impulse towards violent themes is not just a matter of ancient history but is persistently reactivated and reimplanted by the synergistic linkages and revolving doors between military simulation and interactive entertainment that we described in chapter 8.

In our view, however, military foundations and relationships are not in themselves sufficient to explain the role violent games have assumed in the industry's trajectory. The path was by no means set from the start. The early days of interactive gaming saw the emergence, alongside war and shooting games, of various other genres; one has only to think of *Pong* and *Pac-Man*. Home video games were quickly defined as child's play, more specifically, as play for boys under twelve. Though game developers were willing to cater to the aggressive excitement of young males, especially in an arcade setting, the fact that domestic games were largely marketed as toys moderated their content. Technological problems further constrained the intensity of virtual violence; the low graphics capacity of early consoles and computers meant that the holocausts of alien death in *Space Invaders* could only be abstract orgies of geometric disintegration.

The era did see some notoriously violent games, such as Exidy's 1976 *Death Race*, in which drivers mow down pedestrians, and Mystique's 1982 *Custer's Revenge*, where the objective is to rape captive indigenous American women. The outcry generated by these early extreme games, not to mention the sheer repetitiveness of numerous rudimentary point-and-shoot games, probably played some part in the Atari crash that wiped out the US industry in the 1980s. But when Nintendo revived the US video game market in the 1990s, it appeared alert to the fact that extreme content could be a problem and promised quality-controlled family entertainment. These assurances were not to be fulfilled, however, in the overall development of the industry. What tipped the balance in favour of escalating virtual violence was the emergence of new digital design practices arising from the intersection of maturing consumers, improved technology, and, especially, increasingly specialized and competitive marketing.

MARKETING MAYHEM

That violence-filled games are commodities, that their violence is intended to increase their market value, and that commercial competition and promotion create pressures to intensify violence may seem like obvious points. Yet they bear repeating. It is not simply that violence sells but that it is a way to precision-target strategically important market segments. As James Hamilton observes in his study of television, "Where consumers face a variety of entertainment options, violence is an element of product differentiation."[14]

Product differentiation by violence has been crucial to the evolution of video gaming. As we have seen, Nintendo's successful expansion into the home console market was guided by a marketing research apparatus that monitored the tastes, preferences, and buying power of core customers to predict which games would be hits. The research confirmed that "the principle player is a boy eight to seventeen years-old."[15] Once the action–adventure, sports, racing, fighting, and shooting genres adapted from the arcade proved popular with this youthful male audience, the company had a strong economic incentive to continue and amplify the genres, rather than risk breaking ground with new content.

Increasingly, its dialogue with consumers revolved around genres preferred by the most loyal and frequent buyers of games – those whom video game designer Celia Pearce calls "the world's most innocent and maligned victim of demographic opportunism, the ever-vulnerable, ever-receptive, ever-predictable adolescent male."[16]

The dynamism and creativity that characterized the early game design teams diminished as a set of tried and true genres, based on replicating the success of hit games, was defined as what the marketers wanted. In spite of Nintendo's declared family orientation, its games soon attracted criticism for relying on stereotypical narratives of conquest and combat, especially in the martial arts fighting games.[17] In the late 1980s Eugene Provenzo's study of Nintendo's top ten hits highlighted the pre-eminent role of the active boy-warrior hero and the invisibility or passivity of female characters.[18] But even these games would soon appear extraordinarily mild.

By the mid-1990s, the "Nintendo generation" was passing into adolescence. A new entrant to the market, Sega, rapidly grasped that one way of breaking its rival's monopolistic hold on the market was to target these older players and draw the younger ones in their wake. Marketing for older boys meant introducing "adult" or "mature" content in games, an imperative Sega unhesitatingly interpreted as meaning "more violence." Applied to the gaming world, this logic informed much of Sega's overall campaign of "testosterone marketing" and its promotion of provocatively violent games – such as the heart-ripping versions of *Mortal Kombat* – as a way to win the young male niche that was perceived as the key to dominance in the whole business.

Although Nintendo publicly cried outrage, the success of such tactics drew it and many subsequent interactive game companies into a competitive spiral of intensifying graphic violence – violence that was being rendered ever more compelling and immersive by the accelerating power of successive generations of consoles and computers. The process generated more public outcry, calls for government censorship, and, in order to pre-empt such an outcome, the industry's institution of self-administered ratings codes. What game developers quickly learned, however, was that the ratings could themselves be incorporated into the marketing campaigns, an R-rating becoming simply one more audacious component in the promotion of a "wicked" game.

By the mid-1990s, when computer games re-entered the market in a big way, they did so in a setting where the test and touchstone for their emergent 3-D capacities was in the representation of violent action. This was a challenge John Romero and John Carmack of id Software rose to with nightmare élan in their definitive series of first-person shooters, *Wolfenstein*, *Doom*, and *Quake*. The power of genre now created a momentum not merely to perpetuate but to *intensify* game violence. First-person shooters – *Duke Nukem, Unreal, Half-Life, Daikatana* – all have more or less the same plot: run through labyrinthine settings, evade death, hunt down enemies, and kill them at high speed in lavishly detailed ways. The same holds for the equally formulaic martial arts games with their killer moves and horror-film gore. The only significant way to differentiate these games from their competitors is by elaborating and intensifying speed and violence. In this context, the perpetual-innovation economy works through its logic with multiplying rates of "fragging" and ever-more vivid splatterings of "giblets."

Software development is a risky business. Most products fail. There are fortunes to be made with pioneering games that break new cultural ground. But for each successful experiment, scores crash and burn, taking with them companies and careers. This creates a powerful incentive to stick with the tried and true and ride on the coattails of proven success. The repetitive pattern is reinforced by the fact that game developers are recruited from the ranks of game players. Such asexual reproduction gives game culture a strong tendency to simple self-replication, so that shooting, combat, and fighting themes, once established, repeat and proliferate.

The cycle is further amplified by a profit-driven industry's pressures to reduce production costs. As Celia Pearce observes, "The *real* reason why video game designers create violent games is not because the market wants it, but because it is easier."[19] Like the makers of film and television, game developers understand that violence is easy to plot, requiring a minimum of creative scripting and design. From the point of view of marketers, violence is a cultural idiom that requires no translation within increasingly transnational entertainment markets: martial arts games, for example, can cross the Pacific from Japan to the US and back again very easily. All these factors give virtual violence momentum and staying power.

COEVOLVING CARNAGE

The issue of making violent games for young male players raises sharply the issue of the interactive game industry's relationship with its player-customers. Industry spokespeople and digital futurists argue that consumers set the directions of game development. They claim that in an information age environment players enjoy unique opportunities for feedback, codevelopment, and participation. Writing of digital industry in general, Kevin Kelly claims that in a networked environment, "customers are trained and educated by the company, and then the company is trained and educated by the customer. Products in a network culture become updatable franchises that coevolve in continuous improvement with customer use. Think software updates and subscriptions. Companies become clubs or user groups of coevolving customers."[20] There are elements of truth in these arguments. Like all sensible marketers, gaming companies entered into a protracted negotiation with their player-customers and especially strive to enter into dialogue with and cater to their hard-core customers.

To say that cultural intermediaries like marketers and designers "dialogue" and "negotiate" with the gaming consumer may seem perverse. But from the point of view of capital, it makes good sense to open up channels to consumers, respond to their criticisms, adapt to their ideas and interests, and translate the information into new products. We call this mediated-marketing nexus a negotiation in recognizing that cultural industries especially have been at the forefront of audience and market segmentation research, forging a reflexive circuitry of audience surveillance and an acute awareness of, and responsiveness to, changing preferences, tastes, and subcultures.

But in the historical context of a dependence on young males with a penchant for consuming violent virtualities, these processes can work to further amplify the intensity of game violence. One example of dialogue between developers and players involves the games *Half-Life*, *Counter-Strike*, and *Gunman Chronicles*. *Half-Life*, created by the developer Valve and published by Sierra Games, was released in 1998. It follows the classic first-person shooter formula: the player adopts the subject-position of a hero trapped in a labyrinthine laboratory where monstrous experiments have gone wrong; he must shoot his way past mutant creatures and security forces attempting to eliminate all traces of the accident. The game's superior artificial intelligence – which gave the monsters peculiar perspicacity in pursuing the protagonist – made it an immediate hit.

A year and a half after its release a twenty-one-year-old computer science student at Simon Fraser University in Vancouver, Minh Le, adapted the code of the PC game to produce a new creation, also a shooter, but with a different premise. *Counter-Strike* is an online game for multiple players featuring one team in the role of terrorists and the other as counterterrorists. It became so popular that Valve began helping Le and his now-considerable band of collaborators to write code and later arranged for Sierra to publish the "mod." *Counter-Strike* is sold as a packaged addition to *Half-Life*. In 2000 it was named action game of the year by several leading gaming magazines, driving the *Half-Life* game server to the top of the list of online game venues, outstripping rivals such as *Quake 3* and *Unreal Tournament*.

Using the software-development kit that Valve provides, fans continued to create new scenarios, weapons, and characters for *Counter-Strike*. Meanwhile, the logic of modifications continues. A handful of gamers connected through the Internet from sites as diverse as Utah, South Africa, and Germany took Valve's *Counter-Strike* development kit and built a single-player version called *Gunman Chronicles* on top of the *Half-Life* engine. Valve again helped this ad hoc development team perfect the code, then put them in touch with Sierra. Published in 2000, *Gunman* quickly became a top game.[21]

From one perspective, this pattern of successive consumer-led game modifications is an inspiring story of participatory and democratic design, with developers facilitating a series of player-led initiatives in a mutually beneficial manner. But it is not coincidental that the participants come from a young male technoculture fascinated by scenarios of violence – for it is precisely there, as we have seen, that the game industry has cultivated its most devoted and technically adept consumers. And it is to the intensification of this niche that the *Half-Strike/Counter-Strike/Gunman Chronicles* codevelopment saga contributes.

MILITARIZED MASCULINITY

Although it is easy to find spectacular examples of violent games such as *Quake*, *Soldier of Fortune*, or *Counter-Strike*, determining the importance of such content in the game industry as a whole is more difficult. Industry groups and market research companies issue a number of genre breakdowns of interactive software sales. (See Table 1.) These are often drawn on to argue that the explicitly violent games such as "shooters" and "fighting" account for only a minority of industry production, about fifteen percent of the total.

This benign portrait is then often reinforced by statistics breaking down sales according to the official ratings categories of the Entertainment Software Rating Board: in 1998, for example, some seventy-two percent of games sold were deemed appropriate for the "everyone" category, nineteen percent were deemed suitable for "teenagers," and only nine percent fell into the "adult" category. The Interactive Digital Software Association likes to push its claims further by pointing out that the vast majority of top-selling games are rated as appropriate for everyone.

This apparently irrefutable statistical evidence disguises a number of problematic issues. Categorizing games by genre involves complex judgments and, in some surveys, may actually be done with an eye towards fending off the "violence" criticism.[22] Genre accounting obfuscates if high-tech tank warfare and aerial combat are defined as simulation games, or *Mortal Kombat* and *Tekken* are classified under

Table 1
Genre by Unit Sales 1998, percentages*

Genre	Total	Computer	Video Game
Action	24	8	32
Fighting	7	–	11
Racing	13	7	16
Shooters	9	8	9
Simulation	4	10	–
Strategy/RPG	24	43	15
Sports	16	17	16
Other	3	7	1

* NPD, "Video and Computer Games: Genre by Unit Sale."

the sports category because karate and fencing are in the Olympics, or the highly popular *World Wrestling Federation* games are separated from fighting games, or role-playing games whose pivot is the annihilation of opponents are tucked away in their own innocuous-sounding category. In the battle of surveys, a recent study by Children Now calculates that, of the games rated "E", more than seventy-five percent "contained some type of violent content."[23]

Both genre analysis and the industry's self-administered ratings also occlude the degree to which representations of violence have become normalized in game design and in the media in general. IDSA, for example, likes to cite the blockbuster success of Nintendo's *Pokémon* franchise as an example of the industry's wholesome orientation, without mentioning that the key event in this virtual saga is the combat of specially trained fighting creatures. Industry denials of the charge that it actively promotes violent content lost credibility when in 2000 a US Federal Trade Commission investigation found that gaming companies regularly violate the spirit of self-regulation by aiming marketing campaigns for ultraviolent "adult" games at precisely the young players that the industry's own rating system recommends shouldn't buy or play them.[24]

In our view, an account of violent representations in interactive games that focuses only on the notorious "shooters" and "fighters" grossly understates its pervasiveness. Such games, while extreme, are only one manifestation of a larger thematic complex that focuses gaming culture on the subject-positions and discourses of what we term "militarized masculinity." This complex interweaves ingredients that range from shooting and fighting skills to magical spells of destruction, strategic and tactical war games, espionage, and scenarios of exploration and progress culminating in the ability to conquer alien civilization. The elements are dispersed across a very wide variety of genres of gameplay – "shooters," "action," "strategy," "role playing" – and are often combined in "metagenre" syntheses – "role playing plus strategy," "sports plus shooting." But taken together they constitute a shared semiotic nexus revolving around issues of war, conquest, and combat that thematically unites games ranging from *Soulblighter* to *Shogun* to *SpecOps*.

Within this complex there are many modulations and gradations in the verisimilitude and intensity of virtual violence. At its fringes is a netherworld of psychopathologically sadistic games such as *Postal* (in which the character assumed by the player shoots unarmed bystanders), *Carmageddon* (which involves lethal road rage), *Duke Nukem* (in which semi-naked women are blown away) or *Kingpin* (in which people are beaten to a pulp) that in their extreme brutality exceeds anything readily available in popular cinema or television. Seen from this perspective, *Soldier of Fortune* and its ilk are only one component in a hegemonic strain of gaming culture that mobilizes fantasies of instrumental domination and annihilation, tracing out a virtual refrain that chants again and again, though in tones ranging from a muted whisper to a scream of rage, "Command, control, kill."

As Williams reminds us, hegemony never completely exhausts creativity. Though "militarized masculinity" is a dominant theme within game culture it is not the whole story. Within traditions of boy's play and male culture, there are lines leading in different directions. Sports games are a notable instance. Although some of them overlap with a game culture of violence, it is important to recognize that games such as the *Tony Hawk* skateboarding series are as much a part of the gaming scene as *Quake*. There are also recurrent elements in digital game culture that go well beyond both sports and violence. Best-selling puzzle games such as *Tetris* and *Myst* – which have attracted large followings of female as well as male players – are one example. So too are certain "god game" simulations, many of which, like *Civilization* or *Age of Empires*, still have fundamentally militarist subtexts of conquest and imperialism. But others, such as the Maxis series of *Sim* games, are oriented towards the more or less peaceable orchestration of complex urban, ecological or domestic scenarios: *The Sims* (which we discuss at length in chapter 12) can be seen as a culmination of this tendency.

Thus, the picture painted by critics who characterize the interactive game industry according to its bloodiest products is a misrepresentation, as are the whitewashes of its apologists. *Ghost Recon*, *Red Faction*, and *Half-Life* share shelf space and best-seller lists with *The Sims* and *Tony Hawk* and *Roller Coaster Tycoon*. The resulting configuration is more

complex than either side in the virtual violence debate usually acknowledges. But the nexus of war-, combat-, and conquest-oriented games
enjoys a pervasiveness that overlaps genre distinctions, is far more
widely diffused than the industry likes to admit, and is perpetuated
by a number of feedback loops that gives "militarized masculinity" a
persisting centrality in interactive game culture.

TOYS FOR THE BOYS

As the designation "militarized masculinity" highlights, representations
of violence in digital play are closely bound up with the medium's
gender bias. In the great semiotic ledger of culture that divides the
boys from the girls, interactive gaming has, at least until very recently,
been entered firmly on the male side.

In a major survey of the field, published in 2000, Elizabeth Buchanan
offers a schema for considering the subject-positions offered to women
and girls in video games.[25] It lists such issues as invisibility ("do women
appear at all in games?"); agency (do they have "a significant role and
are they able to make decisions and take action that affects the world
of the game," or are they relegated to passive positions as "objects of
desire or detestation ... the victim or the vixen"?); point of view (do
games offer the possibility of aligning oneself with a female character?);
intent ("What is it games are teaching to girls and young women?");
and address (does the video game industry as a whole "speak" to girls
and women so that they can "carve their own niche in arcades, pick
up a gaming magazine and connect with others?").

Buchanan suggests that in most cases the answers reveal that interactive entertainment is deeply sexist and male centred. Many others
agree. In the games examined in the Children Now report, female
characters appeared in only sixteen percent.[26] But it is not just the
number of characters but also the context of representation. The same
report found that of 874 player-controlled characters, only twelve
percent were female; when female characters appeared it was most
often as a "prop."[27] In their analysis of games from a feminist perspective, Nola Alloway and Pam Gilbert argue that video gaming is
"a discursive field within which constructions of hegemonic masculinity dominate," one that produces and circulates versions of masculinity,
femininity, and gender relations that are "narrow, restrictive and
regressive with respect to contemporary moves to encourage more
expansive identities and democratic relationships."[28] Heather Gilmour
notes, "For the most part women and girls are entirely left out of the
spectatorial or productive equation."[29]

This is not to say that there have not been games, such as *Myst* or *Tetris*, that have attracted a strong female following. But, as Buchanan reminds us, one of the crucial factors in estimating the gendering of games is simply "quantity": whatever the achievement of individual games, there is an issue of "plain numbers" in terms of volume and ratio of female-friendly games on the shelves.[30] Once one looks at interactive gaming on this scale and considers these commentators' remarks, it is hard to avoid the verdict that in the video game industry, what teenage boys want, teenage boys get.

This situation, like the industry's preoccupation with scenarios of violence, tracks back to the military origins of interactive play. The game industry, conjured into being by technologically adept and culturally militarized men, made games reflecting the interests of its creators, germinating a young male subculture of digital competence and violent preoccupations. The industry then recruited new game developers from this same subculture, replicating its thematic obsessions and its patterns of female exclusion through successive generations. As Romero put it, "Men design games for themselves because they understand what they know is fun. They don't understand what women find fun."[31] This self-propagating process was amplified as game marketers, recognizing that the critical niche was made up of boys and teenagers, launched powerful and calculatedly gender based appeals to consumers. Sega's marketers, for example, readily admit their ads were "in-your-face, aggressive young male. We were all about testosterone."[32] All this, spliced with the powerful pre-existing divisions that already segregate so many forms of entertainment – from film genres to TV programs to sports activities – along gender lines, and with the overall male predominance in high-tech activities, has (at least until very recently) relentlessly constructed the game-playing subject as male.

This sexist structuring of the industry and its market has perturbed educators, parents, and feminists for several reasons. In most games, either females are invisible (hence invalidated) or they appear in a limited set of stereotypical roles, ranging from passive "prize" to evil threat – what Buchanan calls the "virgin or vixen" syndrome.[33] In a more extreme subsection of gaming culture, female characters are the targets of more or less explicitly sexualized violence. Such a culture seems to provide boys with (yet another) vivid source of instruction in how to ignore, objectify, or even abuse women, while unmistakably informing girls that digital space is not for them. Moreover, such gender representation also entails a politics of technological exclusion. If, as game enthusiasts often argue, interactive entertainment familiarizes children with computer skills that are vital for success in a high-technology world, their being marked as a male domain must significantly disadvantage girls, barring them from "the technological equivalent of a 'head start' program."[34]

GIRL GAMES

At first glance, it might be thought that undoing this exclusionary pattern would be a priority of game companies, for reasons of self-interest if not social conscience. After all, a cultural coding of games as "male" reduces their market to just fewer than fifty percent of the population. Making interactive play a truly mass entertainment medium would seem to demand that girls and women as well as men and boys glue themselves to screen and joystick. There are, however, important economic barriers to this transformation. Making a game – and making a market for a game – is an expensive and protracted venture: today, perhaps five million dollars in development costs, and two years in production time. Although there have been serendipitous successes in developing games that women play, generating a consistent and extensive female gaming culture would require investment: research into what sort of games girls and women might want to play; the recruitment of developers – perhaps scarce female developers – capable of imagining, scripting, and coding such games; and advertising to alert, excite, and persuade potential female consumers. In other words, it would require that digital designers and entrepreneurs take up that perennial question "What do women want?" a conundrum that men, at any rate, have always found time consuming, psychologically challenging, and sometimes expensive to address.

Moreover, there is no guarantee of success: on the contrary. The binary cultural coding of gender means that male and female products are often not just different but actively opposed. Experiments in feminizing interactive entertainment, or just moderating its "testosterone" elements, may actually reduce the appeal of a game, a genre, or a company to reliable male purchasers or even mark the product and its producers as "sissy" or "girly." A turn towards "girl-games" (however they may be defined) is therefore a gamble that bets the proven track record of the masculine niche for the uncertain benefits of a potentially wider cross-gender market, or perhaps for the unexploited but also largely unexplored terrain of female-specific gaming. It demands a substantial economic commitment, without the promise of fast payback.

This calculus of investment strategies is of course affected by changing social and cultural conditions. By the mid-1990s a number of factors made the issue of girls and games exceptionally volatile. These factors included the increasingly saturated and competitive nature of the gaming market; the growing familiarization of women and girls with computers in school and work; the controversy over school shootings, which inevitably raised issues about youthful male aggression and its links to popular culture; and the emergence of a handful of female game designers, such as Brenda Laurel, Celia Pearce, Roberta Williams, and Theresa Duncan, who mounted a critique of sexist stereotyping from within the industry. This changing constellation of influences generated at least four major strategies or responses, all aiming at altering the gender politics of interactive gaming but mobilized by different actors, often in contradiction with one another. We term these four strategies the "Barbie," "Purple Moon," "Psycho Men Killer," and "Lara Croft" responses.

Barbie

The toy company Mattel had been involved in the early growth of us video gaming but largely withdrew after the Atari crash. In 1996 it re-entered the game market with a virtual incarnation of its famous female doll. The computer game *Barbie Fashion Designer* allowed players to design, and then print out, fashion outfits for the famous fashion doll. To the horror of both hard-core gamers and digital artistes, it sold more than five hundred thousand copies in its first two months, outdoing such games as *Doom* and *Myst*. Mattel's answer to the "girl game" issue was simplicity itself: transfer to video games the "pink and blue" gender divide deeply embedded in other realms of commodified children's culture. If in the material world boys played with guns and trucks and girls with dolls and clothes, repeat the formula in cyberspace.

Mattel followed up its initial success with *Barbie Photo Designer*, *Riding Club*, *Hair Styler*, and *Detective*. Other companies have attempted similar strategies; Hasbro, for example, makes *Easy Bake Oven* and *My Little Pony* video game spin-offs. But Mattel remains pre-eminent. In 1998 it commanded 64.5 percent of the commercially designated girl-game market and was soon to acquire the firm that occupied second place, Learning Company, which controlled another 21.6 percent.[35]

It need hardly be said that from the point of view of critics of gender stereotypes, Barbie is a dubious route to the digital empowerment of girls. In any case, despite the stunning success of the Mattel games, their formula may well not be generally applicable. Mattel's software is arguably more an accessory for play with physical Barbies than a video game proper. In the absence of such "corporeal" support, other attempts at virtualizing stereotypically girlish play activities – dolls, ponies, make-up, cooking – have had far less success.

Purple Moon

Contrasting with Mattel's recapitulation of traditional roles has been the attempt to develop "girl games" with less stereotypic content – combining new media with new images of gender. By the late 1990s this project seemed to be gathering some momentum, but then it suffered a major setback. The first part of the story has been documented in a fascinating collection of essays edited by Justine Cassell and Henry Jenkins.[36] They describe this technocultural movement as the result of "an unusual and highly unstable alliance" between female high-technology entrepreneurs and feminist researchers and activists interested in making better opportunities for girls and women in cyberspace, and various industry leaders and venture capitalists who were keen to crack the female market.[37] These forces came together momentarily in an attempt to create games that were attractive to girls, commercially viable, and at least somewhat experimental in their gender representations.

Although there were a variety of academic and commercial strands to this movement, the most famous example was Brenda Laurel's Purple Moon Company. Laurel had been a game designer with Atari in the 1980s and had then gone on to research girls and games at Interval Research Corp, financed by Microsoft's cofounder, Paul Allen. On the strength of more than a thousand interviews, Laurel concluded that girls were not necessarily uninterested in video gaming but had a set of preferences quite distinct from those of boy players. They liked games that stress collecting, creating, and constructing rather than destroying, shooting, and defeating; they had little interest in beating the machine in one-on-one games with the computer but enjoyed multiplayer options; they were not necessarily turned off by violence but became frustrated by games in which players are repeatedly "killed" or penalized (while boys actually enjoyed this process); they placed more emphasis on character, story, and relationships than on the achievement of set goals. Such data suggested that it was possible to make games that would appeal to girls, and that the parameters for doing so were wide enough to introduce content that was far more adventurous than Barbie.

Purple Moon, which was also financed by Allen, attempted commercially to realize the fruits of Laurel's research. Staffed eighty percent by women, a sharp reversal of the usual gender balance in its Silicon Valley habitat, it set out to create "friendship adventures for girls." Its most famous character was Rockett Movedo, a heroine with spiky orange bangs who negotiated tricky situations – such as going to a new school – in relatively commonplace settings. The games included such devices as "emotional navigation" and "relationship hierarchies." Purple Moon's products included the *Rockett* and *Secret Paths* series and a sports game for girls, *Starfire Soccer.* The company had a popular Web site with over 250,000 registered users, spin-off merchandising, and wide publicity; *Time* featured it in an article entitled "A ROM of their Own."[38]

Despite the popularity and publicity, however, Purple Moon ran into catastrophic trouble. It rapidly found itself in a desperate war with Mattel for the scant shelfspace allocated to "girl games." In 1998 it was pulling in $4.7 million against Mattel's $53.3 million.[39] Its best game, *Rockett's New School*, made just over a tenth as much as *Barbie Photo Designer* – $12.9 million versus $1.6 million. Purple Moon won only an estimated 5.7 percent of the girls' market and probably lost about $30 million in 1998 alone, with accumulated losses well above $45 million.[40] While Laurel continued to pursue an active research agenda, investors were looking for financial returns. In 1998 the board decided a public stock offering – the Grail of dot.com companies – was out of reach. Shortly thereafter they pulled out and offered the company for sale. It was snapped up for an undisclosed amount – by Mattel.

This was a bitter pill for the proponents of progressive "girl games." Brenda Laurel "thought it through better than anyone else, and it still failed," Jenkins observed. "That's very frustrating and very worrisome." Laurel herself said, "It is a shame to be hammered by a monolithic model. We have a hegemony issue and it will be interesting to see who takes it on. My hope is that diversity in products for girls will express itself."[41] What this episode demonstrates is the powerful market influences that maintain traditional structures of gender-coded play: on the one hand, the enormous commercial power of behemoths like Mattel, which thrive on stereotypical representations; on the other, limitations on the resources and time that commercial ventures are prepared to commit in seeking out alternatives that are slow to turn a profit.[42]

Psycho Men Killers

It is tempting to offer the demise of Purple Moon simply as a story of enlightenment killed by corporate greed and power. But there is another possible explanation for its failure – that its project was misconceived. Feminist cultural critics have by no means universally applauded Purple Moon or similar ventures. Others have criticized girl-gaming research projects for "conform[ing] to assumptions about gender that are created and reinforced by existing market pressures" – hence duplicating, even if at a more sensitive and female-friendly level, many of the binary oppositions that underpin commercial play commodities.[43] Though the "stereotypical" Barbie and "feminist" Purple Moon strategies might seem sharply opposed, both share the assumption that girls want games that are distinctly different from – indeed, in many ways the polar opposite of – those played by boys. The subtext of what Laurel and her fellow researchers write seems to be that girl play is "nice" – cooperative and peaceable – whereas male games are "nasty" and violent. As Gilmour observes, "The binary structure of this list essentializes women as neo-Victorian subjects."[44]

Other researchers have suggested, however, that when girls *do* play interactive games, at least some of them enjoy the "nasty" male-coded (e.g., violent) games.[45] This point was brought home with a bang by the dramatic appearance in cyberspace of the female *Quake* clans. *Quake* is a big-gun-toting virtual gorefest, often played collectively by online competing tribes or clans. In the midst of this – the ultimate macho gaming scene – a number of clans began identifying themselves as female and declaring a gaming politics of Amazonian militancy. With names like Psycho Men Killers, Die Valkyrie, Crack Whores, Clan PMS, and Riot Grrls, the clans not only sought to prove that they could "frag" opponents as well or better than boys but also in online forums often sharply challenged the idea that it was liberating to make "girl games" based on assumptions about female pacifism.

Such organized female collectivities appear to have had an influence on game designers and the subject-positions they create for players. Girl players hacked the original generically male body of *Quake* avatars to make it assume a female "skin." Anatti Autio has suggested that more "official recognition in the Quake subculture" came when the developer, id Software, included female warrior protagonists in *Quake 2* – an example of the coevolution dynamic we discussed earlier actually working to open games to a wider inscription of subject-positions than had been originally intended.[46]

Female clans appear not only in *Quake* but also in other networked games, such as *Everquest* and *Descent*. "Grrl gamers" – as opposed to "girl games" – make a salutary neobrutalist corrective to "pink and blue" cultural differentiations, however ostensibly female friendly, in the gaming field. At the same time, however, the Psycho Men Killers and their sisters raise issues that are the virtual analogues of debates about female combat roles in the armed forces. Is it a fulfilment or betrayal of feminist aspirations for girls to tote a railgun in virtual space? Is the best answer to "militarized masculinity" a "militarized femininity"? Is it too cynical to suggest that for many game developers who do not care a whit about these questions, the female *Quake* clans have been a welcome way to get off the hook over the issue of game sexism? "Grrl games" can, after all, be easily cited to prove that violent games are not inimical to women; throw in one or two female game developers; add a few she-warrior characters, and the same games can be cranked out in all good conscience – with ever-improved graphics resolution on the flying "giblets" of equal-opportunity virtual flesh: in short, military-industrial business as usual.

Lara Croft

The fullest expression of the game industry's ambiguous attitude towards women is Lara Croft, the fearless and curvaceous heroine of Eidos's *Tomb Raider*. Initially made for the computer, the game is now one of the major software attractions for Sony's PlayStation. Lara is a neocolonialist archaeologist-adventurer, a female Indiana Jones who sleuths, shoots, and seeks hidden treasure in exotic locations sporting a tank-top-and-shorts outfit that displays a digitally crafted body that would make Barbie envious. Since *Tomb Raider* was first launched on the PlayStation in November 1996, Lara Croft has attained cult status. The third sequel is now out, and the game has sold over three million copies. A *Tomb Raider* movie was released in 2000, with multitudinous spin-off and accessory products.[47]

It might be argued that Lara is gaming's best shot at female emancipation. *Tomb Raider* is a game that invites players of either gender to identify themselves as an active and autonomous woman; hence it could be a creation that makes digital play affirming and attractive to girls. But critics have pointed out that Lara's pin-up proportions and titillating outfits invite spectatorship as much as identification. There is a subject/object ambivalence implanted in the game software, which allows both a first-person view through Lara's eyes and a third-person view, looking at Lara as she fights, exercises, and leaps over obstacles. Lara Croft's creator, twenty-one year-old Toby Gard, declares that "strong, independent women are the perfect fantasy girls – the untouchable is always the most desirable."[48]

The part that this voyeurism plays in Lara's popularity is shown by a recent double-page ad for *Tomb Raider* III in *Total Control* magazine.[49] It features Lara sunbathing in the nude, a small towel just barely covering her backside. Above her is the caption: "It's hard to believe I just get better and better." That there are real-person models for the Lara Croft persona – first the Australian twenty-two-year-old actress Rhona Mitra, later Nell McAndrews – helps the promotion of the game. Recently, McAndrews posed for the UK magazine *For Him* in her Lara persona. But where the sexual spectatorship aspect of Lara is most explicit is on the Net, where there is a pirate "Nude Raider" Web site with a patch that disrobes Lara. The site receives a staggering number of visits – up to twenty thousand a day. If the digital version isn't what you want, gamers can get a twenty-four-dollar action figure from Playmates. According to a reviewer with a video gaming Web site, "Playmates is going to charge you $24 for the pleasure of owning and posing Lara, but compared to the street rate for such things, it's a steal."[50]

Croft's adventurousness (not to mention sexiness) may make her attractive to many female players. Just as many are probably alienated by yet another airbrushed idealization of female eroticism. Martial combat games like *Street Fighter* and many role-playing games have always had their quota of lithe and lethal female figures, with slit dresses, leather suits, and deadly moves designed to appeal to male sexual fantasies and fears; indeed, as Gilmour points out, the irony is that "there are more women in fighting games than in any other genre."[51] But *Tomb Raider* marks the first time a successful game series has been centred on such a protagonist.

Lara thus represents a revised approach to game design that more prominently incorporates women into the game world but in a way that intensifies appeal to the male market. This strategy probably owes something to recognition of the growing participation of women and girls in digital culture. But it is also driven by another dynamic – the maturing of the "Nintendo generation" from boys to men. As we have seen, one way the industry can continue developing the male niche is by a constant "up-aging" of the market. Introducing sex to games in the form of shapely heroines is one way to manage the process. Of course, not all the erotic desire Lara provokes necessarily comes from heterosexual men – but we bet a lot of it does. Her success, and that of the many other game heroines that have followed in her footsteps, such as Joanna Dark of Nintendo's *After Dark*, seems to represent the emergence of an ambivalent digital design practice that aims both at encouraging the creation of female markets and cashing in on male libido.

CONCLUSION: GAMING BY DESIGN

It is not hard to imagine the immense sense of possibility that accompanied the invention of interactive multimedia. All the more reason, perhaps, to feel depressed as one surveys the widespread depictions of violent carnage on the boxes displayed in interactive game stores, and to want to decipher the complexes of cultural practice that have led to the industry's continuing preoccupation with representations of militarized masculinity.

As we have seen, interactive game designers and marketers, starting from an intensely militarized institutional incubator, forged a deep connection with their youthful core male gaming *aficionados* but failed or ignored other audiences and gaming options. "Choice" is the very heart of digital design because play must be programmed as a flow of imaginary choices at every point in the game. Different genres are constructed with varying degrees of openness or closure, realism or fantasy, and exploration or immersion, depending on to whom they are sold. The kinds of games that can be computerized are immense. But the kinds of choice offered to players become constrained – reduced to carefully designed selections of what tank to drive, what plane to pilot, which enemy to confront, and which weapons to use next – according to what narratives, characters, and genres of gaming can be promoted and sold profitably. When audiences are small, and resistive, or just unfamiliar and unpredictable, which is the case with women and girls, their choices and tastes may be ignored or met with very limited resources. Thus, the vaunted generational participation in the wired communion of interactive players is largely seen in the empowerment of a male search for an imagined mastery of their world. Within this pattern of hegemonic masculinity there are many gradations. Yet the trained fighters of *Pokémon*, the samurai warlords of *Shogun*, and the mercenary killers of *Soldier of Fortune* are all points on a range of interactive game culture that tends to concentrate around virtual fantasies of violent domination.

This hegemonic bias of experience is not unchangeable. Today, the interactive game industry is at a watershed, where conflicting cultural, economic, and technological vectors press in contrary directions. The culture of militarized masculinity is deeply entrenched, but there are also forces that may diversify the industry away from testosterone-blasted aggression. These include an interest in expanding out of an extremely saturated and competitive "young male" niche, which provides some incentive to respond to the critique of industry sexism made by feminist researchers and entrepreneurs, a critique to which game publishers are partially responsive because of their commercial interest in reaching female consumers.

Industry organizations, anxious to promote the image of an expanding consumer market and repudiate accusations of sexism, now generate rosy estimates of steep rises in female gameplay. Thus, a 1999 survey conducted by the Interactive Digital Software Association claims to "explode the myth that the videogame domain is a boys-only club."[52] According to the survey, approximately thirty-five percent of frequent console gamers and "a whopping 43 percent of frequent PC gamers" are female. The survey also claimed that there was a slight preponderance of female Internet gameplayers (53 percent) over male counterparts (46 percent), a reversal of the game world's typical gendering. However, for online role-playing and strategy games males were still the heaviest users. The survey concludes that "more girls and women are coming on board, as their comfort levels with the technology and the software rise through familiarity with the Internet or products aimed squarely at female players."[53]

Although there is probably truth to the claim that more women and girls are becoming involved in interactive games, there are reasons to doubt that the shift is as great as IDSA likes to represent. Studies of earlier generations of media have shown enormous gender-related differences in what constitutes "use" of entertainment technologies. "Watching" television, for example, has historically meant very different things for men and women. Men tend to watch on a more sustained basis, commanding the remote and selecting programs, while women watch more sporadically, interspersing TV viewing with the performance of domestic duties. The IDSA study does not mention any of the comparable factors that structure digital gaming: duration of play, ownership of systems, choice and purchase of games.

Another survey of video game play conducted in 2001 by the market research company IDC outlines some of the complexities that underlie such aggregate figures. It focused on video games and on the "primary gamer" – that is, "the individual spending the most time playing games on the videogame console" – regardless of age.[54] The survey found that with an average age of just over twenty-one, the primary gamer population was three-quarters male. It attributes what it terms "this male-oriented trend" to three factors: "marketing efforts, game content and perception."[55] There was a growing proportion of female primary gamers on the latest video game consoles – at the time of the survey, the Sony PlayStation 2 and Sega's Dreamcast. Female teenagers made up about ten percent of the primary gamers on these platforms, a higher proportion than with older platforms. The authors concluded that "the gaming industry appears to be making slight headway into this traditionally non-gaming segment." They added that "to adequately penetrate the female gaming segment (as well as older adults, both male and female), the gaming industry still has quite a bit of work to do in terms of marketing efforts and game content."[56] Unfortunate phrasing aside, IDC's conclusion seems more nuanced than the boosterism of the IDSA reports.

There *has* recently been a wider inscription of girls and women into gaming, in terms both of characters and the appearance of women in game magazines. But with the exception of the narrow shelfspace devoted specifically to "girl games" – now effectively monopolized by Mattel – much of this "feminization" seems to follow the lines of the

Lara Croft strategy: that is, putting women into games in a way designed to appeal to young men. The message is that one doesn't have to be a geek to game: you can have the girl too. That the "testosterone zone" ethos remains strong is attested to by one participant who observed of the main industry "E3" conference, "Nearly every company, from the mom-and-pop developers to the holy trinity – Nintendo, Sega, and Sony – populated its booths with models, beach bunnies, and moonlighting strippers."[57] Episodes such as the recent release by Simon and Schuster of *Panty Raiders*, a game based on space aliens stripping nubile women down to their underwear, do not support the industry's newly "gender sensitive" pose.[58]

While the public relations side of the industry strenuously affirms its commitment to gender-equity gaming, market analysts are more cynical. The equation, according to one, is that "catering to boys is much more fun. Video game companies are very good at it, and it makes them rich. And they don't want to mess with a winning formula."[59] In this view, "the video game industry is so huge it can afford to ignore women," and "one side effect of the eruption of game-playing into the mature male consumer market is the continual shelving of plans to design games for females."[60]

Making the game world female friendly is not just a matter of recomposing the cultural discourses coded into software programs, or of the growing familiarity of women and girls with digital technology. Although it involves all these processes, it is also and simultaneously a "bottom line" question of the capitalist market. Breaking down the historically sedimented barriers that fence women out of the cultural and technological circuits of gameplay will not necessarily be financially rewarding to game developers and publishers. Experimentation with marketing to female players faces the risk-versus-repetition syndrome familiar throughout the entertainment industries. This sets the possibility of huge profits – or losses – from audacious innovation against safer income from a steady stream of clones, sequels, and knock-offs of tried-and-true formulas. While the former route may open interactive entertainment to greater participation by girls and women, the inertial momentum and constantly renewed feedback loops of the latter course keeps it in a tight orbit around male players. The current conjunction shows a collision of influences: it is by no means sure that inclusiveness and gender equity will prevail.

Another pressure comes from the mobilizations of movements against violent representations in media. Campaigns against game violence episodically coalesce, calling variously for government censorship, intensified rating systems, industry self-restraint, and a "deglamorization" of virtual bloodshed.[61] Historically, film and television have shown considerable ability to shrug off such pressures. But in the wake of Columbine and other spectacular youth killings, the possibility of civil actions launched by victims of violence caused the interactive game industry some concern. Such attempts have so far been unsuccessful. But they hold in the back of the game developers' minds the kind of situation faced by the tobacco industry: makers of ultraviolent games may yet find themselves liable for gigantic damages. To all the preceding factors we add the undoubted interest of many developers in producing entertainment experiences that are interesting and vitalizing, rather than formulaically destructive and violent.

Such a shift in game culture would be important for the industry and for the larger society it inhabits. As we have already said, the debate about the effects of virtual violence in particular, and media representations of violence in general, remains unresolved. It is a debate that involves not only direct causal connections between virtual experiences and real life events such as Columbine but also more subtle issues about the relation of a culture of violence to psychological well-being and to an open perception of political and social possibility. Furthermore, the "moral panic" approach to video game violence runs the risk not simply of generating a fear of young people but also of demonizing them as a group rather than encouraging social and historical understanding of a cultural practice. Because we know little for sure about how interactive technologies shape us as social subjects, prudence and variety in the construction of virtual culture is desirable. To interpret an unknown as a mandate to forge ahead in generating ultraviolent virtual realities is disingenuous and irresponsible.

Several writers have spoken of "media ecology," comparing the complexity and ubiquity of our contemporary media sphere to that of the biological environment.[62] Ecological activists dealing with issues of species life and death often recommend the adoption of the "precautionary principle" when the uncertain effects of powerful industrial practices are themselves sufficient reason for caution and moderation. Just as ecologists have taught us to be careful about monocultural practices that stifle biodiversity, so too we should be wary of gaming monocultures such as the blood-red flowers of militarized masculinity that so easily crowd out a wider potential of playful designs.

12

Sim Capital

Even if your Sims have a lot of Simoleans in the bank, they are not as happy as they could be. Yes, yes, we all have heard those axioms about thrift, and a penny saved and all, but in the Sims' world, that means you are not buying stuff. And not buying stuff is a problem. Not buying stuff means you're not upgrading the Sims' environment so that it is easier to make them happy. They're consumers, you know, and they are most happy when they can have a choice between pinball or the piano, computer games or the plasma TV. They want it all.

The Sims Game Manual

There is a short story by Stanislaw Lem about some programmers who create a virtual world like *The Sims*. And there is a debate among the creators of the world about whether or not they should tell these simulated people that they just live inside a computer and that they created them. Eventually a law is passed that says they're not allowed to tell the simulated people anything; it's considered unethical. But these people deduce that there are creators watching them, they notice that information is leaking out of the environment and coming back in with feedback, so obviously someone is changing the environment based on what they're doing. Anyway, the moral implication of all that is pretty interesting and I'll just leave it at that.

Will Wright, creator of *The Sims*[1]

INTRODUCTION: PLUGGING IN *THE SIMS*

The interactive gaming industry's first big hit of the new millennium was *The Sims*, released in February of 2000. It was the latest in the highly successful Sim series, launched by the developer Maxis in 1989 with *SimCity*, an urban planning scenario that became one of the best-selling games ever, and continued with such titles as *SimLife*, *SimEarth*,

SimFarm, *SimCopter*, and *SimAnt*, as well as *SimCity* 2000 and 3000. These are what are popularly referred to as "God games," in which the player oversees the development of an entire city or civilization from a near-deific vantage point. *The Sims*, however, drops down a notch on the scale of magnitude. The player directs the day-to-day life of tiny humanoid suburban inhabitants, or Sims, in what appears to be a microcosm of affluent suburban middle-class North America. To an even greater extent than Maxis' earlier creations, *The Sims* is open-ended, with no scenarios, no explicit goals, and no winning objectives other than to keep one's Sims alive and flourishing. Without proper handling or adequate attention to seven basic parameters of existence – "Bladder," "Fun," "Hunger," "Hygiene," "Social Comfort," "Energy," and "Room" – Sims can starve to death, collapse from exhaustion, degenerate into infantile squalor, or collapse into depression. But with careful micromanagement to juggle the demands of work and recreation, Sims prosper, making friends, finding partners, raising children, climbing career ladders, building, designing, furnishing, and decorating their homes, and engaging in happy hours of leisure activities. From these apparently simple elements, Will Wright, the original designer of the whole Sim series, built a commercial and critical triumph: *The Sims* was the best-selling PC game for the first half of 2000, while Ted Friedman, one of the most thoughtful writers on video game theory, said, "I think it will transform computer game makers' ideas of what a successful game can be."[2]

It seems appropriate, therefore, to consider how the line of analysis running throughout this book might illuminate *The Sims* phenomenon. Much writing about digital gaming focuses on the interaction between player and game. We, however, argue that the moment of gameplay is constructed by and embedded in much larger circuits – technological, cultural, and marketing – that in turn interact with one another within the system of information capital. The three-circuits model involves the cultural circuit, which links the player through the game text to its designers; the technological circuit, which ties the computer or console user through his or her machine to its developers; and the marketing circuit, which connects game consumers through the game commodity to its corporate promoters. Each of these circuits constitutes a distinct moment in the gaming process, but they all intimately affect each other, so that the three circuits are superimposed on or interpenetrate each other, producing very complex, dynamic effects either of synchronization and reinforcement or dissonance and interference within and between the various elements. To decontextualize the possibilities of digital gaming – be they promises or perils – from these constituting forces is fundamentally to distort or misrepresent its realities. So let us look briefly at some of the processes as they work out in *The Sims*.

THE SIMS IN THE MARKETING CIRCUIT: BUILDING SIM EMPIRES

The Sims sells for US$49.99. An appropriate computer system to play it on would run close to US$1,000, while an Internet connection (which, as we shall see, is becoming an integral part of the Sim experience) involves regular charges that vary according to provider, location, and quality of connection. Sim players are thus not constituted merely by attraction to the game content and the capability to master it; they must be "consumers" (piracy, for the moment, aside) with a certain level of disposable purchasing power. The game is a commodity; it is surrounded by an elaborate web of marketing practices exhorting people to part with their money to enter its world, and implicitly urging them to attain, sustain, or raise a certain level of income in order to support the computer game playing habit, of which *The Sims* is just one of the most current and alluring instances.

While *The Sims* marketing exploited the cachet of the Maxis name, the hand behind its commodity success was that of the game-publishing giant Electronic Arts, which had bought up Maxis a few years earlier in the round of publisher acquisitions that has done much to consolidate and concentrate the structure of the industry. EA, which posted revenues in 1999 of $1.2 billion, markets under several brand names apart from its own Electronic Arts and EA Sports labels: Origin, Bullfrog, Westwood Studios, Gonzo Games, and Jane's Combat Simulations.[3] When Maxis was added to this stable, the Sim series became a major EA "gaming franchise," and the development and launch of *The Sims* was supervised under the eye of an EA-installed management team.

The strategy, in the words of Maxis's public relations director Patrick Buechner, was to create one or two "event" titles each year. An event game is one that "will be highly anticipated, is capable of drawing the attention of both the gaming and mainstream press, and stays in the Top Ten for an extended period of time."[4] Building *The Sims* as an event title, Buechner said, was in large part a matter of promoting word of mouth, capitalizing "on the game's unique subject matter and Will Wright's status as one of the industry's few celebrities" to "generate feature-length coverage in magazines ranging from *Wired* and *Entertainment Weekly* to daily newspapers like the *Wall Street Journal* and the *New York Times*."[5] Electronic Arts also busily worked the news value of the allegedly high number of female players of *The Sims* to place stories in journals such as *Mademoiselle*, *Working Woman*, and *Cosmopolitan* during the crucial Christmas period. Electronic Arts launched a major television advertising campaign for *The Sims*. It began with a spring ad campaign on MTV and included sponsorship of the popular "Wanna' Be a VJ" contest, the first time EA had advertised a PC-only (rather than console-based) game on TV, followed by a series of eight advertisements running on MTV, TNT, Comedy Central, and other popular television channels throughout the Christmas period of 2000.

But as Buechner makes clear, while launching the game was critical, the more sustained challenge was "to retain customers in an extremely competitive climate."[6] In that regard the whole "virtual community" constituted around TheSims.com and the array of associated Sim fan sites were vital. Running TheSims.com site, with its eighty thousand unique visits per day, provides EA with a sure-fire market outlet to promote further elements of *The Sims* franchise such as *The Sims Living Large* expansion pack, *SimsVille*, which will blend elements of *SimCity* and *The Sims*, and the long-awaited *Sims 2*, which at time of writing is veiled in secrecy. Equally important, it gives an opportunity to gather customer registration information and monitor responses and discussion about games. Earlier in the history of the Sim series, Maxis had culled ideas for new variants from players' comments on the phone hotline for *SimCity*. Buechner's comments on the emergence of follow-ups to *The Sims* make it clear that a similar process of consumer monitoring, feedback, and reinscription is occurring in the Net environment. He refers, for example, to players' manifest interest in creating "more extreme" situations for Sims – houses of torture are apparently a big download favourite on fan sites.

The maker of *The Sims*, Will Wright, suggested that in the long run the game's future probably lay on the Net as a massively multiplayer online game – "I'm thinking 100,000 players."[7] This speculation meshed perfectly with EA's intensifying involvement in Internet gaming, a role that it had already explored through its acquisition of Origin, with its enormously popular multiplayer game *Ultima*. At the end of 1999 EA paid AOL $81 million to become the exclusive game provider for AOL's Game Channel, and for all game content on AOL-brand Net services, including AOL.com, CompuServe, Netcenter, and ICQ. A new AOL/EA gamesite would offer customers the ability to download and play EA's proprietary games, obtain information, and join chat rooms to discuss games. The titanic merger between Time/Warner and AOL in 2000 further boosted the significance of the deal. At the same time that it concluded the agreement with AOL, Electronic Arts announced the acquisition of Kesmai, a developer of multiplayer online entertainment and a unit of another global media giant, Rupert Murdoch's News Corporation. While Kesmai became a wholly owned subsidiary of Electronic Arts, EA formed a nonexclusive distribution agreement with News Corporation to deliver its Fox Interactive online games through its Web sites. These intricate manoeuvrings were central to EA's corporate strategy: its chief financial officer, Stan McKee, announced that the company aimed at getting twenty percent of its sales from online revenues within three years. EA signalled the seriousness of these intentions by releasing a new class of stock tied directly to the performance of its new Internet business unit. AOL purchased ten percent of the new shares and options on an additional five percent as part of the two companies' five-year strategic alliance. Looking at EA's intensifying Net activities, estimated to represent a $350 million investment in Web-based business, analysts suggested that *The Sims* might prove a centrepiece: "This game is not some kind of medieval fantasy world. This could be like real life online."[8] Real life in such a case would clearly be organized and circumscribed by the commercial cyberempires, such as AOL/Time Warner, in whose interlocking corporate webs EA is so carefully positioning *The Sims*.

THE SIMS IN THE TECHNOLOGICAL CIRCUIT: PLAY HACKER

Sim characters live virtually, as digital creatures, on-screen. Sim players are of course corporeal, but they nevertheless inhabit a social sphere within which many vital interactions, from workplace to recreational activities to personal relations, happen "on-screen." They are denizens of a society in which "electronic selves" proliferate. *The Sims* is intimately involved in this process. At the most basic level, a player must have access to a computer, be able to boot up, load a disc, navigate a keyboard, and survive Windows. But more importantly, to really play in the Sim world they have to go online.

The interactive gaming business has been a major force disseminating digital machines and technohabits. Gaming, as industry advocates often suggest, has probably given entire generations hands-on education in "digital technology 101," familiarizing young people, especially boys, with hardware/software differentiations, keyboard skills, screen navigation, bugs, crashes, and system incompatibilities – while contributing significantly, according to several major health studies, to a plague of obesity by its cultivation of sedentary mouse potatoes. The process is now being extended by Internet gaming into the world of online connections, Web pages, patches, downloads, freeware, e-commerce transactions, and chat-rooms.

The Sims is intimately involved in this process. But the process of digital socialization goes a lot further than that. Like many game companies, Maxis has added continuing interest and involvement to its product by harnessing the players' own creativity, adding considerable editing capacities to its Sim series so that players can "customize" their games. In later versions of *SimCity* players could plan their cities using precreated buildings, but they could also use a free architectural tool, initially downloadable from the Net, later included in the game itself, that enabled them to design their own. In *The Sims* editing capacities allow the creation of new character "skins" – "from Frankenstein to the Flash to the entire Kiss band" – as well as thousands of interior decorating options.[9]

Moreover, from the time of *SimCity 2000* Maxis realized that "many people want to share that experience with others, which obviously adds another dimension to the games," and set out to "really take advantage of this form of empowerment" by permitting the exchange of cities and user-created buildings over the Internet and encouraging the appearance of fan sites that enable users to trade and import game elements.[10] For *The Sims* the process of creating game-centred virtual communities was extended into the creation of a Web site, TheSims.com, that not only offers players the opportunity to trade houses and families, "skins" and wallpaper, online, but also "gives them a forum to tell the stories behind their creations," a place where thousands of "scrapbooks" – or virtual novels – about Sim characters can be posted.

TheSims.com also offers players ways to renew and enrich the game, whether from other gamers' creations or from the new household objects released every few weeks by EA on the site, including everything from a cuckoo clock to plants, moose heads to hang on the wall, light fixtures, and a pet guinea pig. In 2000 more than half a million players downloaded the guinea pig alone. Many, however, were not pleased to discover that developer Will Wright had infected the pet with a computer virus. Under certain conditions – if the virtual cage was not cleaned, if the digital rodent bit a Sim character – it could communicate a "disease" that lingeringly killed Sim characters so lovingly tended by their masters. The virus could then be communicated to members of their virtual family, exterminating whole households. Needless to say, such a plague became a topic of irate discussion on TheSims.com and fan sites. The discussions involved issues such as the permissible limits of digital creativity (and destruction) by game developers, the larger problems of viruses, worms, and other malign software agents on the Net, and the proper steps to take to achieve security against such contagions.[11] In a sense, Sim players themselves became "guinea pigs" in a digital experiment, one that, as with many other games, provided an occasion within which users and developers could discuss, negotiate, test, and experiment with the conditions of the digital environment that was so rapidly being expanded by e-capital.

THE SIMS IN THE CULTURAL CIRCUIT:
VIRTUALIZED CONSUMERISM

Neither skilful marketing nor technological pedagogy can explain the attraction of *The Sims* to millions of players. It is especially interesting to consider the issue of role-playing in games, an approach that "emphasizes the opportunity for the gamer to identify with the character on screen."[12] We see games functioning as machines of "interpellation," devices of semiotic address that invite players to take up certain subject positions and exercise certain options, widely or narrowly defined, within those positions, positions that in turn replicate, reverberate with, or revise ideologies embedded in a wide variety of cultural discourses.

In this respect, the most remarked upon aspect of *The Sims* is the way it breaks the dominant code of masculine gender positioning effected by digital gaming – not simply in that it allows players to identify with female characters but, more significantly, because it does so in a conventionally "feminine" domestic setting. The issue is not that there are women Sims (there are now female-warrior roles available in shooters like *Quake* or *Unreal*, while victimized princesses have long been a stock in trade for Nintendo) but that even male Sims operate in a world where the crucial decisions involve domestic design and relationships, rather than "fragging" or ogre decapitation. In this aspect, and in its allowance of same-sex relationships (Sims can have gay partnerships, though not marriages) the game has been seen as groundbreaking. Indeed, much of its success has been attributed to this expanded range of subject-positions. According to Patrick Buechner, *Sims*' buyers departed significantly from the main market of adolescent males and "brought a whole new audience to 'Sim' gaming, attracting older gamers, teenage girls, and many women" (allegedly "somewhere in the range of 30 percent to 40 percent") – although the game was also widely applauded in computer game magazines that were far more attuned to hard core gaming.[13]

Thus, in contrast to the "gender rift" observed in chapter 11 as a pattern in gamer culture, we must see *The Sims* as a significant deviation from the dominant game formula of "militarized masculinity." But this raises a related but distinct issue – the class roles the game makes available to its players. *The Sims* invites its gamer-subjects to identify themselves with the daily lives of middle-class home-owning professional North Americans: "All the Sims, even the poorest, live in a pleasant suburb with well-kept streets and good public schools."[14] In terms of its content, this demonstrates an uncanny mirror-world effect, since its inhabitants are a simulated section of the very computer-owning demographic bracket to which the game is predominantly marketed. In inviting gamers to involve themselves with the details of Sim careers, leisure, and domesticity, the game interpellates or addresses players who are already engaged in a multitude of social discourses identified as precisely the subjects of such career choices, lifestyle decisions, design, purchasing, and domestic decisions. In doing so it not only reflects but also reinforces and reproduces these identities, preoccupations, and roles. The idea that military simulations provide training for soldiers is familiar; what *The Sims* does is provide civilian simulator training for yuppies.

The lesson that is taught, or at least reinforced, is that one must negotiate the daily events and crises occasioned by a life in which commodity consumption is the *raison d'être*. Although the game is open-ended and has no explicit definition of winning or losing, it is not devoid of structure. That structure is provided by getting and spending: "money motivates and frames a Sim's behaviour. Money buys and furnishes homes; feeds Sims; pays their utility bills, their gardeners and their maids."[15] They earn cash by following career paths such as medicine, entertainment, or crime. Progress on these paths depends on improving skills (mechanical aptitude, charisma, physical fitness), building social networks, and going off in high spirits to work – where success ensures an income that allows more material goods or more time for the cultivation of skills and social relationships. "The only obvious objectives are the acquisition of consumer goods and the enlargement of one's home."[16]

As J.C. Herz observes in a brilliant review, the world of *The Sims* is a totally instrumental one "where everything is an object that yields a measurable benefit when some action is performed upon it" and "the only form of success is the acquisition of more and better objects." This is "formally engineered into the game-play" insofar as higher-quality, hence more expensive, objects satisfy Sim needs more efficiently: "The bar graph labelled fun goes up faster if you are watching a high definition television than if you are watching a black-and-white television." This logic structures interaction with humans and non-humans alike: a Sim roommate and the couch he or she is sitting on are analogous, since one is a "conversational object that bumps up the Social bar-graph" while the other yields "Comfort if you sit on it and Energy if you nap on it." Even having children is a means to an end, since it is through the interaction of your Sims' kids with the neighbours that adult Sims get to know each other, and it is only by entering into social networks that one gets the professional advancements that lead to career promotion – and more income. (Other commentators have observed that, although *The Sims* is culturally liberal in terms of allowing homosexual partnerships, having a stay-at-home mate of any gender to take care of housekeeping chores seems fairly indispensable for Sim success.) Herz says, "The Sims live in a perfect consumer society where more stuff makes you happier, period. There is nothing else. So your goals in SimLife are purely material. Work your way up the job ladder so you can earn more money, so you can buy more furniture, a bigger house and more toys."[17]

As the game manual quotation at the beginning of this chapter reveals, the makers of *The Sims* are perfectly explicit about the value structure of the game. Indeed, their avowal of consumerist ideology is so unabashed that it can be interpreted as ironic, a tongue-in-cheek distancing from the world-view it presents. Many intellectual admirers of *The Sims* claim that the manifest limitations of the game's ethos constitute an invitation to critical reflection on consumerism. Herz, for example, concludes that while *The Sims* is "disturbing in its crudeness," by "building a window into the Sim's souls, it prompts us to consider our own."[18] Henry Jenkins reportedly says of *The Sims* that "by simplifying a complex real world into a 'microworld' the game leads players to examine their own lives."[19]

But the games industry, like the rest of popular culture, has learned that irony is a no-lose gambit, a "have your cake and eat it too" strategy whose simultaneous affirmation/negation structure can give the appearance of social critique and retract it in the same moment – thereby letting everything stay just as it is while allowing practitioners to feel safely above it all even as they sink more deeply in. What Naomi Klein describes as "irony's cozy, protected, self-referential niche" is in fact the preferred mode of "cool" marketers who know well how to work with "pre-planned knowing smirks, someone else's couch commentary, and even a running simulation of the viewer's thought patterns."[20] It is a stance that allows hours of immersion in virtual decisions about whether to build one's Sims a hot tub or a patio while still believing oneself immune from psychic possession by these preoccupations. In this sense, it is the ultimate sophisticated expression of a culture that "wants it all," the perfect ploy for the construction of the consumer-subject.

THE GOD-GAME: SIM CAPITAL

This construction is a multilevelled one in which, at times, all the wheels – technology, culture, and marketing – of our three-circuits model synchronize and interlock relentlessly. To buy *The Sims* you have to be a consumer – which is, of course, the subject-position the game discourse invites one to adopt. You have to be a consumer not just of houses and hot tubs but also, and especially, of digital technologies. Playing computer games all night is in fact one of the leisure activities that can be selected in the game for your on-screen Sim (with effects that are positive in terms of mood but deleterious in terms of fatigue). So "Sims R'Us" not just in the content of their imaginary daily lives but also in their virtual condition, as "digital subjects" – a position to which Sim gamers recognize themselves as at least tendentially directed, and one that is affirmed and deepened by the very act of sitting and playing *The Sims* for hours. But to do so one must possess a copy of *The Sims*, a computer to play it on, and an Internet connection to TheSims.com, which requires an income, which in turn requires a career and involves difficult choices about juggling work, relationships, and game-playing leisure ... just as in *The Sims*.

Once one looks at the array of cultural, technological, and marketing dynamics that converge on the purchaser-player of *The Sims*, a final metalevel of identification between the virtual Sims and the real Sim-gamer emerges. Sims are one of what are often termed "God games" in which the gamer acts as the director of subordinate yet recalcitrant minions: one reviewer says that in *The Sims* the player adopts the position not so much of an individual character as of a "household deity."[21] But the relentlessly secular world of the game makes this sort of theological metaphor inappropriate. Nor, given its civilian and contemporary nature, are some of the more militarist or archaic variations common in interactive gaming – commander to troops, emperor to citizens – any more relevant. The more contemporary analogy would seem to be that of a manager, controlling and predicting and directing the behaviour of a very finely tuned market niche, a "segment of one," attempting to nudge and cajole and manipulate in a certain direction, all the while responding to their often unpredictable and unforeseen and unwelcome initiatives in a series of constantly adjusting feedback loops.

But this is of course precisely the way that the game industry is positioned in relation to the Sim game player. Most immediately Electronic Arts, more generally the whole digital gaming complex, and ultimately the larger synergetic webs of corporate media within which this complex is increasingly embedded are striving to create, steer, and shape the lives of their consumer subjects. Up to this point in our narrative we have referred to the digital market system of which the interactive game industry is such an important part by a variety of terms – "post-Fordism," "information capitalism," the "perpetual innovation economy." Now, however, we are tempted to call it simply "Sim Capital." Sim Capital designates the accumulatory regime emerging from the dynamic interplay of transnational enterprise, convergent communication technologies, and postmodern culture, in which increased reliance on simulations both as work tools and as consumer commodities, escalating surveillance and synergistic management of segmented markets, and the cultivation of an increasingly symbiotic relation between production and consumption is mediated through the feedback loops created by ever more sophisticated digital media and virtual technologies. As the virtual Sims are to the Sim player, so the Sim player is to Sim Capital. Playing *The Sims* is, in short, a process in which the player takes up – but could also subvert symbolically – digital capital and learns to elaborate its logic – a logic to which she or he is already subject.

All of this opens up the paradox suggested in the Lem story cited by *Sim* creator Will Wright. In *The Sims* the virtual agents are programmed and the real player is autonomous: the Sim characters are the played, the Sim gamer the player, the one object to the other's subject. But the complexity of the programming of *The Sims*, like that of any good game, yields an apparent independence to its digital creations – so that the Sim characters can seem, in their unpredictability, recalcitrance, crises, collapses, and resistances, more subject than object. On the other hand, once one looks at the convergent economic, technological, and cultural forces shaping *The Sims* gamer – not merely as the participant in a particular scripted and designed play scenario but also as a member of a population amongst which certain levels of technological familiarity are increasingly normalized, required, and rewarded, and as the target of a high-intensity marketing regime designed to elicit certain levels of consumption activity – much of their apparent autonomy and empowerment evaporates. The player reappears as object, not subject, the product of a system, an at least partially programmed and subordinated "subject," as much played upon as player.

But this is not the whole story. For if the Sim player is to Sim Capital as the Sims are to the "real" player, then we should also admit that the complexities of the programming do give rise to unexpected and unpredictable results. Indeed, the point of our "three circuits" model is precisely to highlight this possibility. Although sometimes the circuits do move together, reinforcing and meshing with one another, they also generate friction and contradictions. These can exist in regard to a single game, or more generally throughout the entire game culture. So, emphasizing now the potential ruptures in the circuits rather than their continuities, dissonances rather than harmonies, and turbulence rather than smoothness, we will now look at some other aspects of *The Sims* insofar as they suggest wider tensions and problems within interactive gaming. We see three such major contradictions: in the cultural circuit, between violence and variety, in the technological circuit, between

osures and access, and in the marketing circuit, between commod-
ation and play.

Contradiction One: Violence and Variety in the Cultural Circuit

One reason for the many accolades given to *The Sims* is undoubtedly
the relative absence of violence from its scenarios: the game stands in
striking contrast to the predominance of violence in video games, which
we discussed in chapter 11. Sims may get terminally depressed, break
up relationships, slap each other's faces or even immolate themselves
and their families in carelessly started housefires. But they do not
slaughter each other in large numbers with state-of-the-art automatic
weaponry. The game therefore seems to stand as encouraging evidence
of interactive gaming's capacity to emerge from the shadow of a persistent accusation that has stalked it from its very origins – the charge
that it is obsessed with killing and destruction. It is the beacon of
variety in what is otherwise a tendency towards masculinized violence.
This was particularly welcome to the gaming industry since *The Sims*
release coincided with the damaging report of the US Federal Trade
Commission showing that gaming companies regularly aim marketing
campaigns for ultraviolent "adult" games at young players.[22]

As we saw in chapter 11, there are forces pushing the industry to
diversify away from its traditions of "militarized masculinity," and
recurrent elements in digital game culture that depart from its violent
repertoire. In fact, Will Wright's entire series began as an almost
accidental deviation from the typical scenarios of warfare that have
been so central to the development of digital gaming. In the course of
designing a military game for Nintendo, *Raid on Bungling Bay*, "a
typical shoot-'em-up starring a helicopter that bombed everything in
sight," he included an option for players to generate the island territories that would be attacked and discovered that constructing things
could be as much fun as destroying them.[23] *The Sims*, with its lavish
attention to scenarios of domestic suburban life, can be seen as a
culmination of that tendency.

Such diversification may be in the interests of the industry as a whole:
although the testosterone niche can be mined for a long time, a wider
market could be more lucrative. If, as many feminist critics suggest, the
long-term effect of violent game content is to exclude the majority of
girls and women from gaming, hence from a crucial form of digital
socialization and training, then failure to escape the niche of militarized
masculinity could have serious implications not just for the game industry itself but for the whole so-called "knowledge economy." But though
breaking out of the violence groove might be beneficial to digital industry

as a whole and is part of the reason for Maxis' success with *The Sims*, how far this route will prove to be attractive to other gaming companies is uncertain given the real risks and costs involved in departing from well-established formulas and design practices. And as we shall see at the end of this chapter, within a year of the release of *The Sims*, global crisis was to unleash terrifying forces that threw into doubt any possibility that virtual culture might take a peaceful route.

Contradiction Two:
Enclosures and Access in the Technological Circuit

In 2001 Electronic Arts released an expansion pack, or supplement, for *The Sims*, entitled *The Sims Livin' Large*. Amongst many additions, it offered a new career path for Sims: hacking. Inhabitants of the virtual world could now increase their supply of Simoleans by following a life of digital crime, providing they succeeded in evading the forces of law and order. In real life, however, EA took a far less whimsical attitude towards hacking and piracy, which it claimed cost it some four hundred million dollars in 1999 alone. In that year EA and Sony combined forces to launch civil suits against a games "warez" group named Paradigm, the target of a raid by US marshals that had allegedly uncovered evidence of its operations in the United States, Canada, United Kingdom, Germany, Netherlands, Denmark, Norway, Portugal, Sweden, Russia, and other countries. "Putting an end to software piracy is a top priority for our industry," said Ruth Kennedy, senior vice-president and general counsel to EA in the aftermath of the raid.[24] The crusade led EA into conflict not only with clandestine pirates but also with other corporate giants of the digital world. In 2000 EA, along with Sega and Nintendo, filed a lawsuit against Yahoo, the search engine portal site, accusing it of ignoring sales of counterfeit video games at its auction and in mall areas that it leases to outside merchants, and sued for copyright and trademark infringement, unfair competition, and offering illegal devices for sale. The lawsuit asked the court to order Yahoo to cease sales and sought compensatory damages of up to $100,000 per copyright, and up to $2,500 for each sale of hardware devices such as "Mod Chips."[25] Undoubtedly, one of the games EA was most concerned would be pirated was *The Sims*, whose software includes elaborate anticracking devices, and whose launch was accompanied by an extensive antipiracy campaign launched throughout Sim fan sites.

The contrast between *The Sims*' playful endorsement of hacking and the stern real-life antipiracy stance of its makers nicely demonstrates a second major contradiction facing the interactive entertainment industry – that between enclosure and access. This tension arises in the technology circuit, and in the relation between users and producers of digital technology. But it also involves the contradictions between this technology circuit and the marketing circuit. Although young people have for millennia creatively amused themselves with very simple toys that were often self-made or home-made, virtual gaming requires a sophisticated machine apparatus. Because digital games as a form of play are dependent on high technology, people can be sold the hardware, software, and Net access necessary to enjoy it. Those who ultimately control the technology – those who hold the patents and copyrights to hardware and software, or control wired and wireless networks – profit on this basis, selling people equipment and services, and it is this profit-driven logic that has ignited the game industry's explosive growth. But digital technologies are also subversive of commercial ownership and control. "Piracy" represents a real threat to business profit. Because of the two-faced characteristic of digital technology, the interactive games industry has found itself on the horns of a dilemma. With one hand, it pushes to expand access to digital machines on which its market empires depend. With the other, it strives equally hard to police, contain, and constrain the use of such machines to keep it within the boundaries of commercial profit, and to wipe out hacker practices. Sometimes it seems that the right hand does not know what the left hand is doing.

As we saw in chapter 9, commentators like Peter Lunenfeld argue that capitalism's widespread dissemination of digital tools and skills gives a "new character" to virtual creation that significantly breaks down conventional relations between corporate producers and their consumers.[26] What we would add, however, is that as long as this new character is contained within the shell of the old commercial relationships, digital industries will continue to be subject to the constant leakage of "pirated" value, which is in fact an expression of the unacknowledged surplus capacities for reproduction and circulation they have created. Pirate practices may compel major restructurings in the game industry. The increasing availability of emulator technologies could force gaming companies to operate on an "open box" system whereby all systems operate on a universal standard based on and compatible with the PC, differentiated by speed, graphics, and sound capabilities. Peer-to-peer networks may be commercially assimilated, the exchanges charged according to a carefully monitored system of micropayment. But this would require an array of intellectual property management systems, using encryption, digital watermarks, and other highly intrusive forms of software surveillance. Such arrangements could so complicate and encumber interactions, and subject users to such an unwelcome regime of panoptic scanning, as to be self-destructive. Nor is it likely that they would do more than escalate the war of technological measures and countermeasures between owners and hackers. Whether over time this will be merely annoying or seriously ruinous to Sim Capital is impossible to say. But virtual gaming, like other digital businesses, will continue to be dogged by the spectre of piracy – a haunting apparition that the industry has itself conjured up and beckoned to its own door.

Contradiction Three: Commodification and Play in the Marketing Circuit

Perhaps the deepest paradox of interactive entertainment such as *The Sims* is that between commodification and play. As we have seen, an elaborate marketing apparatus deploying highly sophisticated methods of marketing and surveillance surrounds *The Sims* – and most other digital games. The aim of this apparatus is to create game players who will also, simultaneously and of necessity, be game consumers. The television advertising, promotional stunts, retail displays, synergistic tie-ins, brand envelopment, and viral marketing techniques that saturate digital play are already among the most potent vectors of commercialization in contemporary youth culture. But they may soon seem rudimentary.

In a recent online article entitled "Sim Merchants," digital pundit Steve Johnson points out that in multiplayer Internet game communities, bartering or purchase of fictitious goods and services – armour, weapons, magical powers, buildings, "skins," even characters – is now a standard and accepted part of play. Why not take advantage of an already existing in-game e-commerce opportunity? "The gaming world is wired for play money," he observes: "sooner or later someone is going to start dishing out the real thing." Johnson approvingly suggests that this commercialization will increase the "reality" of virtual experience. "If there is one law of modern commerce," he remarks, it is that "wherever communities of people come together, merchants usually find a way to sell things to them." Following this logic, he concludes that "if these stores start appearing in the virtual cities of our video games, that will just be one more reason to start taking them seriously."[27]

The argument that it is a good thing for our games to be deeply suffused and integrated with real consumer practices – so that we take them "seriously" – nicely demonstrates what Jeremy Rifkin calls "the dialectics of play."[28] Rifkin argues that the commodification of culture is at the centre of contemporary "hypercapitalism." The most dynamic processes of accumulation no longer involve the manufacture of material goods but rather the production and marketing of experiences.

Various types of games and play assume an intense economic importance in this regime, because "play is what people do when they create culture" and "the commodification of cultural experience is, above all else, an effort to colonize play in all of its various dimensions and transform it into purely saleable form."[29]

Unlike many other media, interactive game makers have not, to date, relied substantially on advertising revenues; instead they draw their profits primarily from game sales. But this may soon change. A handful of games already feature ads, typically as banners that appear on hoardings in football and racing games. Online game sites promote affiliate deals, with level maps and short cuts flanked by banner ads for the latest games. In 1999, however, a company named Conducent announced that it had developed software capable of placing dynamic updateable advertisements and product placements, linked to the sponsor's Web site, directly in games. Conducent's marketing director, Robert Regular, described the project as introducing "a model as old as God itself" – that of advertisement support – to gaming.[30]

One obvious use of such software is to advertise forthcoming games – using demos, for example, to count down days until a game's release, provide a link to order copies, and switch to promoting other games from the same vendor. But Conducent also looks forward to cross-selling products other than games, and foresees a future when games and other software will be supported in whole or part by advertising, sponsorship, and product placement.[31] It also promises that "as a by-product of our message delivery system" its software will give publishers and developers "meaningful data about how the game is played" by "discretely retrieving information on a user by user basis" about frequency of play, use of download sites, and area codes of players.[32] At the time of writing, Conducent was in negotiation with games companies such as Gathering of Developers, Eidos, and eGames for adoption of its technology.

As we have seen, digital analysts such as Steve Johnson who are sympathetic to marketization hope this model can be extended to a situation in which networked game environments actually contain e-stores within their virtual worlds. In fact, the process is already under way and involves not just deals with commercial affiliates but also the for-profit vending of the basic components of game play. Characters and possessions in time-consuming multiplayer games such as *Ultima* and *Everquest* can be bought and sold for real money. The main site for these transactions is the online auction site eBay. Tellingly, this phenomenon recently warranted a column in *The Economist*. It reports that one hundred thousand virtual pieces of *Ultima* gold translates into about forty dollars' real cash; that a well-situated castle can change hands for over five hundred; and that *Everquest* players can buy a "Froglock Bonecaster's Robe" for more than seven hundred dollars. So lucrative is this business that some experienced players have reportedly turned mercenary-professional and "given up their jobs to play full-time, making a living by selling their spoils to other players." They even buy items for resale. Such entrepreneur-gamers are viewed with contempt, derisively referred to as "campers" or "farmers" because of their formulaic professional playing style and shunned and excoriated by true *aficionados* for commercializing the game. The *Economist* reporter concludes that, attractive as gaming-for-profit may sound, it requires missing out on the convivial problem-solving aspects of gaming and incurring the opprobrium of fellow players, while "the demands of running a business must also seem rather mundane after a day spent slaying dragons."[33]

If the mounting importance of marketing threatens to stifle the digital game industry's variety and creativity at the point of production, it also creates problems on the consumption side of the business. Many young players, even as they participate in gaming culture, are at least partially aware that they are the targets of the highly calculated marketing strategies of a large relentless corporate machine. The appeal of "virtual worlds" such as those offered by interactive games is the possibility of experiencing something different. Video games, even those that most often evoke social criticism, such as first-person shooters, involve utopian visions – of archaic or futuristic worlds outside capital, of being a renegade "bladerunner" opposing evil corporations and their technoscientific lackeys, of being part of heroic fellowship instead of just another targeted unit in a predictable demographic marketing sector, of being a master architect or a military commander-in-chief rather than just another disposable member of post-industrial labour power.

In *The Sims*, this utopian dream expresses itself in the positioning of the player as the director of a potentially perfectible consumer world, rather than as the manipulated target of massively powerful institutions – as controller, not controlled, inhabiting a "processed world" perhaps, but as the processor, not the processed. Despite the constraints of the preprogrammed nature of games, it is the experience of freedom – an autonomous, anarchic, even antagonistic alternative to the disciplinary power of the surrounding social system – that makes play so exciting. But the more obviously, instrumentally, and cynically the utopian desire is subordinated to the imperatives of branded synergistic capital, the more it exhausts the dissident energy on which it draws. The more the player knows that as they plug in and log on they are being played on by a vast technomarketing apparatus, the more disenchanting the virtual experience risks becoming. To date, game marketers have been successful in making this infinite risk regress. In the long term they may be sowing the dragon's teeth of exhaustion, indifference, and rejection. Such possibilities were brought home in the opening years of the new century as Sim Capital was wracked with, first, crisis and turmoil, and then terror.

SIM CRISIS

In 2000 the Internet economy crashed spectacularly. The game industry in general and the publisher of *The Sims* in particular were caught in the turbulence. In 2001 Electronic Arts reported its first quarterly loss in three years in a general contraction of digital investment that, as we discussed in chapter 8, dampened some of the enthusiasm building around investment in interactive play. We do not want to make too much of the great "dot.com" flare-out, which may be nothing more than the familiar shakeout in a conventional cycle of business consolidation: no one should underestimate Sim Capital's recuperative resilience.

Nonetheless, the bursting of the Internet bubble occurred against a background of growing scepticism about the scope and depth of the Net's popular appeal and economic potential. In 2000 a major multinational study sponsored by the Economic and Social Research Council in Britain involving seventy-six researchers from Britain, the US, Denmark, and Holland found that, contrary to portraits of constantly burgeoning Net use, there was widespread evidence of "drop off and saturation among many groups of users." In particular, many young people were turning away from the online world. "Teenagers' use of the Internet has declined," said Steve Woolgar, director of the research. "They were energized by what you can do on the Net but they have been through all that and then realized there is more to life in the real world and gone back to it." Giving reasons for the exodus, the study said that "some were bored, some were frustrated by the amount of advertising, and others would not pay for Internet access after leaving university."[34]

Beyond this, however, there was the chance that youth cynicism about Sim Capital might transform into criticism and dissent, a possibility highlighted by the resurgence of social movements opposing corporate power in the late 1990s and early 2000s. Ecological protesters, antisweatshop campaigns, and international solidarity movements targeted the labour, environmental, and human rights practices of companies such as McDonald's, Nike, the Gap, Shell, Starbucks, Disney, and Wal-Mart. During mass protests in Seattle and Washington in 1999 these issues converged in a much wider street critique of corporate globalization.[35] Viewing those events, authors such as Naomi

Klein and Kalle Lasn suggested that they represent a growing rejection of the branded consciousness of youth culture.

The interactive game industry was not a specific target of protest. But the scepticism of young activists towards corporate media put them on a collision course with Sony, Microsoft, Nintendo, and Sega. In 2000 an internal Sony memo, "NGO Strategy," laid out the corporation's strategy for countering eco-critics in the wake of protests in Seattle, Washington, and Prague. Sony was especially concerned about European campaigns to force the electronic industry to take responsibility for the environmental and health hazards of product disposal – campaigns that could, for example, result in legislation obliging it to recycle or reuse millions of obsolete PlayStations. Campaigners also aimed at phasing out toxic chemicals commonly used by the electronics industry. Sony's memo highlighted the threat to its profits posed by such organizations as Greenpeace, the Northern Alliance for Sustainability, Silicon Valley Toxics Coalition, and Friends of the Earth and advocated employing "Web investigation agencies," such as the London-based Infonic, that specialize in tracking social movements, countering and discrediting anticorporate activists and managing "spin control" to protect their clients.[36]

Given the tension between media corporations and social activists, nothing could more tellingly illustrate the avidity of the game industry's cool hunters than the release of the game State of Emergency. Published by that most hip of game developers, Rockstar, it offers an "urban riot" scenario based on the Seattle demonstrations against the World Trade Organization.[37] The player, put in the subject-position of an anarchist complete with balaclava, succeeds by causing as much chaos as possible, damaging property, attacking police officers, etc. The game was widely denounced, not only by conservative politicians but also by counterglobalization activists – understandably so, since to date the most serious casualties in such events have been demonstrators shot by police. It says a great deal about the game industry's frantic desire to renew its constantly exhausted stock of symbolic value and keep up with youth culture that it could generate simulation training for "black bloc" anticapitalists.

This is only one of many paradoxes that surround the relation of the new activism to Sim Capital. Ironically, one of the features of the anticorporate movements that most disturbs corporations such as Sony is their mastery of digital networks. It was members of a generation socialized in the use of digital technologies by interactive gaming who today hack corporate Web sites and use the Internet to organize their protests. If this trend persists, then alongside the commercial transformation of play into a sphere of commodification may appear another movement by which information capital generates the networked subjects who will challenge the limits of the world market. Playing *The Sims* may teach us that managing consumers is not always easily done: what it cannot encompass is the possibility that one day youthful Sims might put their hands through the screen and reach for the game controls for themselves.

SIM TERROR

Within a very short time, the threat posed by counterglobalization movements was dwarfed by a much starker danger. When Will Wright's first game, *SimCity*, was released in 1987, the distributor, Broderbund, fearing it might be perceived as too "educational," took special steps to make it more obviously entertaining by adding special "disaster" options – earthquakes, nuclear meltdowns, even an attack from Godzilla. On 11 September 2001, just such a disaster scenario hit Sim Capital as terrorist attacks destroyed the World Trade Center in New York and damaged the Pentagon, killing some three thousand people and triggering war and instability across the planet.

Several commentators saw the catastrophe as a horrific wake-up call to a social system cocooned from global realities by an envelope of simulatory consumerist entertainment and the remote transactions of electronic stock exchanges. Noting that the World Trade Center towers were a centre of "virtual capitalism," the cultural theorist Slavoj Zizek remarked that "we should recall the other defining catastrophe from the beginning of the twentieth century, that of the Titanic." In facing the "raw deal of catastrophe," he suggested, North Americans encountered "what goes on around the world on a daily basis." "The shattering impact ... can be accounted for only against the background of the borderline which today separates the digitalized First World from the Third World 'desert of the real.' It is the awareness that we live in an insulated artificial universe."[38] If this is so, then interactive gaming is an integral part of that artificial universe.

America had watched unphased as the smart bombs and televised coverage of the 1992 Persian Gulf war made the destruction of military and civilian infrastructures seem a "Nintendo war" devoid of real suffering. Few blinked when in 1998 NBC accompanied its broadcast reportage of real US air attacks on Sudan and Afghanistan with images of cruise missiles extracted from *Jane's Fleet Command*, a strategy simulation game.[39] Shortly after 11 September Naomi Klein wrote that "the era of video game war in which the US is at the controls has produced a blinding rage in many parts of the world, a rage at the persistent asymmetry of suffering ... A blinking message is on our collective video-game consoles: game over."[40]

In the interactive game business, as throughout the entertainment complex in general, there was some hand wringing and conscience searching about blithe representations of virtual violence. Activision announced that it would postpone the release of *Spider-Man 2* because it featured fight scenes on top of buildings that resembled the World Trade Center. UbiSoft pulled its latest Tom Clancy terrorism game. Electronic Arts was embarrassed about *Command and Conquer: Red Alert 2*, a military strategy game in which one player must obliterate the Pentagon before moving on to take down the World Trade Center. Despite having released the game a year before the terrorist attacks, it decided to replace the box, on which structures such as the Statue of Liberty and the Empire State Building appear, all damaged or destroyed: consumers who had already purchased the game could get a new box. Electronic Arts also delayed the release of *Majesty*, in which players could receive real-life telephone calls alerting them to in-game terrorist attacks. Perhaps more to the point, Microsoft removed from flight simulators the option of flying into tall buildings.

But it would be unwise to conclude that the disaster of 11 September meant a real interruption for the culture of virtual play, even, or especially, in its most violent variants. On the contrary, the "war against terror" announced by the George W. Bush regime recalled the industry to its origins – as a spin-off of the military-industrial complex. Indeed, it is hard to shake off a sense that the industry's long-standing fascination with "militarized masculinity" constituted a sort of protracted ideological preparation for this very moment, an informal in-house socialization of an entire North American generation so that it would support an armed-to-the-teeth fight to maintain global dominance against those who resent advanced capital's astounding wealth, technological dominance, and consumerist culture.

Almost immediately after 11 September, modified scenarios for the shooter game *Counter-Strike* featuring antiterrorist operations in Afghanistan were circulating on the Net, appearing concurrently with actual US special forces landings around Kandahar.

While the young male *aficionados* of virtual war practised informally, the military-entertainment complex went into high gear. At the Naval Postgraduate School in Monterey, California, programmers in the Modelling, Virtual Environments and Simulation Institute were working on a computer simulation of Osama bin Laden's Al Qaeda network, attempting to "prepar[e] to conjure deserts, communication networks and an army of terrorists cunning enough to design plots of mass destruction," as well as "millions of potential victims," using virtual scenarios and artificial intelligence agents.[41] Engineers working on the program reportedly suggested that "the same simple yet unpredictable interactions that make *The Sims* so lifelike have the potential to illuminate the unpredictable methods of terrorists." So the project is known as "Sim Osama."[42]

CONCLUSION: SIM DREAMS, SIM NIGHTMARES

We make no attempt to predict the outcome of the current situation in which gamelike technologies are harnessed to the task of managing a sliding world market and looming prospects of global war. But it is remarkable to consider how quickly such convulsions can make even the recent promises of techno-utopians look jaded and tarnished. In 2000, for example, Jason Lanier, inventor of the term "virtual reality" and a thoughtful commentator on the wired world, suggested that gaming is an important key to eliminating the digital divide. Looking towards the emerging generation of Internet-connected video game terminals, he wrote: "One of the most promising trends in technology is the opening up of video game machine architectures. If kids can connect to the Net with cheap machines and create their own content and services, they will grow into a new generation of empowered, productive, technically skilled citizens, even if they have to endure crummy schools and bad neighbourhoods."[43] To realize this dream would require that gaming companies provide systems that give users the capacity to act not only as digital consumers but also as digital producers: "If you work with a company in the video game business," Lanier pleads to his business audience, "please consider advocating open architectures. Include a modem. Create wonderful authoring tools so that kids can build their own content additions on top of your titles. Allow open Web access."[44]

Such visions seem remarkably antiquated even a few months later, as media corporations consolidate and cut their Net ventures to weather the approach of economic crisis and global war. More seriously, they do not take into account the long-term contradictions that characterize the circuits of Sim Capital. Thus, amazing technology is often a vehicle for banal or violent content; the prospects for an "open architecture" are contradicted by intellectual property rights; and the military-entertainment complex promises to be the driving force behind interactive innovation.

Our analysis is much closer in spirit to that of Heather Menzies, who argues against abstract techno-utopian discourses that celebrate the hypothetical possibilities of virtuality while ignoring the concrete forces shaping its realization. Drawing on the work of Harold Innis, she suggests that to understand the "systemic biases" of any media we have to look not just at its technical capacities but also at the content it conveys and the communities it constitutes.[45] Citing James Carey's discussion of Innis, she argues for recognition that changes in communications systems affect social change in a multidimensional way, by "altering the structures of interest (the things thought about), by changing the character of symbols (the things thought with) and by changing the nature of community (the area in which thought developed)."[46]

Once virtual media are seen from this holistic perspective rather just as a set of pristine technological potentialities, their deep implication and imbrication in an era of capitalist restructuring becomes immediately apparent. The speed and spatial scope of digital networks, their heavy focus on advertising, consumer entertainment, and business-to-business transaction, and their deployment in a global market converge to create networks in which "everything becomes an information-management business, a continuous loop of marketing/market feedback communication within which a lot of what had been considered culture is subsumed."[47] Although Menzies does not foreclose on the possibility of extracting alternative digital logics from this situation, the thrust of her argument is to suggest that the convergence of virtual technologies, privatized consumer culture, and market logic is powerful enough to create a "bias" in new media that will be extremely difficult to disturb or deviate from.

In many ways, we agree with this analysis. In the operations of the interactive game industry, cultural, marketing, and technological circuits are coordinated to reinforce each other in the creation of the "ideal commodity" of the post-Fordist, postmodern world market – Sim Capital. Culturally, games are repeatedly constructed to allot their players subject-positions of acquisitive householders, market speculators, or planet-dominating warriors – ideological roles that are indispensable to the ascendant centres of the world market system. Technologically, computer and console games feed off and into a dynamic of innovation and expansion of digital systems on which the dominance of these centres depends and in which they are massively invested. Both the culture and technology of interactive games are propelled and organized by a dynamic of expanding media empires in which the principal position is that of the technocultural "consumer" relentlessly solicited and identified by high-intensity marketing techniques. Rhapsodies about the empowering nature of interactive play that ignore these systemic vectors are mystifying and deceptive. They make it impossible for young people, in particular, to understand the forces that are in fact playing upon them, incessantly beckoning to and shaping their lives in accordance with a relentless logic of market demographics. By obscuring or glamourizing these processes, or denying their contradictions, digital idealists actually contribute to confusion and cynicism. It is against such mystification that our analysis is mainly directed.

But if our model of the game industry's circuits aims to show the commercial, technological, and cultural forces constraining possibility, it equally points to openings for alternatives. Here we break from the bleakly determinist tendency of Menzies's analysis and emphasize, with Williams, that no hegemonic system ever completely exhausts creativity and alternatives. But it is only by understanding the systemic determinants of social lives in the age of global capitalist technoculture that the real possibilities for freedom, imagination, and transformation can be assessed.

Paradoxically, games such as *The Sims* may offer us some insight. In the era of fascism Walter Benjamin scanned the film and radio technologies of his age for emancipatory possibilities. In his seminal essay "The Work of Culture in the Age of Cybernetic Systems," Bill Nichols follows this logic into the digital era. Recognizing the formidable subordination of digitalization to commodification, he argues that even in such a powerfully hegemonic setting we should be alert for contesting capabilities.[48] In a passage that directly addresses the issue of games like *Sim City*, *Civilization*, or *Age of Empires*, he suggests that "what falls open to apperception" in such play "is ... the relativism of social order ... the set of systemic principles governing order itself, its dependence on messages-in-circuit, regulated at higher levels to conform to predefined constraints." Such simulations "refute a heritage that celebrates individual free will and subjectivity."[49] Nichols remarks that:

if there is a liberating potential in this, it is clearly not in seeing ourselves as cogs in a machine or elements in a vast simulation, but rather in seeing ourselves as part of a larger whole that is self-regulating and capable of long term survival. At present this larger whole remains dominated by parts that achieve hegemony. But the very apperception of the cybernetic connection, where system governs parts, where the social collectivity of mind governs the autonomous ego of individualism, may also provide the adaptive concepts needed to de-centre control and overturn hierarchy.[50]

The game industry is a powerful and propelling subset of information capitalism – what we have called Sim Capital. In this system, cultural entrepreneurs, technological managers, and corporate marketers attempt to "play" people to maximize profits. Just as one needs to look behind the screen to understand the digital coding that produces the play of video game consoles and computers, so too we need to look behind the scenes to understand the social processes – marketing, technological, cultural – that constitute the game subject. Such understanding is a prerequisite if we are to attain real autonomy and social choice, to truly play rather than be played upon. So in the coda to this book we conclude our examination of Sim Capital with a summary overview of the interactive game industry's paradoxical limits and possibilities.

Coda: Paradox Regained

Digital games are interactive media *par excellence* because their entertainment value arises from the loop between the player and the game,
as the human attempts by the movement of the joystick or keyboard
or mouse to outperform the program against and within which he or
she, with or without networked coplayers, competes. This interactive
feedback cycle is often represented as a dramatic emancipatory
improvement over traditional one-way media and passive audiences –
a step up in cultural creativity, technological empowerment, and consumer sovereignty. In the view of the digerati and silicon futurists,
video and computer games herald a brave new world that has broken
completely with the constraints and compulsions of the mass media.

Our theoretical model of the mediatized marketplace challenges this
simplistic construction in favour of an approach that recognizes some
of the contradictions at play in the interactive game. We offer a critical
media analysis of the interactive game that challenges the celebratory
version of technological determinism – which loses the nuance and
complexity of media theorists such as Harold Innis and Marshall
McLuhan – that claims the commanding heights of popular and public
discourse. In our view, the moment of game play cannot be abstracted
from its historical and social context but has to be seen within the
overarching and constraining cycles of post-Fordist information capitalism, with its three mutually constituting moments of technology,
culture, and marketing. In insisting on the interplay of these three
circuits in the formation of interactive gaming as a social practice, we
have attempted to forge our critique of the deterministic technological
optimism promoted by the global media industries.

We do not deny that interactive media enable the user a degree of
control over the flow of information, the consumer a choice of a new
form of entertainment, and the player a new structure of playful social

interaction. But we insist these interactive potentialities are historically constrained and structured by the processes of game design, technological innovation, and product marketing. The production of play occurs within the three reverberating circuits – technological, cultural, and marketing – in which the game industry sets out to manage the flow of play to gamers. Here, our model highlights three subjectivities through which the gaming audience has been conceived and articulated: in the cultural circuit, inviting a "player" into the flow of meanings in a game narrative; in the technological circuit, researching the expectations of console "users" with regard to the graphics and processing power of the machine; and in the marketing circuit, researching and addressing gamers as a "consumer segment" within a mediated cultural market. It is because the interactive game so clearly crystallizes within itself the convergent logic of these processes that it deserves to be considered an "ideal" – in the sense of exemplary – product of post-Fordist, postmodern promotional capitalism.

Our historical study therefore locates the construction of the interactive gaming subject within the emerging oligopolies of knowledge of the games industry, amidst the corporate webs of Atari, Nintendo, Sega, Sony, Microsoft, and a multitude of other commercial game enterprises, supported by complex and continuous exchanges with a massive military-industrial-entertainment complex. It reveals that gamers are hailed as potential users of digital technologies, as primary customers for entertainment products in a global market, and as pioneering players enmeshed in the warp and woof of a burgeoning market-driven virtual culture. In this sense, the game player has been constantly prefigured by the "x on the wall," that is, the commercial potentiality on which game designers and marketers always have their eye. This "x on the wall," we have seen throughout our historical account, gets rearticulated at each stage of the digital product development, marketing, and sales cycles that characterize the negotiations between digital designers and their potential consumers.

The loyal and habitual gamer therefore describes a composite subjectivity (user, player, customer) whom the game makers imagine throughout the cycle of creative construction of gaming systems and their audiences. The practices of game design are structured by a layered series of market-driven negotiations that take place between game producers and consumers around the world; between software and system designers and the users who are drawn to technology for play; and between marketers and the communities of youthful players entering cyberspace in search of immersive mediated experiences.

What unfolds in the managed dialogue of commercialized digital design is a process in which commodity form and consumer subjectivity circle around each other in a mating dance of mutual provocation and enticement. The digital industry's mediated relationship with consumers is linked to an expanding potential to communicate those very needs and desires to which it replies. Detailed monitoring of changing tastes and market patterns allows potential emerging hits to be identified quickly or even pre-emptively designed and then pushed very hard through high-intensity advertising campaigns and integrated into synergistic marketing webs. These promotional efforts in turn shape gamers' future tastes and expectations, cutting thematic grooves, such as that of militarized masculinity, into the entire terrain of digital play. Seen in this light, the portrait of high-technology companies as so-called cozy "clubs" of "coevolving customers" is revealed as a hopeless idealism, for it conveniently overlooks the considerable power of giant corporations, perhaps not to determine customer preferences absolutely but certainly to cull, tilt, skew, and incrementally construct the direction of "coevolving" tastes.

Guiding the direction of coevolving tastes, game companies have expanded the scope and range of their communication with targeted audiences – but only where such negotiation promotes those very needs and desires to which it can profitably respond. As companies become more sensitive to consumers' tastes, they must render their knowledge useful for targeting and branding decisions. In this construction of gaming audiences, digital designers have accumulated cultural knowledge – or more precisely, engaged in interpretation, analysis, judgment, strategy, and creativity in relation to gaming experiences – because understanding gaming audiences is critical to the smooth operation of the interactive entertainment industry. Yet this communication effort is directed less at informing and empowering than at anticipating and capitalizing on gamers' future tastes and expectations. It therefore seems fantastical to describe the gamer as an entirely free agent and an equal partner in the process of the game's construction. Although viewed as subjects by the game makers, the positions that have been constructed for players in contemporary mediated markets do not imply self-awareness, consumer sovereignty, or cultural democratization.

We want to reiterate that in our proposed three-circuits model we intend that "circuits" be understood in a dialectical Innisian sense that allows for creativity and change within cultural systems. Information capitalism attempts to subsume the cultural, technological, and marketing circuits within an overriding logic of profit accumulation, balancing and synchronizing the multiple circuit flows in ways that reinforce its overarching logic. But as we have seen, circuits can blow out and spin out, go too fast or too slow: commodities fail to find buyers, texts may not connect with audiences; potential users ignore technologies;

marketing strategies provoke consumer resistance. Circuits can interrupt or contradict each other: new technologies may be used, new cultural practices may emerge, and new texts may be read in ways that subvert, rather than support, commodity exchange. Because the subject is situated at the interplay and overlap of different social circuits that move according to a different logic, there is always the possibility of the unpredictable spark and the uncontainable power surge.

The experience of interactive play arises from the interaction amongst these different circuits, at the convergence of which is situated the gamer at his or her console or computer. The various simultaneous subject-positions of the gamer – as player, user, and consumer – may corroborate or contradict each other. In the course of this book, we have argued that player involvement in the storylines of "militarized masculinity," the user's technology-based experiences of immersive and accelerated virtual environments, and consumer identification with synergistic corporate brands all combine to give interactive gaming a powerful bias – one that arises from and reproduces the cultural, economic, and technological structures of globally dominant, heavily militarized, digitally networked transnational information capitalism, which we have dubbed Sim Capital.

Yet at the same time, this bias is subject to various disturbances and subversions. It is challenged – culturally, by experimentation with different types of game worlds, technologically, through hacking and piracy, economically, by the multiple forms of disaffection and dissent from commodified play that are now emerging even within game culture. Facing these paradoxes, digital empires must confront a multitude of contending voices and interests – from brand-loyal gamers, to dissident hackers, to concerned parents, to other media industries and beyond – who will also want to shape the direction our digital culture will take. The point of our model of the game industry's circuits is precisely that their control loops can be broken or come into contradiction with one another. Although the feedback cycles and complex interplays of interactive game practice most often work to cut established routes more deeply, they also contain contrary possibilities. There are subordinate but real tendencies that might work on the game industry to expand its thematic diversity from its historical preoccupations with "militarized masculinity," tendencies that can push it away from closed and proprietorial technologies and towards more accessible and open lines of development; and that may disenchant players with the logic of programmatically commodified entertainment. The realization of such options may create more diversity within digital play and more alternatives to digital play and thus lead to more free play all around.

On this point, we again draw inspiration from Raymond Williams, who in his last major work, *Towards 2000*, published in 1983, addressed the topic of "new media." Williams maintained his sceptical perspective towards the trajectory of a marketized media system but also rejected "an unholy combination of technological determinism with cultural pessimism" that looked to the future only with gloom.[1] Rather, he urged critical media analysts and public intellectuals of all sorts to put their energies at the service of a "politics of hope," saying that "once the inevitabilities are challenged, we begin gathering our resources for a journey of hope."[2] The fate of the digital world market, as we see it, is profoundly problematic and very uncertain. The positive side of this prediction is that the future of interactive gaming culture and the fate of Sim Capital itself remain open.

Notes

CHAPTER ONE

1 Emmanuel Mesthene, cited in Leiss, *Under Technology's Thumb*, 25.
2 Robert Forbes cited in ibid.
3 Negroponte, *Being Digital*, 231.
4 Ibid., 229.
5 Ibid., 6.
6 Bell, *The Coming of Post-Industrial Society*, 27–30.
7 Toffler, *The Third Wave*.
8 Ibid., 154.
9 Ibid., 158.
10 Ibid., 349.
11 Leiss, 27–9.
12 Ibid., 33.
13 Ibid., 27.
14 Robins, "Cyberspace and the World We Live In," 136.
15 Ibid., 135.
16 Leiss, *Under Technology's Thumb*, 34.
17 Adam Smith, *An Inquiry into the Nature and Causes of the Wealth of Nations*.
18 For a scathing critique of the "new economy," see Frank, *One Market under God*.
19 Gates, *The Road Ahead*, 171.
20 Ibid.
21 Cairncross, *The Death of Distance*, 1.
22 Ibid., 119.
23 Ibid., 118.
24 Browning and Reiss, "Encyclopedia of the New Economy."

25 For a scathing critique of *Wired*'s position, see Barbrook and Cameron, "The Californian Ideology."
26 GTE Entertainment, "Titanic Adventure out of Time."
27 "Video games" are played on dedicated consoles, usually attached to a television set, or on hand-held game sets. "Computer games" – bought on floppy disk or CD or delivered direct through Internet link – are played on personal computers that can also be used for other purposes, such as word processing. The terms "interactive games" and "digital games" embrace both video games and computer games. We try to keep these categories distinct, but industry-generated statistics and journalistic reports often do not.
28 Hayes and Dinsey, with Parker, *Games War*, 5.
29 Takahashi, "Games Get Serious," 66. These figures are cited as the industry's best recent performance. In the wake of the Internet crash, sales for 2000 flagged or at least flattened. Calculations of revenues from digital play vary enormously, depending on who is counting and what they count. According to Informa Media Group, worldwide revenues amounted to $50 billion in 2001 and were predicted to reach $86 billion by 2006. These figures included hardware for consoles and hand-held games, software for both and for games played on personal computers, arcades, online game revenues, game rentals, wireless games, and interactive TV (cited in Takahashi, "The Game of War," 60). The research firm IDC puts US electronic gaming revenues at around $15 billion, about half of it from video game software, another twenty-five percent from console hardware, most of the rest from PC game software, and less than five percent going to online gaming (cited in Mowrey, "Let the Games Begin," 30).
30 IDSA, *State of the Industry: Report 2000–2001*.
31 See Kline, with Banerjee, "Video Game Culture."
32 IDSA, *State of the Industry: Report 2000–2001*. For discussion of these statistics see chapter 11.
33 IDSA, *State of the Industry: Report 2000–2001*.
34 Stone, *The War of Desire*, 27.
35 Lewis, *The Friction Free Economy*, 2.
36 Negroponte, *Being Digital*, 62–3.
37 Garnham, "Constraints on Multimedia Convergence," 115.
38 Toffler, *The Third Wave*, 163.
39 Ibid.
40 See Toffler, *The Third Wave*, 265–88.
41 Negroponte, *Being Digital*, 231.
42 Ibid., 204.
43 Ibid.
44 See Ohmae, *The Borderless World*.

45 Ohmae, "Letter from Japan," 161–2.

46 Ibid., 162

47 Ibid., 161.

48 Ibid.

49 Rushkoff, *Playing the Future*, 181.

50 Ibid., 180.

51 Ibid., 180–1.

52 Rushkoff, *Media Virus!*, 31.

53 Rushkoff, *Playing the Future*, 181.

54 Ibid., 269.

55 See Davis, Hirschl, and Stack, eds., *Cutting Edge*; Menzies, *Whose Brave New World?*; McChesney, *Rich Media, Poor Democracy*; Robins and Webster, *Times of Technoculture*.

56 See Brook and Boal, eds. *Resisting the Virtual Life*; Noble, *The Religion of Technology*; Sale, *Rebels against the Future*.

57 "A few years ago, Douglas Rushkoff earned road trips of media mileage with the revelation that he had undergone a techno-conversion: Once an Internet-worshipping cyber-evangelist, a willing consultant to translate Net culture to Corporate America, Rushkoff became convinced that the digital revolution had been taken over by corporate interests, who used it to coerce us into gratuitous purchases. The product of this conversion was a book, *Coercion: Why We Listen To What They Say.*" Shulgan, "Open Source Everything."

58 See Kline, "Pleasures of the Screen."

59 Anonymous, personal interview.

60 Ibid.

61 Takahashi, "The Game of War."

62 Robins, "Cyberspace and the World We Live In," 137.

63 The best popular account of the industry is Herz, *Joystick Nation*.

PART ONE

1 Lee, *Consumer Culture Reborn*.

CHAPTER TWO

1 Innis, *The Bias of Communication*, 33–60.

2 Ibid., xxvii.

3 Ibid., 3–32.

4 Ibid., 31.

5 Innis, *Empire and Communications*, 117–69.

6 McLuhan, Foreword, v.

7 McLuhan, *The Mechanical Bride*, v.

8 McLuhan, *The Gutenberg Galaxy*, 1.

9 Ibid., 5.

10 McLuhan, *Understanding Media*, 41–7.

11 Ibid., 8.

12 Ibid., 24.

13 Kroker, *Technology and the Canadian Mind*; Menzies, "Digital Networks." Quotes are from McLuhan, *Understanding Media*, 3–4, 46.

14 McLuhan, *Understanding Media*, 234–45.

15 Ibid., 242.

16 Ibid., 241.

17 Ibid., 238.

18 Ibid., 235.

19 For an overview of the political economy approach, see Mosco, *The Political Economy of Communication*.

20 Adorno and Horkheimer, *Dialectic of Enlightenment*.

21 Marcuse, *One-Dimensional Man*.

22 Herbert Schiller, *Communication and Cultural Domination*.

23 Marx, Preface to *A Contribution to the Critique of Political Economy*.

24 Marx's most sustained exposition of the "circuit" comes in *Capital*, Vol. 2.

25 Garnham, *Capitalism and Communication*, 45.

26 Ibid., 47.

27 This is what the political economist Dallas Smythe famously termed the "audience commodity." See Smythe, *Dependency Road*.

28 This phrase comes from Leiss, *The Limits to Satisfaction*.

29 Leiss, Kline, and Jhally, *Social Communication in Advertising*, 123.

30 Herman and Chomsky, *Manufacturing Consent*.

31 Schiller, *Digital Capitalism*; Herman and McChesney, *The Global Media*.

32 McChesney, *Rich Media, Poor Democracy*, 3.

33 Grossberg, *Bringing It All Back Home*, 242.

34 Hall, "Encoding/Decoding."

35 Among the better-known texts in the reception studies work in cultural studies, see Morley, *The Nationwide Audience*, and Ang, *Watching Dallas*.

36 Poole, *Trigger Happy*.

37 See the collection edited by Cassell and Jenkins, *From Barbie to Mortal Kombat*; McNamee, "Youth, Gender and Video Games"; Alloway and Gilbert, "Video Game Culture."

38 See Bleeker, "Urban Crisis: Past, Present and Virtual"; Stephenson, "The Microserfs Are Revolting."

39 Ferguson and Golding, "Cultural Studies and Changing Times," xxi.

40 McGuigan, *Cultural Populism*.

41 According to Mosco, cultural studies often seems to suggest that "media diversity is not a substantial problem because" the texts, themselves

open to multiple interpretation, "create their own diversity, whatever the number of formal producers and distributors." Mosco, *The Political Economy of Communication*, 259.

42 Slater and Tonkiss, *Market Society*, 168.

43 Lawrence Grossberg, cited in Murdock, "Across the Great Divide," 91.

44 Ferguson and Golding, "Cultural Studies and Changing Times."

45 Mosco, *The Political Economy of Communication*, 262.

46 See "Colloquy"; Clarke, "Dupes and Guerillas"; Ferguson and Golding, eds., *Cultural Studies in Question*; McGuigan, *Cultural Populism*; Morley, "So-Called Cultural Studies."

47 Williams, *Television, The Long Revolution*, and *Communications*.

48 Williams, *Television*, 124, 122.

49 Ibid., 120.

50 Spigel, Introduction to Williams, *Television*, xv–xvi.

51 Williams, *Television*, 122.

52 Williams, *Marxism and Literature*, 136.

53 Williams's classic engagement with "determinism" is in the essay "Base and Superstructure in Marxist Cultural Theory."

54 Williams, *Television*, 7.

55 Williams, *Towards 2000*, 146.

56 Williams, *Television*, 8.

57 Ibid., 124.

58 Ibid., 129.

59 Ibid., 14.

60 Ibid., 80, 85.

61 Ibid., 18

62 Ibid., 17.

63 Ibid., 115.

64 Ibid., 13–25.

65 Ibid., 114.

66 Spigel, Introduction, xii–xiii.

67 Williams, *Television*, 4.

68 See Johnson, "What Is Cultural Studies Anyway?" See also the "Culture, Media and Identities" series published by Sage in association with the Open University. Each book addresses a "moment" in the circuit of culture: representation, identity, production, consumption, and regulation.

69 du Gay, "Introduction," 3.

70 Cockburn, "The Circuit of Technology," 32–47.

71 Althusser, *Lenin and Philosophy*, 170.

72 Cockburn, "The Circuit of Technology."

73 Latour, *Science in Action*.

74 McRobbie, *In the Culture Society*, 71.

75 Ibid., 23, 24.

76 The notion of "mutual constitution" is developed by Mosco in *The Political Economy of Communication*. This conception, Mosco notes, "broadens the knowledge process from simple determination to multiple, dynamic interactions" (137). "Interactions among elements that are themselves in the process of formation and definition, the term *constitution* foregrounds the *process of becoming* within all elements of the social field. No thing is fully formed or clearly defined, but one can specify processes at work within and between them that define the nature of the constitutive process and the relationships among the elements" (138).

CHAPTER THREE

1 Williams, *Television*, 122.
2 Harvey, *The Condition of Postmodernity*.
3 The founding text of Regulation School analysis is Aglietta, *A Theory of Capitalist Regulation*. The most accessible introduction is Lipietz, *Mirages and Miracles*.
4 Lipietz, "Reflections on a Tale," 32–3.
5 This concept derives from Antonio Gramsci's far-ranging if fragmentary essay of the 1930s, "Americanism and Fordism." Gramsci proposed that the introduction into Europe of American assembly-line methods, typically associated with the factories of Henry Ford, brought with it profound changes in the socialization of the workforce, patterns of urbanization, the role of the state, artistic and intellectual movements, and the cultural definition of gender roles. See Gramsci, *Selections from the Prison Notebooks*.
6 A number of social historians have argued that Fordist mass production was predicated on this new set of practices in demand management of mass consumption, which makes "marketing communication" key to the analysis of capitalism's expansion in the twentieth century. See Ewen, *The Captains of Consciousness*; Leiss, Kline, and Jhally, *Social Communication in Advertising*; Marchand, *Advertising the American Dream*.
7 Our account largely follows that of Lee, *Consumer Culture Reborn*, 101–19.
8 Aglietta, *Capitalist Regulation*, 123–4.
9 The "industrial paradigm" is introduced as a distinct element late in the Regulation School's work. See Lipietz and Leborgne, "New Technologies, New Modes of Regulation," and also Perez, "Structural Change and the Assimilation of New Technologies."
10 See Bell, *The Coming of Post-Industrial Society*; Toffler, *The Third Wave*.
11 See Levidow, "Foreclosing the Future"; Parker and Slaughter, "Management by Stress." For a powerful critique of the nebulousness of the

"flexibility" concept see Pollert, "Dismantling Flexibility"; Sivanandan, "All That Melts into Air Is Solid"; Rustin, "The Trouble with 'New Times'"; Clarke, *New Times and Old Enemies*.

12 For example, Pelaez and Holloway, in "Learning to Bow," dismiss most Fordist/post-Fordist analysis as resting on a blunt technological determinism whereby it is the sheer force of new technologies that produces the new era. Graham, "Fordism/Post-Fordism, Marxism-Post-Marxism," suggests that discussions of post-Fordism, by implicitly accepting the success of capital's restructuring, accept the "vitality and uncontested hegemony" of capital's reproduction but "[obscure] the weaknesses and instabilities in that process; it hides the failures and unevenness that make non capitalist alternatives an existing and future option" (49). As Barbrook argues in "Mistranslations," spin-offs from the Regulation School's work such as the New Times analysis in some ways distort their source.

13 This includes, pre-eminently, Harvey's magisterial examination of "flexible accumulation" and its cultural implications in *The Condition of Postmodernity*, Lee's discussion of post-Fordist consumption patterns in *Consumer Culture Reborn*, Tony Smith's analyses of the consequences of lean production for workers and consumers in *Technology and Capital in the Age of Lean Production*, and Robins and Webster's *Times of Technoculture* on the significance of technological change in domestic and educational spaces.

14 Lee, *Consumer Culture Reborn*, 112.

15 Lee reminds us that the consolidation of Fordism spanned the convulsions of the Great Depression and the cataclysm of the Second World War; the appearance of its successor, if it is accomplished at all, may be equally protracted.

16 Dohse, Jurgens, and Malsch, "From 'Fordism' to 'Toyotism'?"; Wark, "From Fordism to Sonyism"; Tremblay, "The Information Society."

17 Morris-Suzuki, *Beyond Computopia*, and also "Robots and Capitalism."

18 Morris-Suzuki, *Beyond Computopia*, 76.

19 Ibid.

20 Ibid.

21 Ibid.

22 Kundnani, "Where Do You Want to Go Today?" 57.

23 Morris-Suzuki, *Beyond Computopia*, 88, 77.

24 Ibid., 76.

25 Morris-Suzuki, "Robots and Capitalism."

26 Kundnani, "Where Do You Want to Go Today?" 58.

27 Ibid.

28 Lee, *Consumer Culture Reborn*, 133. See also Harvey, *The Condition of Postmodernity*, and Hall, "The Meaning of New Times."

29 Lee, *Consumer Culture Reborn*, 133.

30 Ibid., 135.

31 Ibid.

32 Ibid., 133–5. See also Robins and Webster, "Cybernetic Capitalism"; and Garnham, *Capitalism and Communication*.

33 Winston, *Media, Technology and Society*, 232.

34 See for example the works by Harvey, Hall, Lee, and du Gay cited in the bibliography, as well as Heffernan, *Capital, Class and Technology*.

35 Harvey, *The Condition of Postmodernity*, 171.

36 Lyotard, *The Postmodern Condition*.

37 Baudrillard, *Simulations*.

38 In *Postmodernism and the Other*, 10, Sardar suggests that video gaming provides an ideal metaphor for the symbolic landscape:

> Postmodernism posits the world as a video game: seduced by the allure of the spectacle, we have all become characters in the global video game, zapping our way from here to there, fighting wars in cyberspace, making love to digitized bits of information. All social life is now being regulated not by reality but by simulations, models, pure images and representations. These in turn create new simulations, and the whole process continues in a relentless stream in which the behaviour of individuals and societies bears no relationship to any reality: everything and everyone is drowned in pure simulation.

39 Jameson, "Postmodernism," 87.

40 Video game advertisements quoted in David Brown, *Cybertrends*, 25–6.

41 Baudrillard, *The Consumer Society*, 77–8, emphasis in the original.

42 Ibid., 125.

43 Ibid.

44 Wernick, "Sign and Commodity," 157.

45 Some of the classic studies of advertising include Ewen, *Captains of Consciousness*; Leiss, Kline, and Jhally, *Social Communication in Advertising*; Marchand, *Advertising the American Dream*; Ohmann, *Selling Culture*.

46 Wernick, *Promotional Culture*, 185.

47 Bell, *The Cultural Contradictions of Capitalism*, and *The Coming of Post-Industrial Society*.

48 On cultural intermediaries see Bourdieu, *Distinction*; du Gay, ed., *Production of Culture/Cultures of Production*.

49 Lee, *Consumer Culture Reborn*, 115.

50 Ibid.

51 See Nixon, "Circulating Culture."

52 Lee, *Consumer Culture Reborn*, 125.

53 Ibid., 120, emphasis in the original.

54 Ibid., 119.
55 Ibid., 129.
56 Ibid., 130.
57 Ibid., 130–1.
58 Ibid., 119.
59 Ibid., 128.
60 Ibid.
61 Garnham, "Constraints on Multimedia Convergence," 115.
62 Herz, *Joystick Nation*, 2–3.

PART TWO

1 Williams, *Television*, 8.
2 Williams, "Base and Superstructure in Marxist Cultural Theory," 48.
3 McRobbie, *In the Culture Society*, 29.
4 Williams, *Television*, 8.

CHAPTER FOUR

1 Pearce, "Beyond Shoot Your Friends," 220.
2 Title to the "first video game" is disputed, the main contenders being *Spacewar* and a primitive tennis game – the ancestor of *Pong* – invented in 1958 by William Higinbotham, a Manhattan Project atomic engineer working at a US nuclear research facility, which we discuss later. See Herman, *Phoenix*, 1–9; see also Poole, *Trigger Happy*, 29–34. As Pearce notes ("Beyond Shoot Your Friends," 220), *Spacewar* is perhaps more accurately described as the "first multiplayer, realtime computer game." Pearce suggests that "the first record we have of computer gaming was the attempt by MIT's John McCarthy, inventor of the term 'artificial intelligence,' to write a program that would enable the IBM 704 computer (known as the Hulking Giant) to play chess, roundabout 1959" (219). Among other early games played on this IBM machine, she claims, were *Bouncing Ball* and *Mouse in the Maze*.
3 Schumpeter, *Capitalism, Socialism, and Democracy*, 81–6.
4 Abernathy and Utterback, "Patterns of Industrial Innovation."
5 Barbrook and Cameron, "The Californian Ideology."
6 See Carter, *The Final Frontier*.
7 For a discussion of these institutional arrangements, see Haddon, "The Development of Interactive Games," and "Electronic and Computer Games."
8 See Castells, *The Network Society*, vol. 1; De Landa, *War in the Age of Intelligent Machines*; Edwards, *Computers and the Politics of Discourse in Cold War America*.

9 Levy, *Hackers*, 41; the quote is Steve Russell, cited in Herz, *Joystick Nation*, 6.

10 See "The Online Hacker Jargon File."

11 Haddon, "The Development of Interactive Games," 307.

12 Levy, *Hackers*.

13 Ibid., 27.

14 Ibid., 29, 33.

15 Stone, *The War of Desire*, 13–15.

16 Ibid., 14.

17 Henderson and Clark, "Architectural Innovation," 12.

18 See Brand, "Fanatic Life and Symbolic Death."

19 See Van de Ven, *Central Problems in the Management of Innovation*.

20 Hefferman, *Capital, Class and Technology*.

21 Haddon, "Electronic and Computer Games," 56.

22 See Stone, *The War of Desire*, 66–9; see also Herz, *Joystick Nation*, 11.

23 Herz, *Joystick Nation*, 11.

24 Stone, *The War of Desire*, 67.

25 Haddon, "Electronic and Computer Games," 53.

26 An interesting history of the video game is also provided in Burnham, *Supercade*.

27 Cohen, *Zap! The Rise and Fall of Atari*, 17.

28 Dolan, "Behind the Screens."

29 Sheff, *Game Over*.

30 Burnham, *Supercade*, 336.

31 Leonard Herman et al., "The History of Video Games."

32 Burnham, *Supercade*, 52–6; Cohen, *Zap! The Rise and Fall of Atari*, 18–9.

33 Burnham, *Supercade*, 52.

34 Ibid., 55.

35 Haddon, "The Development of Interactive Games," 311.

36 Friedrich, "Machine of the Year."

37 For an interesting account of US computer gaming culture in the 1980s, see Bennahum, *Extra Life: Coming of Age in Cyberspace*.

38 Cohen, *Zap! The Rise and Fall of Atari*, 36.

39 Haddon, "The Development of Interactive Games," 309.

40 Leonard Herman et al., "The History of Video Games."

41 Haddon, "The Development of Interactive Games," 68.

42 Cohen, *Zap! The Rise and Fall of Atari*, 78.

43 Haddon, "The Development of Interactive Games," 310–11.

44 Burnham, *Supercade*, 216.

45 Ibid., 234.

46 Ibid., 182.

47 Cited in Dolan, "Behind the Screens."

48 Stone, *The War of Desire*, 127, emphasis in the original.

49 Ibid.

50 Burnham, *Supercade*, 210.

51 Sawyer, Dunne, and Berg, *Game Developer's Marketplace*, 35.

52 Leonard Herman et al., "History of Video Games."

53 Crawford, *The Art of Computer Game Design*.

54 David Myers, "Computer Game Genres," 286.

55 Crawford, *The Art of Computer Game Design*.

56 Ibid.

57 Ibid.

58 Ibid.

59 Ibid.

60 Ibid.

61 Ibid.

62 Ibid.

63 Ibid.

64 Ibid.

65 Ibid.

66 Ibid.

67 Burnham, *Supercade*, 216.

68 Katz, "Networked Synthetic Environments," 119.

69 Ibid.

70 For a discussion of the linkages between Atari and military research, see Stone, *The War of Desire*, 129.

71 Moody, *The Visionary Position*, 235–7. Warren Katz is also a founder of MÄK Technologies, which develops games for both military and commercial markets. We discuss MÄK in chapter 8.

72 Weisman, "The Stories We Played," 463.

73 Cohen, *Zap! The Rise and Fall of Atari*, 59–60.

74 Ibid., 57.

75 Ibid., 61.

76 See Kline, *Out of the Garden*.

77 Cited in Stern and Schoenhaus, *Toyland*, 95.

78 Cited in ibid.

79 Siegel, "What Corporate America Must Learn," 86.

80 Ibid.

81 Sheff, *Game Over*, 149.

82 Burnham, *Supercade*, 277.

83 Cohen, *Zap! The Rise and Fall of Atari*, 79.

84 Stone, *The War of Desire*, 131.

85 Bushnell, cited in Dolan, "Behind the Screens."

86 See Stern and Schoenhaus, *Toyland*, 92–109.

87 Harvey, *The Condition of Postmodernity*.

CHAPTER FIVE

1 See Provenzo, *Video Kids: Making Sense of Nintendo*, and Kinder, *Playing with Power in Movies, Television, and Video Games: From Muppet Babies to Teenage Mutant Ninja Turtles*. The titles attest to how closely Nintendo was identified with the entire video game phenomenon.

2 Sheff, *Game Over*.

3 Sheff, *Game Over*; Katayama, *Japanese Business*, 171.

4 See Morris-Suzuki, *Beyond Computopia*.

5 Leonard Herman, *Phoenix*, 103.

6 Herz, *Joystick Nation*, 20.

7 Katayama, *Japanese Business*, 161.

8 Leonard Herman, *Phoenix*, 103–4.

9 Sheff, *Game Over*, 158–9.

10 Kinder, *Playing with Power*, 89–90.

11 Ibid., 90; Sheff, *Game Over*, 172.

12 Shapiro and Varian, *Information Rules*, 204.

13 Sheff, *Game Over*, 31.

14 Clapes, *Softwars*, 248.

15 Sheff, *Game Over*, 181.

16 Ibid., 161.

17 Clapes, *Software*, 250.

18 See Sheff, *Game Over*; Hayes and Dinsey, with Parker, *Games War*.

19 Kinder, *Playing with Power*, 91.

20 Stiles, "Home Video Game Market Overview."

21 Hayes and Dinsey, with Parker, *Games War*, 32.

22 Katayama, *Japanese Business*, 172.

23 Hayes and Dinsey, with Parker, *Games War*, 43.

24 Ibid., 43.

25 Sheff, *Game Over*, 370–1.

26 "This strategy was so successful that it was challenged in the courts. In December 1988, Tengen Inc., a former licensee based in California, brought an antitrust lawsuit against Nintendo, which responded with a countersuit for patent infringement. In 1989 Atari brought a $100 million lawsuit against Nintendo for preventing competitors from making game cartridges that would work on Nintendo's hardware." Kinder, *Playing with Power*, 92.

27 Sheff, *Game Over*, 276.

28 Shapiro and Varian, *Information Rules*, 245.

29 Sheff, *Game Over*, 71.

30 Sawyer, Dunne, and Berg, *The Game Developer's Marketplace*, 498.

31 Sheff, *Game Over*, 259.

32 Ibid., 48–9.

33 Ibid., 117.
34 Ibid., 124.
35 Ibid., 126.
36 Ibid., 116.
37 Cited in Katayama, *Japanese Business*, 178.
38 Sawyer, Dunne, and Berg, *The Game Developer's Marketplace*, 167.
39 Hayes and Dinsey, with Parker, *Games War*, 34.
40 Katayama, *Japanese Business*, 178.
41 Leonard Herman, *Phoenix*, 117.
42 Ibid.
43 Ibid., 118.
44 Cited in Katayama, *Japanese Business*, 175–6.
45 Katayama, *Japanese Business*, 175.
46 Ibid., 178.
47 Kinder, *Playing with Power*, 89.
48 Hayes and Dinsey, with Parker, *Games War*, 67.
49 Sheff, *Game Over*, 174.
50 Hayes and Dinsey, with Parker, *Games War*, 13.
51 Ibid., 71.
52 Sheff, *Game Over*, 167.
53 Ibid., 175–7.
54 Ibid., 178.
55 Jenkins, "'x logic,'" 67.
56 Sheff, *Game Over*, 179.
57 Ibid., 180.
58 Ibid., 179–80.
59 Ibid., 182.
60 Ibid., 183.
61 Pargh, "A Prisoner of Zelda."
62 Cited in Sheff, *Game Over*, 181.
63 Cited in ibid.
64 Cited in ibid., 183.
65 Sheff, *Game Over*, 183.
66 Tom Forester, *Silicon Samurai*.
67 Morley and Robins, *Spaces of Identity*, 168.
68 Ibid., 1, 170.
69 Ibid., 149.
70 Sheff, *Game Over*, 262.
71 Ibid., 281.
72 Ibid., 281–2.
73 Ibid., 195.
74 Ibid., 261–2. The quote is from p. 282.
75 Jenkins, "'x logic,'" 56.

76 Provenzo's discussion includes seven games produced by licensees but also three of Nintendo's platform games: *Zelda II – The Adventures of Link*, *Super Mario Bros. 2*, *The Legend of Zelda* (in first, second, and fifth place). Provenzo argues that "the themes of rescue, revenge and good versus evil are found in all ten games" (*Video Kids*, 88). The themes of aggression were "more apparent" in the games by licensees (some of which, such as *Double Dragon*, were "about violence pure and simple") than in the platform games, he claims.

77 Murray, *Hamlet on the Holodeck*, 129.

78 Fuller and Jenkins, "Nintendo and New World Travel Writing," 61.

79 Ibid., 58.

80 Ibid., 70.

81 Ibid.

82 This may well be an answer that Jenkins is unwilling to make himself. In his response to Provenzo's work in "'x logic': Repositioning Nintendo in Children's Lives," Jenkins is at pains to reject what he sees as a romanticized view of childhood innocence "corrupted" by an invasive technocorporation. In place of Provenzo's insistence on the closure of Nintendo narratives, he insists on the latitude available to players in their explorations of Nintendo's digital space – the options, opportunities for improvisational or imaginative extension of the game experience. In many ways he produces a fuller account of the pleasures of gaming culture. But in his desire to avoid a crude reductionism, he risks another form of romanticism, one that overstates the player's autonomy and obscures the overarching commercial design of the context within which this activity occurs. The real pleasures he and Murray describe are dependent on consumerist purchases, purchases that are purposefully designed to exclude competing and alternative activities, and the various spontaneous improvisations and constructions he describes are increasingly calculated into the feedback loop of marketing and monitoring. Through his reluctance to acknowledge this, Jenkins in many ways loses the force of his own "colonization" metaphor. To push a point, his analysis of corporate-child relations suggests an account of colonist-native relations, emphasizing the undoubted agency exercised by indigenous people in the trade for trinkets, firearms, and alcohol, while downplaying the extraordinary dominative thrust of the whole process.

83 Moore, *Crossing the Chasm*, 71.

84 Miles, "Towards an Understanding," 35.

85 Green, Reid, and Bigum, "Teaching the Nintendo Generation?," 24.

86 Herz, *Joystick Nation*, 132.

87 Ibid.

88 Main, cited in ibid., 132–3.

89 Herz, *Joystick Nation*, 134.

90 Fuller and Jenkins, "Nintendo and New World Travel Writing," 70.

CHAPTER SIX

1 See Burnham, *Supercade*, 336.
2 Shapiro and Varian, *Information Rules*, 196.
3 Ibid.
4 Ibid.
5 Stiles, "Home Video Game Market Overview."
6 Johnstone, "Dream Machines," 66.
7 Battelle with Johnstone, "Seizing the Next Level."
8 Ibid.
9 Irina Heirakuji, cited in Battelle with Johnstone, "Seizing the Next Level."
10 "Associate Planning Director Irina Heirakuji spent months in the bedrooms, living-rooms, and homes of America's youth, filming what kids wanted, what kids said, and why kids thought Sega was cool. She found that while most were married to the 8-bit NES system, they lusted in their hearts for Sega. 'There was this base level of dissatisfaction with Nintendo,' she says. 'Not many kids had Sega, but they went to the homes of those that did.'" Battelle with Johnstone, "Seizing the Next Level."
11 Steve Estenazi, cited in Battelle with Johnstone, "Seizing the Next Level."
12 Battelle with Johnstone, "Seizing the Next Level."
13 Ibid.
14 Ibid.
15 Tim Dunley, cited in "How Sega and Sony Try and Get In Your Head."
16 Pearce, "Beyond Shoot Your Friends," 211.
17 Hayes and Dinsey, with Parker, *Games War*, 69.
18 Ibid., 70.
19 Ibid.
20 "How Sega and Sony Try and Get In Your Head."
21 Bill Moyers, cited in Hamilton, *Channeling Violence*, 31.
22 Leonard Herman, *Phoenix*, 198.
23 Ibid.
24 Kinder, "Contextualizing Video Game Violence," 35.
25 Ibid., 28, 32.
26 Herz, *Joystick Nation*, 184.
27 Ibid., 185.
28 Kinder, "Contextualizing Video Game Violence," 25.
29 Leonard Herman, *Phoenix*, 199–200.
30 Leonard Herman et al., "The History of Video Games."
31 Leonard Herman, *Phoenix*, 197.
32 Herz, *Joystick Nation*, 192–3.
33 Leonard Herman, *Phoenix*, 198.

34 Sawyer, Dunne, and Berg, *The Game Developer's Marketplace*, 368.

35 Ibid., 371.

36 Hayes and Dinsey, with Parker, *Games War*, 52–3.

37 Ibid., 53.

38 Ibid., 20.

39 Ibid., 21.

40 Ibid., 73.

41 Doug Glen, cited in Battelle with Johnstone, "Seizing the Next Level."

42 Battelle with Johnstone, "Seizing the Next Level"; Leonard Herman, *Phoenix*, 194.

43 Leonard Herman, *Phoenix*, 272.

44 Ibid., 231.

45 Wylie and Peline, "Back from Hard Knocks U."

46 See Leonard Herman et al., "The History of Video Games."

47 Burstein and Kline, *Road Warriors*, 183.

48 Leonard Herman et al., "The History of Video Games."

49 Leonard Herman, *Phoenix*, 170.

50 Battelle with Johnstone.

51 Stiles, "Home Video Game Market Overview."

52 Leonard Herman, *Phoenix*, 228–30.

53 For an interesting account of US computer gaming culture in the 1980s, see Bennahum, *Extra Life*.

54 Statistics from United States, Department of Commerce, National Telecommunications and Information Administration, "Falling through the Net II."

55 Sawyer, Dunne, and Berg, *The Game Developer's Marketplace*, 385.

56 See Laidlaw, "The Egos at Id."

57 Sawyer, Dunne, and Berg, *The Game Developer's Marketplace*, 45.

58 McCandles, "Legion of Doom."

59 On the relationships and environments behind the making of *Myst* and its successor, *Riven*, see two articles by Carroll, "Guerillas in the *Myst*" and "D(*Riven*)."

60 LaPlant and Seidner, *Playing for Profit*, 146–7.

61 David Miles, "The CD-ROM Novel *Myst*," 4.

62 *Myst User's Manual*.

63 Laidlaw, "The Egos at Id"; Carroll, "Guerillas in the *Myst*."

64 Carroll, "Guerrillas in the *Myst*"; Rothstein, "A New Art Form."

65 David Miles, "The CD-ROM Novel *Myst*," 4.

66 On the emergence of these landscapes and on the class position of their inhabitants, see Harvey, *The Condition of Postmodernity*; Zukin, *Landscapes of Power*; Castells, *The Informational City*.

67 Herz, *Joystick Nation*, 151.

68 Carroll, "(D)*Riven*."

69 Schumpeter, *Capitalism, Socialism, and Democracy*.

CHAPTER SEVEN

1 Flower, "The Americanization of Sony."
2 Cited in Levy, "Here Comes PlayStation 2," 57.
3 "Game Makers to Launch New Generations," c4.
4 Goodfellow, "Sony Comes on Strong in Video-Game Wars," d5.
5 Ibid.; Nathan, Sony: The Private Life, 304–5.
6 Alexander and Associates, "Shakeup at Sony."
7 Ibid.
8 Brandt, "Nintendo Battles for Its Life."
9 George Bulat, cited in "Sony Plays the Video-Game Market," c15.
10 Chip Herman, cited in "How Sega and Sony Try and Get in Your Heads."
11 Lee Clow, cited in "How Sega and Sony Try and Get in Your Heads."
12 Ibid.
13 Brandt, "Nintendo Battles for Its Life."
14 Goodfellow, "Sony Comes on Strong," d5.
15 DFC Intelligence, "Interactive Electronic Entertainment Industry Overview."
16 Ibid.
17 See Goodfellow, "Sony Comes on Strong."
18 Roberts, "Sony Changes the Game," 124.
19 Ibid.
20 Goodfellow, "Sony Comes on Strong."
21 DFC Intelligence, "Interactive Electronic Entertainment Industry Overview."
22 Ibid.
23 Goodfellow, "Sony Comes on Strong."
24 Sales figures from "In Their Dreams," 71; Croal, "The Art of the Game," 80.
25 Alexander and Associates, "Comparing Generations of Console Gaming."
26 Dan Schiller, Digital Capitalism, 91.
27 Paul Saffo, cited in Flanigan, "Whether Apple Grows Anew Depends on Human Factor," d1, d6, cited in Dan Schiller, Digital Capitalism, 91.
28 Taylor, "Pentium II Aimed at Games Market," 13.
29 Cited in ibid.
30 Information in this paragraph from Burstein and Kline, Road Warriors, 189–90.
31 Probst, "A New Opportunity."
32 Sawyer, Dunne, and Berg, Game Developer's Marketplace, 397–8.
33 Probst, "A New Opportunity."
34 Cited in LaPlant and Seidner, Playing for Profit, 148.
35 LaPlant and Seidner, Playing for Profit, 147–51.
36 Dunn, "Finding Art in an Internet Game," c9.
37 Schiesel, "The Ins and Outs of Playing Games Online."

38 Herz, "A Designer's Farewell to His Fantasy Realm," G5.

39 Richard Garriott, cited in ibid.

40 Herz, "A Designer's Farewell to His Fantasy Realm," G5.

41 Kim, "Killers Have More Fun."

42 Ibid.

43 For a hilarious account of the exigencies of *Ultima* play see Lizard, "Kill Bunnies, Sell Meat."

44 Kim, "Killers Have More Fun."

45 "Open Letter from Lord British."

46 Bennahum, "Massive Attack," 163.

47 Cited in ibid., 163.

48 Ibid.

49 Bennahum, "Your Email's Gonna Get You," 105.

50 Bennahum, "Massive Attack," 167.

51 Newman, "From Microsoft Word to Microsoft World."

52 Microsoft, "Directx."

53 Drummond, *Renegades of the Empire*, 12.

54 Sawyer, Dunne, and Berg, *Game Developer's Marketplace*, 398.

55 When the second edition of Windows 98 was released it was found to have an unanticipated incompatibility with the installation programs for Talonsoft's acclaimed *East Front* and *West Front* war games. Unable to persuade Microsoft to issue a remedial patch immediately, Talonsoft was compelled to devise elaborate work-around instructions that made the installation of these games anything but "point and click."

56 Sawyer, Dunne, and Berg, *Game Developer's Marketplace*, 52.

57 Newman, "From Microsoft Word."

58 Sawyer, Dunne, and Berg, *Game Developer's Marketplace*, 314.

59 Newman, "From Microsoft Word."

60 Drummond, *Renegades of the Empire*, 12.

61 Sawyer, Dunne, and Berg, *Game Developer's Marketplace*, 680.

62 Newman, "From Microsoft Word."

63 Drummond, *Renegades of the Empire*, 97.

64 Eng, "Net Games Are Drawing Crowds," 74.

65 Ibid., 73–4.

66 Cited in Bennahum, "Massive Attack," 166–7.

67 Ibid.

68 Herz, "Technology as the Guiding Hand of History," G20.

69 Cited in ibid.

CHAPTER EIGHT

1 Cited in Levy, "Here Comes PlayStation 2," 55.

2 Piore and Sabel, *The Second Industrial Divide*.

3 For an example of this debate in regard to changing structures of Hollywood production see the exchange between Christopherson and Storper (on the side of flexible Fordism) and Aksoy and Robins (on the side of digital empires). The relevant texts are Christopherson and Storper, "The Effects of Flexible Specialization"; Storper, "The Transition to Flexible Specialization"; Aksoy and Robins, "Hollywood for the 21st Century." The quote is from Aksoy and Robins, "Hollywood for the 21st Century," 17.

4 Scally, "PC, Video Game Software Sales Hit Record High in 1997," 65; "Young at Heart," 70.

5 Kapica, "Fun and Games Drive Computer Innovations," C10; Atkinson, "Video Games: In Praise of Folly," D5.

6 Lewis, "Dreamcast Is a Toy."

7 "Sega to Acquire Toy Firm," B6.

8 Carlton and Hamilton, "Infighting Mars Sega Comeback," M2.

9 Takahashi, "Let the Games Begin," 80.

10 Takahashi, "The Game of War."

11 Ibid., 58.

12 Ibid., 60.

13 Ken Kutaragi, cited in Dolan, "Behind the Scenes."

14 McLuhan, *Understanding Media*, 245.

15 Cited in Levy, "Here Comes PlayStation 2," 55–9.

16 See O'Brien, "The Making of the XBOX," 142. See also "In Their Dreams," 71–2; and Sheff, "Sony's Plan for World Recreation," 264–75.

17 Markoff, "Tuning in to the Fight of the (Next) Century."

18 Hayes and Dinsey, with Parker, *Games War*, 43.

19 Interactive Digital Software Association (IDSA), "The State of the Entertainment Software Industry 1998," 4.

20 Ibid., 17.

21 "Babes with Guns," 75.

22 Mosco, *The Political Economy of Communication*, 109.

23 See Sawyer, Dunne, and Berg, *Game Developer's Marketplace*, 395, 431.

24 Sawyer, Dunne, and Berg, *Game Developer's Marketplace*, 431.

25 Cited in IDSA, "The State of the Entertainment Software Industry 1998," 18.

26 Ibid., 17.

27 Sawyer, Dunne, and Berg, *Game Developer's Marketplace*, 430.

28 Ibid.

29 Stoddard, "Sony's PlayStation 2 Could be Used to Launch Missiles, Japan says," B1.

30 See Barbrook and Cameron, "The Californian Ideology"; De Landa, *War in the Age of Intelligent Machines*; Edwards, *Computers and the Politics of Discourse in Cold War America*; Castells, *The Network Society*.

31 See Herz, *Joystick Nation*; Poole, *Trigger Happy*; and Kent, *The First Quarter*.

32 McLuhan, *Understanding Media*, 239.

33 This term is often attributed to McKenzie Wark, "The Information War." See also Lenoir, "All But War Is Simulation"; Leslie, *The Gulf War as Popular Entertainment*.

34 Der Derian, *Virtuous War*, xi.

35 Ibid., 174–5.

36 Ibid., 217.

37 Ibid., 164.

38 Pearce, "Beyond Shoot Your Friends," 224.

39 Stapleton, "Theme Parks," 429.

40 Ibid.

41 The military-video games interplay continues to spiral into many other realms too, such as media buying for the advertising of video games. Explains one industry guide: "Exploring alternative advertising outlets that aren't game oriented, but reach the same target market – especially among special interests like sports, driving, military, and so on – should be your first advertising foray outside of game industry media." Sawyer, Dunne, and Berg, *Game Developer's Marketplace*, 482.

42 "MÄK Technologies Awarded Contract." See also "MÄK Technologies Wins Army Contract."

43 MÄK Technologies, "About MÄK Technologies."

44 "MÄK Technologies Awarded Contract."

45 Ibid.; "MÄK Technologies Wins Army Contract."

46 "MÄK Technologies Awarded Contract."

47 It is worth noting that Warren Katz and John Morrison, who founded MÄK Technologies, were both MIT graduates and "original members ... of the SIMNET project team, which developed the first low-cost, networked 3D simulators for the Department of Defense." MÄK Technologies, "About MÄK Technologies."

48 "MÄK Technologies Awarded Contract."

49 Ibid.

50 Ibid.

51 Lee, *Consumer Culture Reborn*, 133–5. See also Robins and Webster, "Cybernetic Capitalism"; and Garnham, *Capitalism and Communication*.

52 Alexander and Associates, "Game Platforms and the Future of the Internet."

53 IDSA says household penetration of next-generation game consoles grew from six million in 1996 to twenty-five to thirty million at the end of 1998 and was projected to reach thirty-five to forty million by the end of 1999. The Yankee Group (Press release) estimates an installed base of 35.9 million consoles in 1999, growing to 43.5 million by the end of

2003, with some eighty-five percent being next-generation machines, e.g., PlayStation, Nintendo 64, or Saturn. Henry and Hause ("Videogame Consumer Segmentation Survey 1999") claim that in 1999 thirty-eight percent of all US households had game consoles.

54 United States, Department of Commerce, National Telecommunications and Information Administration, "Falling through the Net IV."
55 Yankee Group, Press release.
56 Interactive Digital Software Association (IDSA), "State of the Industry: Report 2000–2001."
57 Alexander and Associates, "Comparing Generations of Console Gaming."
58 Ibid.
59 IDSA, "State of the Industry: Report 2000–2001."
60 PC data cited in Frauenfelder, "Death Match," 153.
61 Alexander and Associates, "Comparing Generations of Console Gaming."
62 "Electronic Games Market Expands as Players Mature", C22.
63 Yankee Group, Press release; IDSA, "The State of the Entertainment Software Industry 1998."
64 Marriott, "I Don't Know Who You Are," G1.
65 Schiesel, "The Ins and Outs of Playing Games Online," D1.
66 Hatlestad, "Games People Play."
67 United States, Department of Commerce, National Telecommunications and Information Administration, "Falling through the Net IV."
68 Marriot, "I Don't Know Who You Are."
69 Ibid.
70 McNealy, "Let the Online Games Begin."
71 Ibid.
72 Bloom and Takahashi, "Can Games Make Money Online?"
73 Ibid.
74 Bennahum, "Massive Attack," 163.
75 Kharif, "Let the Games Begin – Online."
76 Ohmae, "Letter from Japan," 158.
77 Morley and Robins, *Spaces of Identity*; Forester, *Silicon Samurai*.
78 Reich, *The Work of Nations*.
79 Johnstone, "Video Games: Dream Machines," 66.
80 IDSA, "State of the Entertainment Software Industry 1997."
81 Ibid.
82 Herz, *Joystick Nation*, 170.
83 Hiroshi Imanishi, cited in Sheff, *Game Over*, 414.
84 Cited in Sheff, *Game Over*, 416–17.
85 Stevens and Grover, "The Entertainment Glut," 88–95.
86 Petersen, "Chavez Steps into Pepsi's Ring."

87 "Electronic Arts Does Its First Thai Language Game."
88 Takahashi, "Let the Games Begin"; Brautigam, "Mousing to Megabucks," E4.
89 United Nations, *Human Development Report*, 30.
90 Ibid., 33.
91 Michael Roberts, "Internet Revolution."
92 For an insightful study see Cassidy, *Dot.Con: The Greatest Story Ever Sold*.

CHAPTER NINE

1 IDSA, "The State of the Entertainment Software Industry: 1997 Executive Summary." These figures should be treated with caution, since they are contradicted by a more recent news release from IDSA, "Economic Impacts of the Demand for Playing Interactive Entertainment Software." IDSA attributes the discrepancies to differences in sampling method between the surveys.
2 Sawyer, Dunne, and Berg, *Game Developer's Marketplace*, 233–48.
3 Kraft and Sharpe, "Software Globalization."
4 Katayama, *Japanese Business*, 173–4.
5 Ibid., 179.
6 Ibid., 182.
7 Sheff, *Game Over*, 39–40.
8 Ibid., 40.
9 Cited in Katayama, *Japanese Business*, 189.
10 Cited in Katayama, *Japanese Business*, 177.
11 Herz, *Joystick Nation*, 93, 91.
12 Industry manuals list salaries for video game producers rising from $40,000 upward, passing at the top levels to over $100,000. Sawyer, Dunne, and Berg, *Game Developer's Marketplace*, 233.
13 "Babes with Guns," 74–5.
14 Lessard and Baldwin, *Netslaves: True Tales of Working the Web*; Ross, "Applying the Anti-Sweatshop Model to High Tech Industries"; Thompson, "Why Your Fabulous Job Sucks."
15 In his study of the early days of Atari, Cohen gives an evocative portrait of the company's Silicon Valley employees (*Zap! The Rise and Fall of Atari*). On this ethos see also Hayes, *Behind the Silicon Curtain*. For a more cheery recent account of game programming, see Bronson, *The Nudist on the Late Shift*, 98–137.
16 Thompson, "Why Your Fabulous Job Sucks," 58.
17 Ibid.
18 Ibid.
19 Ibid.

20 Ibid., 62

21 The classic version is Piore and Sabel, *The Second Industrial Divide.*

22 Tapscott, Ticoll, and Lowy, *Digital Capital.*

23 Toffler, *The Third Wave,* 265–88.

24 The blurring of play and work in the gaming sector is a striking illustration of an argument made by the political economist Dallas Smythe that media audiences are in fact engaged in *work* by watching advertisements and commercial content, and by learning to be good consumers of advertisers' products. See Smythe, *Dependency Road.*

25 Ron Bertram, marketing director at Nintendo of Canada, cited in Stafford, *Insert Coin.*

26 Sawyer, Dunne, and Berg, *Game Developer's Marketplace,* 413.

27 Stapleton, "Theme Parks," 432.

28 Ibid.

29 Ibid., 431, 432. See also Moody, *The Visionary Position.*

30 IDSA, "The State of the Entertainment Software Industry 1998," 14.

31 A report on game development in the magazine *Computer Gaming World* ("Action: Building the Perfect Game," 172) notes that "across the board, designers seem to take the input of their testers seriously, implementing changes and suggestions every step of the way." In Pandemic's *Battlezone,* tester suggestions led to a major redesign of the user interface. In the case of Sierra Studio's enormously successful *Half-Life,* implemented suggestions from testers reportedly ran the gamut from "Make this ladder more obvious" to "This monster really sucks and needs to be redesigned."

32 Herz, *Joystick Nation,* 118.

33 Ibid.

34 Cited in "Simulations: Building the Perfect Game," 185.

35 Cited in "Action: Building the Perfect Game," 172.

36 McCandles, "Legion of Doom."

37 Ibid.

38 Cited in Herz, "Under Sony's Wing," E4.

39 Herz, "Under Sony's Wing."

40 Ibid.

41 Rogers and Larsen, *Silicon Valley Fever,* 132; Cohen, *Zap!,* 141; "Atari Gets Tough."

42 Weinstein, "Ex-Workers Win Back Pay."

43 "Video Game Company Moves."

44 "First Round of NAFTA Unemployment Benefits Issued."

45 "Small Firms in Japan."

46 Cited in Katayama, *Japanese Business,* 170.

47 Katayama, *Japanese Business,* 169.

48 Muller, "The Best Game Nintendo Built," 30.

49 Herz, *Joystick Nation*, 113.

50 Sheff, *Game Over*, 295.

51 Ibid., 294.

52 All data are from Fischer, "Keyboard Firm Ranks with the Best." For full discussion and documentation see Dyer-Witheford, "The Work in Digital Play."

53 There is an extensive literature on working conditions in the *maquila-doras*. A recent important contribution is Peña, *The Terror of the Machine*.

54 Information on the struggle is from Communications Workers of America, "CWA Files NAFTA Complaint"; Collier, "NAFTA Labor Problems Haunt New Trade Debate"; and "Mexico: Fackligt Stod Fran USA," which says there are four hundred employees at the factory, as opposed to Collier's nine hundred.

55 Asia Monitor Resource Center, "The Hong Kong Takeover of South China," 12.

56 Rizvi, "Toying With Workers," 8.

57 O'Brien, "The Making of the xbox," 142.

58 Business Software Alliance, "Sixth Annual BSA Global Software Piracy Study," 1.

59 McCandless, "Warez Wars." See also Tetzalf, "Yo-Ho-Ho and a Server of Warez."

60 IDSA, "Fast Facts."

61 Cited in MSNBC, "Video-Game Pirates on the Loose."

62 IDSA, "Fast Facts."

63 Tetzalf, "Yo-Ho-Ho and a Server of Warez."

64 See MSNBC, "Video-Game Pirates on the Loose."

65 Cited in MSNBC, "Video-Game Pirates on the Loose."

66 Wen, "Why Emulators Make Video Game Makers Quake."

67 Ibid.

68 Pilieci, "Searching for Sega in All the Wrong Places," D5.

69 Lunenfeld, *Snap to Grid*, 5.

70 Ibid., 5–7.

71 Cited in Harris, *Digital Property*, 162.

72 MSNBC, "Video-Game Pirates on the Loose."

73 McCandles, "Warez Wars."

74 In 1992 FBI agents raided the Massachusetts office of the Davey Jones Locker Bulletin Board, which offered subscribers in thirty-six states and eleven countries downloads of popular business and entertainment software for ninety-nine dollars a year. Warshofsky, 193.

75 McCandles, "Legion of Doom."

76 Ibid.

77 Weber, "Maverick Programmers," B1.

78 Cited in Hansen et al., "Games May Point to Future."
79 "Game Over?" *New Scientist.*
80 For enunciations of this doctrine see Kelly, "New Rules for the New Economy"; Barlow, "The Economy of Ideas."
81 Johnstone, "Pirates Ahoy!" 68.
82 Ibid.
83 Warshofsky, *The Patent Wars,* 2.
84 "Intellectual Property: Bazaar Software."
85 Ibid.
86 Ibid.
87 Ye, "Computer Insects," 292–3.
88 Peckham, "Soft Copy," 20.
89 See Software and Information Industry Association.
90 Peckham, "Soft Copy," 20.
91 IDSA, "Fast Facts."
92 Terranova, "Free Labor," 33.
93 Ibid., 53.
94 Ibid., 46.
95 Ibid., 55.
96 Johnstone, "Pirates Ahoy!" 68.

CHAPTER TEN

1 Bloom, "Nintendo Builds a Better Monster."
2 Wernick, *Promotional Culture,* 187.
3 See Wernick, *Promotional Culture.*
4 Slater and Tonkiss, *Market Society,* 15. Emphasis added.
5 Murdock, "Base Notes," 100.
6 Slater and Tonkiss, *Market Society,* 176; see 176–81.
7 Wernick, "Sign and Commodity," 157.
8 See Goldman and Papson, *Sign Wars.*
9 Yuen, "Games Aren't Just for Kids Anymore."
10 du Gay et al., *Doing Cultural Studies,* 59.
11 Pastore, "Young Americans Take Their Spending Online."
12 Pigeon, "Packaging Up Coolness," 21.
13 Moore, *Crossing the Chasm,* 27.
14 Nixon, "Circulating Culture," 181.
15 Moore, *Crossing the Chasm,* 66.
16 IDSA, "State of the Industry Report: 2000–2001."
17 According to a 2000–01 IDSA study, when asked to rank various entertainment activities in order of fun, the response was: 1. playing video/PC games; 2. watching television; 3. surfing the Internet; 4. reading books; 5. going to the movies. "State of the Industry Report: 2000–2001."

18 Cited in "How Sega and Sony Try and Get in Your Heads."

19 Moore, *Crossing the Chasm*, 156.

20 For a more detailed analysis of late-1990s video game advertisements, see Kline and de Peuter, "Ghosts in the Machine." See also Nixon, "Fun and Games are Serious Business."

21 Ron Bertram, cited in Stafford, *Insert Coin*.

22 Mills, "Chaos Rules," 17.

23 Chip Herman, cited in "How Sega and Sony Try Get In Your Heads."

24 Ron Bertram, cited in Mills, "Chaos Rules," 20.

25 "How Sega and Sony Try and Get in Your Heads."

26 Chip Herman, cited in "How Sega and Sony Try and Get in Your Heads."

27 Haven Dubrul, cited in "How Sega and Sony Try and Get in Your Heads."

28 Kinder, *Playing with Power*, 109.

29 IDSA, "Fast Facts."

30 Klein, *No Logo*, 148.

31 "The Movie," *Tomb Raider 2000* Press release no. 2.

32 Ibid.

33 Paul Baldwin, cited in "Interviewing the Big Guys."

34 "Lara in Nike Commercial."

35 Keighley, "Angelina Jolie: Lara's Splitting Image?"

36 Bloom, "Hollywood Square," 90–2.

37 Koelsh, *The Infomedia Revolution*, 91.

38 Jon Hussman, cited in Myers, "Today Sega City, Tomorrow the World," 18.

39 Ibid., 22.

40 Ibid., 19.

41 Ibid., 18–9.

42 Kinney, "The Virtual Squadrons of Fightertown," 486.

43 Hettema, "Location-Based Entertainment," 375–6.

44 Mayers, "Focusing on Fun," D1.

45 Posner, "A Really Big Show," 38.

46 Roger Harris, cited in ibid.

47 Harvey, *The Condition of Postmodernity*. See also Zukin, *Landscapes of Power*, and *The Cultures of Cities*.

48 For a discussion of cool-hunting, see Klein, *No Logo*, 63–85.

49 Pigeon, "Packaging Up Coolness," 21.

50 Sawyer, Dunne, and Berg, *Game Developer's Marketplace*, 380.

51 Johnson, "What is Cultural Studies Anyway?" 83.

52 Herz, *Joystick Nation*, 107; Spencer Nilson, head of Sega Music Group, cited in Herz, 108.

53 Herz, *Joystick Nation*, 108.

54 "Club iMusic Showcase: *Wipeout XL*."
55 Video games are further integrated into the promotional strategies of the music industry as rock bands design their own video games. For example, in 1999 the rap group Wu-Tang Clan released *Shaolin Style*, a fighting game that featured a soundtrack with previously unreleased songs. (Mernagh and Keast, "Video Games Saved the Radio Stars," 21).
56 Turner, "Rockstar and the Cooling of Video Games."
57 Ibid.
58 Mernagh and Keast, "Video Games Saved the Radio Stars," 21.
59 Herz, *Joystick Nation*, 179.
60 Nintendo, "Nintendo Gives Consumers 1000 More Reasons to Shop at Tommy Hilfiger."
61 Ron Bertram, cited in Dawson, "Brand Games."
62 Dawson, "Brand Games."
63 Herman and McChesney, "The Global Media in the Late 1990s," 190.
64 IDSA, "The State of the Entertainment Software Industry 1997."
65 Baldwin, cited in "Interviewing the Big Guys."
66 IDSA, "The State of the Entertainment Software Industry 1997."
67 George Harrison, cited in Rich, "More Delays Cause Nintendo to Make Line Up Changes."
68 See Walkerdine, Dudfield, and Studdert, "Sex and Violence."
69 Toffler, *The Third Wave*, 161–2.
70 Seiter, "Gotta Catch 'Em All."
71 Celia Pearce, cited in Bloom, "Nintendo Builds a Better Monster."
72 Canadian Press Newswire, "Pokémon Phenomenon Strikes Today."
73 Snider, "'Pocket Monsters' Pocket Sales."
74 Canadian Press Newswire, "Pokémon Phenomenon Strikes Today."
75 Satoshi Tajiri, cited in Larimer, "The Ultimate Game Freak."
76 Deacon, "The Craze That Ate Your Kids," 74.
77 Satoshi Tajiri, cited in Larimer, "The Ultimate Game Freak."
78 Warner Bros., "The Making of Pokémon."
79 McCormick, "Fantastic Franchises," 68.
80 Christopher Byrne, cited in Deacon, "The Craze that Ate Your Kids," 74.
81 Wheelwright, "Game Buy: Deep Pocketed Monsters," E1.
82 Mount, "Shorts Watch Pokémon Frenzy," D3.
83 Warner Bros., "The Making of Pokémon."
84 Ibid.
85 Kirkland, "Pokemon: 2nd Wave."
86 Al Kahn, cited in McCormick, "Fantastic Franchises," 68.
87 "Pokémon Rules TV, Too."
88 Berman, "Japanese Invasion," 46.
89 Peter Moss, cited in Deacon, "The Craze That Ate Your Kids," 74.
90 McGovern, "Childhood's End," 48–50.

91 Berman, "Japanese Invasion," 46.
92 Danielle Giovanelli, cited in McCormick, "Warner Unleashes Massive Campaign," 108.
93 Obayashi, "Strong Yen Batters Nintendo Profit," c16.
94 Hiroshi Imanishi, cited in Schaefer, "Pokémon Boom," c11.
95 Douglas Lowenstein, "Opening Remarks."
96 "Pokémon Fans in Stores and on Wall Street," d2.
97 Zhao and Schiller, "A Dance with Wolves?" 139.
98 Ibid., 140.
99 Seiter, "Gotta Catch 'Em All."
100 See Kline, *Out of the Garden*.
101 Ibid., 347.
102 Ibid., 349.
103 Rifkin, *The Age of Access*, 263.
104 Ibid., 261.
105 Ibid.
106 Ibid., 265.
107 See interview with Satoshi Tajiri in Larimer, "The Ultimate Game Freak."

CHAPTER ELEVEN

1 Kay, "Are Video Games Bad For Your Health?" b4.
2 Whitta, "Soldier of Fortune II," 47.
3 Cited in Whitta, "Soldier of Fortune II," 48.
4 Cited in Poole, *Trigger Happy*, 218.
5 Grossman, *On Killing*; Grossman and DeGaetano, *Stop Teaching Our Kids to Kill*. See also Keegan, "In the Line of Fire."
6 Kushner, "Feed Daily."
7 For the industry perspective, see the report from IDSA, "Video Games and Youth Violence." For a children's advocacy group perspective, see the report by Glaubke et al., "Fair Play."
8 Le Diberder and Le Diberder, *L'Univers des Jeux Vidéo*.
9 For discussion of the equivocal evidence on this score, see Kline, *Out of the Garden*, 316–21.
10 Williams, *Television*, 113.
11 Ibid., 115.
12 Herz, *Joystick Nation*, 194.
13 Pesce, "The Trigger Principle"; Herz, *Joystick Nation*, 183.
14 Hamilton, *Channeling Violence*, 31.
15 Ron Bertram, cited in Stafford, *Insert Coin*.
16 Pearce, "Beyond Shoot Your Friends," 210.
17 Provenzo, *Video Kids*.

18 Ibid., 72–135.

19 Pearce, "Beyond Shoot Your Friends," 213. See also Keegan, "Culture Quake."

20 Kelly, *Out of Control*, 200.

21 Shafer, "Development ... la Mod," 96.

22 The difficulty with this sort of exercise is that the strategy/RPG games include games such as *Pokémon*, *The Legend of Zelda*, *Sim City*, and *Wheel of Fortune*. Another genre breakdown cited by IDSA in 1999 gave 3-D Action/Adventure 44.8 percent; Sports 25 percent; General Strategy 10.6; Card/Casino/Quiz and Board Games 10.97 percent; and another category of Action 9.7 percent (IDSA, "1999 Genre Breakdown by Sales").

23 Glaubke et al., *Fair Play*, 9.

24 United States, Federal Trade Commission, "FTC Releases Report." See also Jenkins, "Professor Jenkins Goes to Washington."

25 Buchanan, "Strangers in the 'Myst' of Video Gaming."

26 Glaubke et al., *Fair Play*, 2.

27 Ibid., 12.

28 Alloway and Gilbert, "Video Game Culture," 99–100.

29 Gilmour, "What Girls Want."

30 Ibid.

31 John Romero, cited in Cassell and Jenkins, "Chess for Girls?" 26.

32 Tim Dunley, cited in "How Sega and Sony Try and Get In Your Heads."

33 Buchanan, "Strangers in the 'Myst' of Video Gaming."

34 Cassell and Jenkins, "Chess for Girls?" 3.

35 Takahashi, "Software Firm Crushed by Barbie," C18.

36 Cassell and Jenkins, *From Barbie to Mortal Kombat*.

37 Ibid., 4.

38 Harmon, "With the Best Research and Intentions, a Game Maker Fails."

39 PC data cited in ibid.

40 Takahashi, "Software Firm Crushed by Barbie," C18.

41 Cited in ibid.

42 Laurel's own reflections on these events can be found in her 2001 book *Utopian Entrepreneur*.

43 Cassell and Jenkins, *From Barbie to Mortal Kombat*, 24.

44 Gilmour, "What Girls Want."

45 Cunningham, "Mortal Kombat and Computer Game Girls."

46 Autio, "Gender Politics of a Virtual Warrior." For another interesting account of gender roles in this virtual environment, see Breeze, "Quakeing in My Boots."

47 Thompson, "Meet Lara Croft, Virtual Sensation," C1.

48 Toby Gard, cited in ibid., C14.

49 *Total Control*.

50 "Hands On Lara Croft."

51 Gilmour, "What Girls Want."

52 IDSA, cited in Sutton, "Special Report."

53 Ibid.

54 Rahmat, "IDC's 2001 Video Game Survey."

55 Ibid.

56 Ibid.

57 Boulware, "Raiders of the Lost Panty."

58 Ibid.

59 McGregor, "Toys for the Boys," 10.

60 Ibid.

61 See Grossman, *On Killing,* and Grossman and DeGaetano, *Stop Teaching Our Kids to Kill.*

62 See Babe, "Understanding the Cultural Ecology Model," 1–23, and Kalle Lasn, *Culture Jam,* 9–29.

CHAPTER TWELVE

1 Cited in "Good Will Gaming," 70.

2 Cited in Segan, "A World of Their Own."

3 Business Wire, "Electronic Arts Ships More Than One Million Copies of *The Sims.*"

4 "Maxis' Patrick Buechner."

5 Ibid.

6 Ibid.

7 Cited in "Good Will Gaming," 70.

8 Farmer, "Electronic Arts Makes Net Push with AOL Deal"; Bloomberg News, "Electronic Arts Poised for Net Play," "Short Take."

9 Lombardi, "Welcome to the Dollhouse," 67.

10 All quotes in this paragraph are from ibid.

11 Markoff, "Something Is Killing the Sims."

12 Friedman, "Making Sense of Software," 78.

13 "Maxis' Patrick Buechner."

14 Lombardi, "Welcome to the Dollhouse," 67.

15 Ibid.

16 Ibid.

17 All quotes in this paragraph are from Herz, "The Sims Who Die With the Most Toys Win."

18 Ibid.

19 Jenkins, cited in Segan, "A World of Their Own."

20 Klein, *No Logo,* 78–9.

21 Lombardi, "Welcome to the Dollhouse," 67.

22 United States, Federal Trade Commission, "FTC Releases Report."

23 Friedman, "Making Sense of Software," 80.
24 Cited in Business Wire, "Electronic Arts."
25 Abreu, "Videogame Makers Sue Yahoo Over Piracy."
26 Lunenfeld, *Snap to Grid*, 5.
27 Steve Johnson, "The Sim Salesman."
28 Rifkin, *The Age of Access*, 260.
29 Ibid.
30 Cited in Janelle Brown, "Coming Soon to Computer Games – Advertising."
31 "Advertising in Videogames: Plug and Play."
32 Woosely, "Conducent Entices Game Companies."
33 "Where There's MUD, There's Brass," 82.
34 Uhlig, "Youth Grow Bored with Internet," A1, A11.
35 See Cavanagh and Broad, "Global Reach," 21–4; Ross, ed., *No Sweat*; and also Dyer-Witheford, *Cyber-Marx*.
36 Wazir, "Eating the Greens."
37 Associated Press, "Those Seattle WTO riots?" See also MacDonald, "Click Here to Smash Capitalism," R2.
38 Zizek, "Welcome to the Desert of the Real."
39 Herz, "War as Art."
40 Klein, "War Isn't a Game After All," A10.
41 Kaplan, "Virtual War Waged on New Terrain," A1, A4.
42 Ibid.
43 Lanier, "Let's Conquer the Divide."
44 Ibid.
45 Menzies, "Digital Networks," 540.
46 Carey, "Canadian Communications Theory," 45, cited in Menzies, "Digital Networks," 550.
47 Menzies, "Digital Networks," 545.
48 See Nichols, "The Work of Culture," 90–116, for a discussion of this case.
49 Ibid., 111.
50 Ibid., 112.

CODA

1 Williams, *Towards 2000*, 129.
2 Ibid., 268.

Bibliography

Abernathy, W., and J. Utterback. "Patterns of Industrial Innovation." *Technology Review* 50, no. 7 (1978): 40–7.

Abreu, Elinor. "Videogame Makers Sue Yahoo Over Piracy." *The Industry Standard*. 29 March 2000. Available online at http://www.thestandard.com/article/0,1902,13465,00.html. Accessed 23 March 2002.

"Action: Building the Perfect Game." *Computer Gaming World* (January 1999): 172.

Adorno, Theodor, and Max Horkheimer. *Dialectic of Enlightenment*. New York: Herder and Herder, 1972.

"Advertising in Videogames: Plug and Play." *The Economist* (24 July 1999). Available online at http://www.economist.com. Accessed 12 May 2000.

Aglietta, Michel. *A Theory of Capitalist Regulation: The US Experience*. London: New Left Books, 1979.

Aksoy, Asu, and Kevin Robins. "Hollywood for the Twenty-First Century: Global Competition for Critical Mass in Image Markets." *Cambridge Journal of Economics* 16 (1992): 1–22.

Alexander and Associates. "Comparing Generations of Console Gaming." New York: Alexander and Associates, 20 June 2000. Available online at http://www.alexassoc.com/games/Gdmemo.shtml. Accessed 16 March 2002.

– "Game Platforms and the Future of the Internet." New York: Alexander and Associates, 1999. Available online at http://www.alexassoc.com/white/indexgames.html. Accessed 12 May 2000.

– "Shakeup at Sony: What's the PlayStation Costing?" New York: Alexander and Associates, 1995. Available online at http://www.alexassoc.com/white/Sony.html. Accessed 17 May 1998.

Alloway, Nola, and Pam Gilbert. "Video Game Culture: Playing with Masculinity, Violence, and Pleasure." In *Wired-up: Young People and the Electronic Media*, edited by Sue Howard, 95–114.

Althusser, Louis. *Lenin and Philosophy*. London: New Left Books, 1977.

Amin, Ash, ed. *Post-Fordism: A Reader.* Oxford: Blackwell, 1994.

Ang, Ien. *Watching Dallas: Soap Opera and the Melodramatic Imagination.* London: Methuen, 1985.

Anonymous. Personal interview, Vancouver, BC, 1997.

Asia Monitor Resource Center. "The Hong Kong Takeover of South China." *Multinational Monitor* 18, no. 6 (1997): 12.

Associated Press. "Those Seattle WTO Riots? It's Just a Game Now, Folks." *Seattle Times*, 29 May 2001. Available online at http://www.seattletimes.com. Accessed 25 September 2001.

"Atari Gets Tough." Newsbytes News Network, 5 June 1984. Available online at http://www.newsbytes.com/pubNews/84/56938.html. Accessed 1 June 1999.

Atkinson, Bill. "Video Games: In Praise of Folly." *Globe and Mail* (Toronto), 18 July 1998, D5.

Autio, Antti. "Gender Politics of a Virtual Warrior: Masculinity and Masculine Discourse in Quake." August 1998. Available online at http://www.student.oulu.fi/~anautio/quake/quake.html. Accessed 18 March 2002.

Babe, Robert. "Understanding the Cultural Ecology Model." In *Cultural Ecology: The Changing Dynamics of Communications*, edited by Danielle Cliché, 1–23.

"Babes with Guns: Britain's Videogame Industry." *The Economist* (22 February 1997): 74–5.

Barbrook, Richard. "Mistranslations: Lipietz in London and Paris." *Science as Culture* 8 (1990): 80–117.

Barbrook, Richard, and Andy Cameron. "The Californian Ideology." *Science as Culture* 6, no. 1 (1996): 44–72.

Barlow, John Perry. "The Economy of Ideas." *Wired* 2, no. 3 (March 1994). Available online at http://www.wired.com. Accessed 1 March 1998.

Battelle, John, with Bob Johnstone. "Seizing the Next Level: Sega's Plans for World Domination." *Wired* 1, no. 6 (December 1993). Available online at http://www.wired.com. Accessed 14 March 2002.

Baudrillard, Jean. *The Consumer Society: Myths and Structures.* London: Sage, 1998 [1970].

– *Simulations.* New York: Semiotext(e), 1983.

Bell, Daniel. *The Coming of Post-Industrial Society: A Venture in Social Forecasting.* New York: Basic Books, 1973.

– *The Cultural Contradictions of Capitalism.* New York: Basic Books, 1976.

Bennahum, David S. *Extra Life: Coming of Age in Cyberspace.* New York: Basic Books, 1998.

– "Massive Attack," *Spin* (14 December 1998): 163–7.

– "Your Email's Gonna Get You." *Wired* 7, no. 5 (May 1999). Available online at http://www.wired.com. Accessed 14 March 2002.

Berman, Marc. "Japanese Invasion." *MediaWeek* 10, no. 11 (13 March 2000): 46.

Bleeker, Julian. "Urban Crisis: Past, Present and Virtual." *Socialist Review* 24, nos. 1–2 (1995): 189–223.

Bloom, David. "Hollywood Square." *Red Herring* (13 April 2001): 90–2.

– "Nintendo Builds a Better Monster." *Red Herring* (6 April 2001). Available online at http://www.redherring.com. Accessed 22 January 2002.

Bloom, David, and Dean Takahashi. "Can Games Make Money Online?" *Red Herring* (6 April 2001). Available online at http://www.redherring.com. Accessed 23 January 2002.

Bloomberg News. "Electronic Arts Poised for Net Play." CNET News.com (30 July 1999). Available online at http://news.cnet.com. Accessed 4 August 1999.

– "Short Take: EA Invests for Web Games, PlayStation2." CNET News.com (21 August 2000). Available online at http://news.cnet.com. Accessed 18 September 2000.

Boulware, Jack. "Raiders of the Lost Panty." *Salon.com* (16 May 2000). Available online at http://www.salon.com. Accessed 18 March 2002.

Bourdieu, Pierre. *Distinction: A Social Critique of the Judgement of Taste.* Cambridge, MA: Harvard University Press, 1984.

Brand, Stewart. "Fanatic Life and Symbolic Death Among the Computer Bums." *Rolling Stone* (7 December 1972). Available online at http://www.wheels.org/spacewar/stone/rolling_stone.html. Accessed 22 April 1998.

Brandt, Richard L. "Nintendo Battles for its Life." *Upside Magazine* (1 October 1995). Available online at http://www.upside.com. Accessed 1 May 1998.

Brautigam, Tara. "Mousing to Megabucks." *Toronto Star*, 8 January 2002, E4.

Breeze, Mary-Anne. "Quake-ing in My Boots: Examining Clan Community Construction in an online Gamer Population." *Cyber-Sociology Magazine* 2 (20 November 1997). Available online at http://members.aol.com/cybersoc. Accessed 2 April 2000.

Brook, James, and Iain A. Boal, eds. *Resisting the Virtual Life: The Culture and Politics of Information.* San Francisco: City Lights, 1995.

Bronson, Po. *The Nudist on the Late Shift and Other Tales of Silicon Valley.* New York: Broadway, 1999.

Brown, David. *Cybertrends: Chaos, Power, and Accountability in the Information Age.* London: Viking, 1997.

Brown, Janelle. "Coming Soon to Computer Games – Advertising." *Salon.com* (21 March 1999). Available online at http://www.salon.com. Accessed 4 April 2002.

Browning, John, and Spencer Reiss. "Encyclopedia of the New Economy." *Wired* (April 1998). Available online at http://www.hotwired.com. Accessed 3 April 1998.

Buchanan, Elizabeth. "Strangers in the 'Myst' of Video Gaming: Ethics and Representation." *Computer Professionals for Social Responsibility Newsletter* 18, no. 1 (Winter 2000). Available online at http://www.cpsr.org. Accessed 20 April 2000.

Burnham, Van. *Supercade: A Visual History of the Videogame Age, 1971–1984*. Cambridge, MA: MIT Press, 2001.

Burstein, Daniel, and David Kline. *Road Warriors: Dreams and Nightmares Along the Information Highway*. New York: Penguin, 1996.

Business Software Alliance. "Sixth Annual BSA Global Software Piracy Study." Washington, DC. May 2001. Available online at http://www.bsa.org/resources/2001–05–21.55.pdf. Accessed 2 April 2002.

Business Wire. "Electronic Arts and Sony Computer Entertainment America Nab Internet Pirate Ring: Companies File Joint Lawsuit against Online Pirates." 6 August 1999. Available online at http://www.retailsupport.ea.com/corporate/pressreleases/piracy.html. Accessed 28 March 2002.

– "Electronic Arts Ships More than One Million Copies of *The Sims*." 12 April 2000. Available online at http://www.findarticles.com. Accessed 15 May 2000.

Cairncross, Frances. *The Death of Distance: How the Communications Revolution Will Change Our Lives*. Boston: Harvard Business School Press, 1997.

Caldwell, John Thornton, ed. *Theories of the New Media: A Historical Perspective*. London: Athlone, 2000.

– *Electronic Media and Technoculture*. New Jersey: Rutgers, 2000.

Canadian Press Newswire. "Pokémon Phenomenon Strikes Today." 9 November 1999.

Carey, James. "Canadian Communications Theory: Extensions and Interpretations of Harold Innis." In *Studies in Canadian Communication*, edited by Gertrude Joch Robinson and Donald F. Theall, 27–59.

Carlton, Jim, and David P. Hamilton. "Infighting Mars Sega Comeback." *Wall Street Journal* (New York), 8 September 1999.

Carter, Dale. *The Final Frontier: The Rise and Fall of the American Rocket State*. London: Verso, 1988.

Carroll, Jon. "D(Riven)." *Wired* 5, no. 9 (September 1997). Available online at http://www.wired.com. Accessed 4 May 2000.

– "Guerillas in the *Myst*." *Wired* 2, no. 8 (August 1994). Available online at http://www.wired.com. Accessed 4 May 2000.

Cassell, Justine, and Henry Jenkins, eds. *From Barbie to Mortal Kombat: Gender and Computer Games*. Cambridge, MA: MIT Press, 1998.

Cassell, Justine, and Henry Jenkins. "Chess for Girls? Feminism and Computer Games." In *From Barbie to Mortal Kombat: Gender and Computer Games*, edited by Justine Cassell and Henry Jenkins, 2–45.

Cassidy, John. *Dot.Con: The Greatest Story Ever Sold*. New York: Harper Row, 2002.

Castells, Manuel. *The Informational City: Information Technology, Economic Restructuring, and the Urban-Regional Process*. Oxford: Blackwell, 1989.

– *The Network Society: The Information Age Vol. 1*. Oxford: Blackwell, 2000.

Cavanagh, John, and Robin Broad. "Global Reach: Workers Fight the Multinationals." *The Nation* (18 March 1996): 21–4.

Christopherson, Susan, and Michael Storper. "The Effects of Flexible Specialization on Industrial Politics and the Labour Market: The Motion Picture Industry." *Industrial and Labour Relations Review* 42, no. 3 (1989): 331–47.

Clapes, Anthony. *Softwars: The Legal Battles for Control of the Software Industry.* Westport, CT: Quorum, 1993.

Clarke, John. "Dupes and Guerillas: The Dialectics of Cultural Consumption." In *The Consumer Society Reader,* edited by Martyn J. Lee, 288–93.

– *New Times and Old Enemies: Essays on Cultural Studies and America.* New York: Harper Collins, 1991.

Cliché, Danielle, ed. *Cultural Ecology: The Changing Dynamics of Communications.* London: International Institute of Communications, 1997.

"Club iMusic Showcase: Wipeout XL." *IMusic.com* (February 1998). Available online at http://imusic.com/showcase/club/wipeo.html. Accessed 16 May 1998.

"Colloquy." *Critical Studies in Mass Communication* 12, no. 1 (1995): 60–100.

Cockburn, Cynthia. "The Circuit of Technology: Gender, Identity and Power." In *Consuming Technologies: Media and Information in Domestic Spaces,* edited by Roger Silverstone and Eric Hirsch, 32–47.

Cohen, Scott. *Zap! The Rise and Fall of Atari.* New York: McGraw-Hill, 1984.

Collier, Robert. "NAFTA Labor Problems Haunt New Trade Debate." *San Francisco Chronicle,* 10 September 1997, A1.

Communications Workers of America Washington, DC. "CWA Files NAFTA Complaint to Aid Mexican Telecom Workers." 17 October 1996. Available online at http://www.cwa-union.org. Accessed 8 November 1998.

Cook, Daniel, ed. *Symbolic Childhood.* New York: Peter Lang Publishing, 2002.

Crawford, Chris. *The Art of Computer Game Design.* Berkeley, CA: Osborne/McGraw-Hill, 1984. Available online at http://www.vancouver.wsu.edu/fac/peabody/game-book/Coverpage.html. Accessed 11 March 2002.

Croal, N'Gai. "The Art of the Game." *Newsweek* (6 March 2000): 80–3.

Cunningham, Helen. "Mortal Kombat and Computer Game Girls." In *Theories of the New Media: A Historical Perspective,* edited by John Thornton Caldwell, 213–17.

Davis, Jim, Thomas Hirschl, and Michael Stack, eds. *Cutting Edge: Technology, Information, Capitalism and Social Revolution.* London: Verso, 1997.

Dawson, Brett. "Brand Games." *Undercurrents.* Canadian Broadcasting Corporation, originally broadcast April 2000.

Deacon, James. "The Craze That Ate Your Kids," *Maclean's* 112, no. 45 (8 November 1999): 74.

Der Derian, James. *Virtuous War: Mapping the Military-Industrial-Media Entertainment Network*. Boulder, CO: Westview Press, 2001.

De Landa, Manuel. *War in the Age of Intelligent Machines*. New York: Zone Books, 1991.

DFC Intelligence. "Interactive Electronic Entertainment Industry Overview." San Diego: DFC Intelligence, December 1997. Available online at http://www.dfcint.com/news/news1old.html. Accessed 21 May 1998.

Dodsworth, Clark, Jr., ed. *Digital Illusion: Entertaining the Future with High Technology*. New York: ACM Press, SIGGRAPH Series, 1998.

Dohse, Knuth, Ulrich Jurgens, and Thomas Malsch. "From 'Fordism' to 'Toyotism'? The Social Organization of the Labor Process in the Japanese Automobile Industry." *Politics and Society* 14, no. 2 (1985): 115–46.

Dolan, Mike. "Behind the Screens." *Wired* 9, no. 5 (May 2001). Available online at http://www.wired.com. Accessed 16 March 2002.

Drummond, Michael. *Renegades of the Empire: How Three Software Warriors Started a Revolution Behind the Walls of Fortress Microsoft*. New York: Crown, 1999.

du Gay, Paul. "Introduction." In Paul du Gay, Stuart Hall, Linda James, Hugh Mackay, and Keith Negus. *Doing Cultural Studies: The Story of the Sony Walkman*.

du Gay, Paul, Stuart Hall, Linda James, Hugh Mackay, and Keith Negus. *Doing Cultural Studies: The Story of the Sony Walkman*. London: Sage/Open University, 1997.

du Gay, Paul, ed. *Production of Culture/Cultures of Production*. London: Sage/Open University, 1997.

Dunn, Ashley. "Finding Art in an Internet Game." *New York Times*, 30 July 1997, C9.

Dutton, William, ed. *Information and Communication Technologies: Visions and Realities*. Oxford: Oxford University Press, 1996.

Dyer-Witheford, Nick. *Cyber-Marx: Cycles and Circuits of Struggle in High-Technology Capitalism*. Urbana-Champaign, IL: University of Illinois Press, 1999.

– "The Work in Digital Play: Video Gaming's Transnational and Gendered Division of Labor." *Journal of International Communication* 6, no. 1 (June 1999): 69–93.

Edwards, Paul. *Computers and the Politics of Discourse in Cold War America*. Cambridge, MA: MIT Press, 1997.

"Electronic Arts Does Its First Thai Language Game." *Bangkok Post*, 5 August 1998. Available online at http://www.bkpost.samart.co.th/news/DBarchive/. Accessed 23 April 1999.

"Electronic Games Market Expands as Players Mature." *Financial Post* (Toronto) 89, no. 17 (29 April 1995): C22.

Eng, Paul. "Net Games Are Drawing Crowds." *Business Week* (31 March 1997): 73–4.

Ewen, Stuart. *The Captains of Consciousness: Advertising and the Social Roots of the Consumer Culture.* New York: McGraw-Hill, 1976.

Farmer, Melanie Austria. "Electronic Arts Makes Net Push with AOL Deal." CNET News.com (22 November 1999). Available online at http://news.cnet.com. Accessed 18 May 2000.

Ferguson, Marjorie, and Peter Golding, eds. *Cultural Studies in Question.* London: Sage, 1997.

– "Cultural Studies and Changing Times: An Introduction." In *Cultural Studies in Question,* edited by Marjorie Ferguson and Peter Golding, xiii–xxvii.

"First Round of NAFTA Unemployment Benefits Issued." *NAFTA Monitor* 1, no. 7 (8 February 1994). Available online at http://www.etext.org/politics/naftamonitor/volume.1/nm01-007. Accessed 3 September 1999.

Fischer, Alan D. "Keyboard Firm Ranks with the Best." *Arizona Daily Star* (Tucson), 22 January 1996, D1. Available online at http://www.azstarnet.com. Accessed 3 September 1999.

Flanigan, James. "Whether Apple Grows Anew Depends on Human Factor." *Los Angeles Times,* 10 August 1997, D1, D6.

Flower, Joe. "The Americanization of Sony: The Untold Story of How Sony Is Rapidly Becoming an American Company." *Wired* 2, no. 6 (June 1994): 94–8, 134–5.

Forester, Tom. *Silicon Samurai: How Japan Conquered the World's IT Industry.* Cambridge, MA: Blackwell, 1993.

Frank, Thomas. *One Market under God: Extreme Capitalism, Market Populism, and the End of Economic Democracy.* New York: Doubleday, 2000.

Frauenfelder, Mark. "Death Match: Your Guide to the Box Wars." *Wired* 9, no. 5 (May 2001). Available online at http://www.wired.com. Accessed 3 January 2002.

Friedman, Ted. "Making Sense of Software: Computer Games and Interactive Textuality." In *CyberSociety: Computer Mediated Communication and Community,* edited by Steven G. Jones, 73–89.

Friedrich, Otto. "Machine of the Year: The Computer Moves In." *Time* (3 January 1983). Available online at http://www.time.com. Accessed 11 March 2002.

Fuller, Mary, and Henry Jenkins. "Nintendo and New World Travel Writing: A Dialogue." In *CyberSociety: Computer-Mediated Communication and Community,* edited by Steven G. Jones, 52–72.

"Game Makers to Launch New Generations." *Financial Post* (Toronto), 8–10 July 1995, C4.

"Game Over?" *New Scientist* (16 January 1999). Available online at http://www.newscientist.com. Accessed 7 March 2000.

Garnham, Nicholas. "Constraints on Multimedia Convergence." In *Information and Communication Technologies: Visions and Realities,* edited by William Dutton, 103–19.

– *Capitalism and Communication: Global Culture and the Economics of Information.* London: Sage, 1990.

Gates, Bill. *The Road Ahead.* New York: Norton, 1995.

Gilmour, Heather. "What Girls Want: Female Spectators and Interactive Games." Available online at http://www-cntv.usc.edu/heim/girlga.html. Accessed 18 March 2002.

Glaubke, Christina R., Patti Miller, McCrae A. Parker, and Eileen Espejo. "Fair Play: Violence, Gender and Race in Video Games." Los Angeles: Children Now, 2002. Available online at http://www.childrennow.org/pubs-media.htm. Accessed 19 March 2002.

Goldman, Robert, and Stephen Papson. *Sign Wars: The Cluttered Landscape of Advertising.* New York: Guilford Press, 1996.

Goodfellow, Kris. "Sony Comes on Strong in Video-Game Wars." *New York Times,* 25 May 1998, D5.

"Good Will Gaming: An Interview With Gaming's Great Constructor: Will Wright." *Computer Gaming World* (May 2000): 70–6.

Graham, Julie. "Fordism/Post-Fordism, Marxism-Post-Marxism." *Rethinking Marxism* 4, no. 1 (1991): 39–58.

Gramsci, Antonio. *Selections from the Prison Notebooks.* Edited by Quintin Hoare and Geoffrey Nowell-Smith. New York: International Publishers, 1971.

Green, Bill, Jo-Anne Reid, and Chris Bigum. "Teaching the Nintendo Generation? Children, Computer Culture and Popular Technologies." In *Wired-up: Young People and the Electronic Media,* edited by Sue Howard, 19–41.

Greenfield, Patricia M., and Rodney C. Cocking, eds. *Interacting with Video.* Norwood, NJ: Ablex, 1996.

Grossberg, Lawrence. *Bringing It All Back Home: Essays on Cultural Studies.* Durham, NC: Duke University Press, 1997.

Grossman, David. *On Killing: The Psychological Costs of Learning to Kill in War and Society.* Boston, MA: Little, Brown, 1996.

Grossman, David, and Gloria DeGaetano. *Stop Teaching Our Kids to Kill: A Call to Action against TV, Movie and Video Game Violence.* New York: Random House, 1999.

GTE Entertainment. "Titanic Adventure Out of Time." Available online at http://www.gamespot.com. Accessed 4 May 1999.

Haddon, Leslie. "The Development of Interactive Games." In *The Media Reader: Continuity and Transformation,* edited by Hugh Mackay and Tim O'Sullivan, 305–27.

– "Electronic and Computer Games: The History of an Interactive Medium." *Screen* 29, no. 2 (1988): 52–75.

Hall, Stuart. "Encoding/decoding." In *Culture, Media, Language,* edited by Stuart Hall, Dorothy Hobson, Andrew Lowe, and Paul Willis, 128–39.

– "The Meaning of New Times." In *New Times: The Shape of Politics in the 1990s*, edited by Stuart Hall and Martin Jacques, 116–36.

Hall, Stuart, Dorothy Hobson, Andrew Lowe, and Paul Willis, eds. *Culture, Media, Language*. London: Hutchinson, 1980.

Hall, Stuart, and Martin Jacques, eds. *New Times: The Shape of Politics in the 1990s*. London: Lawrence and Wishart, 1989.

Hamilton, James T. *Channeling Violence: The Economic Market for Violent Television Programming*. Princeton, NJ: Princeton University Press, 1998.

"Hands On Lara Croft." *Game Informer Magazine*. Available online at http://www.gameinformer.com/news/feb99/020499b.html. Accessed 2 April 1999.

Hansen, Evan, John Borland, and Mike Yamamoto. "Games May Point to Future." CNET News.com. Available online at http://news.cnet.com. Accessed 17 March 2002.

Harmon, Amy. "With the Best Research and Intentions, a Game Maker Fails." *New York Times on the Web*, 22 March 1999. Available online at http://www.idg.net/english/crd_idg_net_763537.html. Accessed 18 March 2002.

Harris, Lesley Ellen. *Digital Property: Currency of the 21st Century*. Toronto: McGraw-Hill, Ryerson, 1998.

Harvey, David. *The Condition of Postmodernity: An Enquiry into the Origins of Cultural Change*. Oxford: Blackwell, 1989.

Hatlestad, Luc. "Games People Play." *Red Herring* (June 1997). Available online at http://www.redherring.com. Accessed 2 April 2002.

Hayes, Dennis. *Behind the Silicon Curtain: The Seductions of Work in a Lonely Era*. Boston: South End Press, 1989.

Hayes, Michael, and Stuart Dinsey, with Nick Parker. *Games War: Video Games – A Business Review*. London: Bowerdean Publishing, 1995.

Heffernan, Nick. *Capital, Class and Technology in Contemporary American Culture: Projecting Post-Fordism*. London: Pluto, 2000.

Henderson, Rebecca, and Kim Clark. "Architectural Innovation: The Reconfiguration of Existing Product Technologies and the Failure of Established Firms." *Administrative Science Quarterly* 35 (1990): 9–30

Henry, Kelly, and Kevin Hause. "Videogame Consumer Segmentation Survey 1999." Fromingham, MA: International Data Corp. Available online at http://www.idcresearch.com/Press/default. Accessed 2 February 2000.

Herman, Andrew, and Thomas Swiss, eds. *The World Wide Web and Contemporary Cultural Theory*. New York: Routledge, 2000.

Herman, Edward S., and Robert W. McChesney. *The Global Media: The New Missionaries of Corporate Capitalism*. London: Cassell, 1997.

– "The Global Media in the Late 1990s." In *The Media Reader: Continuity and Transformation*, edited by Hugh Mackay and Tim O'Sullivan, 178–210.

Herman, Edward S., and Noam Chomsky. *Manufacturing Consent: The Political Economy of the Mass Media*. New York: Pantheon, 1988.

Herman, Leonard. *Phoenix: The Fall and Rise of Videogames.* 2d ed. Union, NJ: Rolenta, 1997.

Herman, Leonard, Jer Horwitz, Steve Kent, and Skyler Miller. "The History of Video Games." *GameSpot.* 2001. Available online at http://gamespot.com. Accessed 4 April 2002.

Herz, J.C. "The Sims Who Die with the Most Toys Win." *New York Times,* 10 February 2000, G10.

– "Technology as the Guiding Hand of History." *New York Times,* 16 December 1999, G20.

– "A Designer's Farewell to His Fantasy Realm." *New York Times,* 15 July 1999, G5.

– "War as Art." *New York Times,* 24 December 1998, G4.

– "Under Sony's Wing: Novel Games Incubate." *New York Times,* 28 May 1998, E4.

– *Joystick Nation: How Videogames Ate Our Quarters, Won Our Hearts, and Rewired Our Minds.* Boston: Little, Brown, 1997.

Hettema, Phil. "Location-Based Entertainment." In *Digital Illusion: Entertaining the Future with High Technology,* edited by Clark Dodsworth, Jr., 373–80.

"How Sega and Sony Try and Get in Your Heads." *Next Generation* (1998). Available online at http://www.next-generation.com/features/marketing/sysmarketing.html. Accessed 10 November 1998.

Howard, Sue, ed. *Wired-up: Young People and the Electronic Media.* London: University College London Press, 1988.

Innis, Harold A. *The Bias of Communication.* Toronto: University of Toronto Press, 1991 [1951].

– *Empire and Communications.* Victoria, BC: Press Porépic, 1986 [1950].

"Intellectual Property: Bazaar Software." *The Economist* (5 March 1997). Available online from http://www.economist.com. Accessed 24 July 1999.

Interactive Digital Software Association (IDSA). "Economic Impacts of the Demand for Playing Interactive Entertainment Software." Washington, DC: IDSA. April 2001. Available online at http://www.idsa.com/pressroom.html. Accessed 4 May 2001.

– "Fast Facts." Washington, DC: IDSA, 1999. Available online at http://www.idsa.com/releases/industry.htm. Accessed 1 November 1999.

– "The State of the Entertainment Software Industry 1998: An IDSA Report." Washington, DC: IDSA, 1998. Available online at http://www.idsa.com/press/execusum.html. Accessed 24 July 1999.

– "The State of the Entertainment Software Industry 1997: Executive Summary." Washington, DC: IDSA, 1998. Accessed 24 July 1999.

– "State of the Industry: Report 2000–2001." Washington, DC: IDSA, 2002. Available online at http://www.idsa.com/releases/SOTI2001.pdf. Accessed 17 March 2002.

- "Video Games and Youth Violence: Examining the Facts." Washington, DC: IDSA, October 2001. Available online at http://www.idsa.com/IDSAfinal.pdf. Accessed 18 March 2002.
- "1999 Genre Breakdown by Sales." Washington, DC: IDSA, 1999. Available online at http://www.idsa.com.charts.html. Accessed 1 November 1999.
- "What Does the Future Hold for the Interactive Entertainment Industry?" Washington, DC: IDSA, 1999. Available online at http://www.idsa.com/releases/industry.html. Accessed 1 November 1999.

"Interviewing the Big Guys." *Video Game Advisor.* Available online at http://www.videogamespot.com/features/universal/hov/index.html. Accessed 14 November 1999.

"In Their Dreams." *The Economist* (26 February 2000): 71–2.

Jameson, Fredric. "Postmodernism: Or the Cultural Logic of Late Capitalism." *New Left Review* 146 (1984): 55–92.

Jenkins, Henry. "Professor Jenkins Goes to Washington." July 1999. Available online at http://web.mit.edu/21fms/www/faculty/henry3/profjenkins.html. Accessed 30 March 2001.
- "'x Logic': Repositioning Nintendo in Children's Lives." *Quarterly Review of Film and Video* 14, no. 4 (1993): 55–70.

Johnson, Richard. "What Is Cultural Studies Anyway?" In *What Is Cultural Studies? A Reader,* edited John Storey, 75–114.

Johnson, Steve. "The Sim Salesman." *Feed* (26 January 1999). Available online at http://www.feedmag.com/column/interface/ci164lofi.html. Accessed 25 April 2000.

Johnstone, Bob. "Pirates Ahoy!" *Far Eastern Economic Review* (24–31 December 1992): 68.
- "Video Games: Dream Machines." *Far Eastern Economic Review* (24–31 December 1992): 66.

Jones, Steven G., ed. *CyberSociety: Computer Mediated Communication and Community.* Thousand Oaks, CA: Sage, 1995.

Kapica, Jack. "Fun and Games Drive Computer Innovations." *Globe and Mail* (Toronto), 18 July 1998, C10.

Kaplan, Karen. "Virtual War Waged on New Terrain." *Toronto Star,* 4 November 2001, A1, 4.

Katayama, Osamu. *Japanese Business: Into the 21st Century.* London: Athlone, 1996.

Katz, Warren. "Networked Synthetic Environments: From DARPA to Your Virtual Neighborhood." In *Digital Illusion: Entertaining the Future with High Technology,* edited by Clark Dodsworth, Jr., 115–27.

Kay, Jonathan. "Are Video Games Bad for Your Health?" *National Post* (Toronto), 22 July 2000, B4.

Keegan, Paul. "Culture Quake." *Mother Jones* 29, no. 6 (November/December 1999): 42–9. Available online at http://www.motherjones.com. Accessed 18 March 2002.

– "In the Line of Fire." *Guardian* (London), 1 June 1999, G2.

Keighley, Geoff. "Angelina Jolie: Lara's Splitting Image?" *GameSpot* PC, 14 June 2001. Available online at http://gamespot.com/gamespot. Accessed January 10 2002.

Kelly, Kevin. *Out of Control: The Rise of Neo-Biological Civilization.* New York: Addison-Wesley Publishing, 1994.

– "New Rules for the New Economy: Twelve Dependable Principles for Thriving in a Turbulent World." *Wired* 5, no. 09 (September 1997). Available online at http://www.wired.com. Accessed 1 May 1998.

Kent, Steven L. *The First Quarter: A 25-Year History of Video Games.* Washington: BWD Press, 2000.

Kharif, Olga. "Let the Games Begin – Online." *Business Week Online,* 13 December 2001. Available online at http://businessweek.com. Accessed 10 January 2002.

Kim, Amy Jo. "Killers Have More Fun." *Wired* 6., no. 5 (May 1998). Available online at http://www.wired.com. Accessed 2 April 2002.

Kinder, Marsha. *Playing with Power in Movies, Television, and Video Games: From Muppet Babies to Teenage Mutant Ninja Turtles.* Berkeley: University of California Press, 1991.

– "Contextualizing Video Game Violence: From Teenage Mutant Ninja Turtles 1 to Mortal Kombat 2." In *Interacting with Video,* edited by Patricia M. Greenfield and Rodney C. Cocking, 25–38.

Kinney, David. "The Virtual Squadrons of Fightertown." In *Digital Illusion: Entertaining the Future with High Technology,* edited by Clark Dodsworth, Jr., 479–87.

Kirkland, Bruce. "Pokémon: 2nd Wave." *Canoe,* 11 March 2000. Available online at http://www.canoe.ca/JamMoviesFeaturesP/pokemon_1.html. Accessed 24 March 2002.

Klein, Naomi. *No Logo: Taking Aim at the Brand Bullies.* Toronto: Knopf Canada, 2000.

– "War Isn't a Game After All." *Globe and Mail* (Toronto), 14 September 2001, A10.

Kline, Stephen. "Pleasures of the Screen: Why Young People Play Video Games." *Proceedings of the International Toy Research Seminar.* Angouléme, France, 9–14 November 1997.

– *Out of the Garden: Toys and Children's Culture in the Age of TV Marketing.* Toronto: Garamond Press, 1993.

Kline, Stephen, and Greig de Peuter. "Ghosts in the Machine: Postmodern Childhood, Video Gaming, and Advertising." In *Symbolic Childhood,* edited by Daniel Cook, 255–78.

Kline, Stephen, with Albert Banerjee. "Video Game Culture: Leisure and Play Preferences of BC Teens." Burnaby, BC: Media Analysis Lab, Simon Fraser University, 1998.

Koelsh, Frank. *The Infomedia Revolution: How It Is Changing Our World and Your Life*. Toronto: McGraw-Hill, 1995.

Kraft, Philip, and Richard Sharpe. "Software Globalization." Paper presented at the "Work, Difference, and Social Change" conference, State University of New York at Binghamton, 8–10 May 1998.

Kroker, Arthur. *Technology and the Canadian Mind: Innis/McLuhan/Grant*. Montreal: New World Perspectives, 1984.

Kundnani, Arun. "Where Do You Want to Go Today? The Rise of Information Capital." *Race and Class* 40, nos. 2–3 (1998–99): 49–71.

Kushner, David. "Feed Daily." *Feed* (25 April 2000). Available online at http://www.feedmag.com/daily/dy042500_master.html. Accessed 13 May 2000.

Laidlaw, Marc. "The Egos at Id." *Wired* 4, no. 8 (August 1996). Available online at http://www.wired.com. Accessed 4 April 1999.

Lanier, Jaron. "Let's Conquer the Divide." *CIO Magazine* (15 January 2000). Available online at http://www.cio.com/archive/011500_diff_content.html. Accessed 15 March 2000.

LaPlant, Alice, and Rich Seidner. *Playing for Profit*. New York: Wiley, 1999.

"Lara in Nike Commercial." *Tomb Raider 2000 World Wide*. Available online at http://www.trww.com/nike.shtml. Accessed 8 October 2000.

Larimer, Tim. "The Ultimate Game Freak." *Time Asia* 154, no. 20 (22 November 1999). Available online at http://www.time.com/time/asia. Accessed 23 March 2002.

Lasn, Kalle. *Culture Jam: The Uncooling of America*. New York: Morrow, 1999.

Latour, Bruno. *Science in Action*. Milton Keynes: Open University Press, 1987.

Laurel, Brenda. *Utopian Entrepreneur*. Cambridge, MA: MIT Press, 2001.

Le Diberder, Alain, and Frédéric Le Diberder. *L'Univers des Jeux Vidéo*. Paris: Éditions La Découverte, 1998.

Lee, Martyn J. *Consumer Culture Reborn: The Cultural Politics of Consumption*. London: Routledge, 1993.

Lee, Martyn J., ed. *The Consumer Society Reader*. Oxford: Blackwell, 2000.

Leiss, William. *The Limits to Satisfaction: An Essay on the Problem of Needs and Commodities*. Toronto and Buffalo: University of Toronto Press, 1976.

– *Under Technology's Thumb*. Kingston and Montreal: McGill-Queen's University Press, 1990.

Leiss, William, Stephen Kline, and Sut Jhally. *Social Communication in Advertising: Persons, Products and Images of Well Being*. 2d ed. Toronto, ON: Nelson Canada, 1990.

Lenoir, Tim. "All But War Is Simulation: The Military-Entertainment Complex." *Configurations* 8, no. 3 (Fall 2000): 289–335. Available online at http://www.stanford.edu/dept/HPS/TimLenoir/MilitaryEntertainmentComplex.htm. Accessed 10 March 2002.

Leslie, Paul, ed. *The Gulf War as Popular Entertainment: An Analysis of the Military-Industrial Media Complex*. Lewiston, NY: Edwin Mellen Press, 1997.

Lessard, Bill, and Steve Baldwin. *Netslaves: True Tales of Working the Web*. New York: McGraw Hill, 2000.

Levidow, Les. "Foreclosing the Future." *Science and Society* 8 (1990): 59–79.

Levy, Stephen. *Hackers: Unsung Heroes of the Computer Revolution*. Garden City, NY: Anchor Press/Doubleday, 1984.

– "Here Comes PlayStation 2." *Newsweek* (6 March 2000): 55–9.

Lewis, T.G. *The Friction Free Economy: Marketing Strategies for a Wired World*. New York: Harper, 1997.

Lewis, Peter H. "Dreamcast Is a Toy, Yes, but It's a PC at Heart." *New York Times*, 9 September 1998.

Lipietz, Alain. *Mirages and Miracles: The Crisis of Global Fordism*. London: Verso, 1987.

– "Reflections on a Tale: The Marxist Foundations of the Concepts of Regulation and Accumulation." *Studies in Political Economy* 26 (Summer 1988): 32–3.

Lipietz, Alain, and D. Leborgne. "New Technologies, New Modes of Regulation: Some Spatial Implications." *Environment and Planning D: Society and Space* 6 (1988): 263–80.

Lizard. "Kill Bunnies, Sell Meat, Kill More Bunnies: Are We Having Fun Yet?" *Wired* 6, no. 5 (May 1998). Available online at http://www.wired.com. Accessed 28 March 2002.

Lombardi, Chris. "Welcome to the Dollhouse." *Computer Gaming World* (May 2000): 67.

Lowenstein, Douglas. Opening Remarks. "Computer and Video Games Come of Age: A National Conference to Explore the Current State of an Emerging Entertainment Medium," at the Program in Comparative Media Studies, Massachusetts Institute of Technology, Cambridge, MA, 10–11 February 2000. Available online at http://web.mit.edu/cms/games/opening.html. Accessed 21 March 2002.

Lunenfeld, Peter. *Snap to Grid: A User's Guide to Digital Arts, Media and Cultures*. Cambridge, MA: MIT Press, 2000.

Lyotard, Jean-François. *The Postmodern Condition: A Report on Knowledge*. Minneapolis: University of Minnesota, 1984.

MacDonald, Gayle. "Click Here to Smash Capitalism." *Globe and Mail* (Toronto), 2 March 2002, R2.

Mackay, Hugh, and Tim O'Sullivan, eds. *The Media Reader: Continuity and Transformation*. London: Sage, 1999.

MÄK Technologies. "About MÄK Technologies." Available online at http://www.mak.com/backgrounder.htm. Accessed 16 March 2002.

– "MÄK Technologies Awarded Contract for First Dual-Use Video Game." 14 April 1997. Available online at http://www.mak.com/pr_dualgame.htm. Accessed 16 March 2002.

– "MÄK Technologies Wins Army Contract to Develop HLA Compliant Sequel to Tank Simulation Game." 9 July 1998. Available online at http://www.mak.com/pr_spearhead2.htm. Accessed 16 March 2002.

Marchand, Roland. *Advertising the American Dream: Making Way for Modernity, 1920–1940*. Berkeley: University of California Press, 1985.

Marcuse, Herbert. *One-Dimensional Man*. London: Routledge and Kegan Paul, 1964.

Markoff, John. "Something Is Killing the Sims, and It's No Accident." *New York Times on the Web*, 27 April 2000. Available online at http://www.nytimes.com. Accessed 15 May 2000.

– "Tuning in to the Fight of the (Next) Century." *New York Times on the Web*, 7 March 1999. Available online at http://www.nytimes.com. Accessed 15 March 1999.

Marriott, Michel. "I Don't Know Who You Are, but (Click) You're Toast." *New York Times*, 29 October 1998, G1.

Marx, Karl. *Capital Vol. 2*. London: Lawrence and Wishart, 1978 [1885].

– *A Contribution to the Critique of Political Economy*. New York: International Publishers, 1970 [1859].

"Maxis' Patrick Buechner." Interview with Marc Dultz. *Game Week.com*. September 2000. Available online at http://www.gameweek.com. Accessed 25 May 2001.

Mayers, Adam. "Focusing on Fun." *Toronto Star*, 13 July 1997, D1.

McCandles, David. "Legion of Doom." *Wired* 6, no. 3 (May 1998). Available online at http://www.wired.com. Accessed 17 March 2002.

– "Warez Wars." *Wired* 5, no. 4 (April 1997). Available online at http://www.wirednews.com. Accessed 4 February 1998.

McChesney, Robert. *Rich Media, Poor Democracy: Communication Politics in Dubious Times*. New York: New Press, 2000.

McCormick, Moira. "Fantastic Franchises." *Billboard* (19 February 2000): 68.

– "Warner Unleashes Massive Campaign for 'Pokémon' Release." *Billboard* (22 January 2000): 108.

McGovern, Celeste. "Childhood's End: Profiteers in the Entertainment Economy Are Turning Kids into the Slaves of Consumerism." *British Columbia Report* 10, no. 14 (17 May 1999): 48–50.

McGregor, Alexander. "Toys for the Boys." *Financial Times* (London), 12–13 December 1998, 10.

McGuigan, Jim. *Cultural Populism*. London: Routledge, 1992.

McLuhan, Marshall. *Understanding Media: The Extensions of Man*. Cambridge, MA: MIT Press, 1995 [1964].

– Foreword to Harold A. Innis. *Empire and Communications*. Revised by Mary Q. Innis. Toronto and Buffalo: University of Toronto Press, 1972 [1950].

– *The Gutenberg Galaxy: The Making of Typographic Man*. Toronto: University of Toronto Press, 1962.

– *The Mechanical Bride: Folklore of Industrial Man*. New York: Vanguard Press, 1951.

McNamee, Sarah. "Youth, Gender and Video Games: Power and Control in the Home." In *Cool Places: Geographies of Youth Culture*, edited by Tracey Skelton and Gill Valentine, 195–206.

McNealy, Paul-Jon. "Let the Online Games Begin." GartnerG2. 15 November 2001. Available online at http://www.gartnerg2.com. Accessed 20 January 2002.

McRobbie, Angela. *In the Culture Society: Art, Fashion, and Popular Music*. London: Routledge, 1999.

Menzies, Heather. *Whose Brave New World? The Information Highway and the New Economy*. Toronto: Between the Lines, 1996.

– "Digital Networks: The Medium of Globalization and the Message." *Canadian Journal of Communication* 24 (1999): 539–55.

Mernagh, Matt, and James Keast. "Video Games Saved the Radio Stars." *Exclaim!* (Toronto), July 1999, 20–1.

Microsoft. "Directx." Available online at http://www.microsoft.com/directx.

"Mexico: Fackligt Stod Fran USA." *Dagens Arbete*. Available online at http://www.dagensarbet/fkl/mex970521a_013.htm. Accessed 11 August 1998.

Miles, David. "The CD-ROM Novel *Myst* and McLuhan's Fourth Law of Media: *Myst* and Its 'Retrievals.'" *Journal of Communication* 46, no. 2 (1996): 4–18.

Miles, Steven. "Towards an Understanding of the Relationship between Youth Identities and Consumer Culture." *Youth and Policy* 51 (1996): 35–45.

Mills, Lara. "Chaos Rules: Today's Media-Savvy Kids Find Traditional Linear Narrative a Bore." *Marketing Magazine* (20/27 July 1998): 17, 20.

Moody, Fred. *The Visionary Position: The Inside Story of the Digital Dreamers Who Are Making Virtual Reality a Reality*. New York: Random House, 1999.

Moore, Geoffrey A. *Crossing the Chasm: Marketing and Selling Technology Products to Mainstream Consumers*. Oxford: Capstone, 1998.

Morris-Suzuki, Tessa. *Beyond Computopia: Information, Automation and Democracy in Japan*. London: Kegan Paul, 1988.

– "Robots and Capitalism." In *Cutting Edge: Technology, Information, Capitalism and Social Revolution*, edited by Jim Davis, Thomas Hirschl, and Michael Stack, 13–27.

Morley, David. *The Nationwide Audience: Structure and Decoding*. London: British Film Institute, 1983.

– "So-Called Cultural Studies: Dead Ends and Reinvented Wheels." *Cultural Studies* 12, no. 4 (1998): 476–97.

Morley, David, and Kevin Robins. *Spaces of Identity: Global Media, Electronic Landscapes and Cultural Boundaries*. London: Routledge, 1995.

Mowrey, Mark A. "Let the Games Begin!" *Red Herring* (15 October 2001): 30.

Mosco, Vincent. *The Political Economy of Communication: Rethinking and Renewal*. London: Sage, 1996.

Mosco, Vincent, and Janet Wasko, eds. *The Political Economy of Information*. Madison: University of Wisconsin Press, 1988.

Mount, Ian. "Shorts Watch Pokémon Frenzy." *Financial Post* (Toronto), 19 October 1999, D3.

MSNBC. "Video-Game Pirates on the Loose." 6 August 1999. Available online at http://www.techtv.com/extendedplay/print/0,23102,2310674,00.html. Accessed 30 March 2002.

Muller, E.J. "The Best Game Nintendo Built." *Distribution* 92, no. 12 (1993): 30–6.

Murdock, Graham. "Base Notes: The Conditions of Cultural Practice." In *Cultural Studies in Question*, edited by Marjorie Ferguson and Peter Golding, 86–101.

– "Across the Great Divide: Cultural Analysis and the Condition of Democracy." *Critical Studies in Mass Communication* 12, no. 1 (1995): 89–95.

Murray, Janet. *Hamlet on the Holodeck: The Future of Narrative in Cyberspace*. New York: Free Press, 1997.

Myers, David. "Computer Game Genres." *Play and Culture* 3 (1989): 286–301.

Myers, Jennifer. "Today Sega City, Tomorrow the World." *Profit* (February–March 1998): 16–22.

Myst User's Manual. San Francisco, CA: UbiSoft Entertainment.

Nathan, John. *Sony: The Private Life*. Boston: Houghton Mifflin, 1999.

Negroponte, Nicholas. *Being Digital*. London: Hodder and Stoughton, 1995.

Newman, Nathan. "From Microsoft Word to Microsoft World: How Microsoft Is Building a Global Monopoly." A NetAction white paper. San Francisco, CA: The Tides Center/NetAction, 1997. Available online at http://www.netaction.org/msoft/world/Msword2world.html. Accessed 4 May 1998.

Nichols, Bill. "The Work of Culture in the Age of Cybernetic Systems." In *Electronic Media and Technoculture*, edited by John Thornton Caldwell, 90–116.

Nintendo of America. "Nintendo Gives Consumers 1000 More Reasons to Shop at Tommy Hilfiger." Available online at http://www.nintendo.com/corp/press/073098. Accessed 30 July 1998.

Nixon, Helen. "Fun and Games Are Serious Business." In *Digital Diversions: Youth Culture in the Age of Multimedia*, edited by Julian Sefton-Green, 21–42.

Nixon, Sean. "Circulating Culture." In *Production of Culture/Cultures of Production*, edited by Paul du Gay, 177–220.

Noble, David. *The Religion of Technology: The Divinity of Man and the Spirit of Invention*. New York: Knopf, 1997.

NPD. "Video and Computer Games: Genre by Unit Sale." Port Washington, NY: The NPD Group. Available online at http://www.npd/corp/tracking/gd/sld026.htm. Accessed 2 July 2000.

Obayashi, Yuka. "Strong Yen Batters Nintendo Profit: Console Sales Slip." *Financial Post* (Toronto), 25 November 1999, C16.

O'Brien, Jeffrey M. "The Making of the xbox." *Wired* 9, no. 11 (November 2001). Available online at http://www.wired.com. Accessed 22 March 2002.

Ohmae, Kenichi. *The Borderless World: Power and Strategy in the Interlinked Economy*. New York: Harper Perennial, 1991.

– "Letter from Japan." *Harvard Business Review* (May–June 1995): 154–63.

Ohmann, Richard. *Selling Culture: Magazines, Markets, and Class at the Turn of the Century*. London: Verso, 1998.

"The Online Hacker Jargon File, version 4.3.3." Available online at http://www.tuxedo.org/~esr/jargon/. Accessed 20 September 2002.

"Open Letter from Lord British." *GameSpot*. Available online at http://www.headline.gamespot.com/news/97_11/18_Ultima/index.html. Accessed 6 May 1999.

Pargh, Andrew. "A Prisoner of Zelda and Two Lost Pilots Are on the Line: Video-Game Makers Face Rash of Christmas Callers Who Can't Run Their Games." *Wall Street Journal* (New York), 30 December 1992.

Parker, Mike, and Jane Slaughter. "Management By Stress." *Science and Society* 8 (1990): 27–58.

Pastore, Michael. "Young Americans Take Their Spending Online." *Cyber-Atlas* (19 September 2000). Available online at http://cyberatlas.internet.com/big_picture/demographics/article/0,,5901_463961,00.html. Accessed 10 January 2002.

Pearce, Celia. "Beyond Shoot Your Friends: A Call to Arms in the Battle against Violence." In *Digital Illusion: Entertaining the Future with High Technology*, edited by Clark Dodsworth, Jr., 209–28.

Peckham, Mathew. "Soft Copy." *Computer Game* 134 (2002): 20.

Pelaez, Eloina, and John Holloway. "Learning to Bow: Post-Fordism and Technological Determinism." *Science as Culture* 8 (1990): 15–27.

Peña, Devon G. *The Terror of the Machine: Technology, Work, Gender, and Ecology on the US-Mexico Border*. Austin, TX: University of Texas Press, 1997.

Perez, C. "Structural Change and the Assimilation of New Technologies in the Economic and Social System." *Futures* 15 (1983): 357–75.

Pesce, Mark. "The Trigger Principle." *Feed* (3 February 2000).

Petersen, Lisa Marie. "Chavez Steps into Pepsi's Ring." *Brandweek* (27 September 1993).

Pigeon, Thomas. "Packaging Up Coolness: What Designers Must Do to Capture the Attention of Teens and Tweens." *Marketing Magazine* (20–27 July 1998): 21.

Pilieci, Vito. "Searching for Sega in All the Wrong Places" *Financial Post* (Toronto), 12 August 2000, D5.

Piore, Michael J., and Charles Sabel. *The Second Industrial Divide: Possibilities for Prosperity.* New York: Basic, 1984.

"Pokémon Fans in Stores and on Wall Street." *Financial Post (National Post),* Toronto, 4 August 1999, D2.

"Pokémon Rules TV, Too." *Canoe* (19 November 1999). Available online at http://www.canoe.ca/JamMoviesFeaturesP/pokemon_1.htm. Accessed 23 March 2002.

Pollert, Anna. "Dismantling Flexibility." *Capital and Class* 34 (1988): 43–75.

Poole, Steven. *Trigger Happy: The Inner Life of Video Games.* London: Fourth Estate, 2000.

Posner, Michael. "A Really Big Show: New Theatres Offer a Total Entertainment Experience." *Maclean's* 110, no. 32 (11 August 1997): 38.

Probst, Larry. "A New Opportunity for PC-Based Entertainment Software." *Red Herring* (December 1995). Available online at http://www.redherring.com/mag/issue26/new.html. Accessed 4 May 1999.

Provenzo, Eugene F., Jr. *Video Kids: Making Sense of Nintendo.* Cambridge, MA: Harvard University Press, 1991.

Rahmat, Omid. "IDC's 2001 Video Game Survey." *Tom's Hardware Guide* (31 August 2001). Available online at http://www.tomshardware.com/business/01q3/010831/. Accessed 25 October 2001.

Reich, Robert. *The Work of Nations: Preparing Ourselves for 21st Century Capitalism.* New York: Knopf, 1991.

Rich, Jason R. "More Delays Cause Nintendo to Make Line Up Changes." In "Industry News for October." *Video Game Advisor* (October 1997). Available online at http://www.vgadvisor.com. Accessed 10 May 1999.

Rifkin, Jeremy. *The Age of Access: The New Culture of Hypercapitalism Where All of Life Is a Paid-For Experience.* New York: Putnam, 2000.

Rizvi, Haider. "Toying With Workers." *Multinational Monitor* 17, no. 4 (April 1996): 12.

Roberts, Michael. "Internet Revolution: A New Paradigm or Another Bubble?" In *Defence of Marxism*, 29 March 2000. Available online at http://www.marxist.com/economy/internet_bubble300.html.

Roberts, Paul. "Sony Changes the Game." *Fast Company* (August–September 1997): 118–29.

Robins, Kevin. "Cyberspace and the World We Live In." *Body and Society* 1, nos. 3–4 (1995): 135–55.

Robins, Kevin, and Frank Webster. *Times of Technoculture: From the Information Society to the Virtual Life.* London: Routledge, 1999.

– *The Technical Fix: Education, Computers and Industry.* London: Macmillan, 1989.

– "Cybernetic Capitalism: Information, Technology, Everyday Life." In *The Political Economy of Information*, edited by Vincent Mosco and Janet Wasko, 44–75.

Robinson, Gertrude Joch, and Donald F. Theall, eds. *Studies in Canadian Communication.* Montreal: McGill Studies in Communication, 1975.

Rogers, Everett M., and Judith K. Larsen. *Silicon Valley Fever: Growth of High-Technology Culture.* New York: Basic Books, 1984.

Ross, Andrew. "Applying the Anti-Sweatshop Model to High Tech Industries." *New Labor Forum* (Spring 1999). Available online at http://www.tao.ca/writing/archives/nettime/0432.html. Accessed 19 May 1999.

Ross, Andrew, ed. *No Sweat: Fashion, Free Trade and the Rights of Garment Workers.* Verso: London, 1997.

Rothstein, Edward. "A New Art Form May Rise from the '*Myst.*'" *New York Times*, 4 December 1995.

Rushkoff, Douglas. *Playing the Future: What We Can Learn from Digital Kids.* New York: Riverhead Books, 1999.

– *Media Virus! Hidden Agendas in Popular Culture.* New York: Ballantine Books, 1994.

Rustin, Michael. "The Trouble with 'New Times.'" In *New Times: The Changing Face of Politics in the 1990s*, edited by Stuart Hall and Martin Jacques, 303–20.

Sale, Kirkpatrick. *Rebels against the Future: The Luddites and Their War on the Industrial Revolution: Lessons for the Computer Age.* Reading, MA: Addison-Wesley, 1995.

Sardar, Ziauddin. *Postmodernism and the Other: The New Imperialism of Western Culture.* London: Pluto Press, 1998.

Sawyer, Ben, Alex Dunne, and Tor Berg. *Game Developer's Marketplace.* Toronto: Coriolis Group Books, 1998.

Scally, Robert. "PC, Video Game Software Sales Hit Record High in 1997." *Discount Store News*, 9 March 1998, 65.

Schaefer, Gary. "Pokémon Boom Helps Push Nintendo Profit to Six-Year High." *Financial Post* (Toronto), 27 May 1999, C11.

Schell, Orville, and David Shambaugh, eds. *The China Reader: The Reform Era.* New York: Vintage, 1999.

Schiesel, Seth. "Is There Real Gold in Online Fantasy Games?" *New York Times*, 7 July 1997, D1.

Schiller, Dan. *Digital Capitalism: Networking the Global Market System.* Cambridge, MA: MIT Press, 1999.

Schiller, Herbert. *Communication and Cultural Domination.* New York: International Arts and Sciences Press, 1976.

Schumpeter, Joseph A. *Capitalism, Socialism, and Democracy.* New York: Harper and Row, 1942.

Sefton-Green, Julian, ed. *Youth Culture in the Age of Multimedia.* London: University College London Press, 1998.

"Sega to Acquire Toy Firm Bandai for $1.09 Billion." *Globe and Mail* (Toronto), 24 January 1997, B6.

Segan, Sascha. "A World of Their Own: Online Game Let's You Play God." ABC.news.com. 22 March 2000. Available online at http://www.abc-news.go.com. Accessed 10 May 2000.

Seiter, Ellen. "Gotta Catch 'Em All – Pokémon: Problems in the Study of Children's Global Multi-Media." Paper presented at "Research in Childhood: Sociology, Culture, and History," University of Southern Denmark, Odense, 28–31 October 1999. Available online at http://www.hum.sdu.dk/center/kultur/bue/ric-papers/seiter-poke.pdf.

Shafer, Scott Tyler. "Development ... la Mod." *Red Herring* (15 April 2001): 96.

Shapiro, Carl, and Hal R. Varian. *Information Rules: A Strategic Guide to the Network Economy.* Boston, MA: Harvard Business School Press, 1999.

Sheff, David. *Game Over: How Nintendo Zapped an American Industry, Captured Your Dollars, and Enslaved Your Children.* New York: Random House, 1993.

– "Sony's Plan for World Recreation." *Wired* 7, no. 11 (November 1999): 264–75.

Shulgan, Christopher. "Open Source Everything." *Shift.com.* Available online at http://www.shift.com. Accessed 4 March 2002.

Siegel, Ritasue. "What Corporate America Must Learn from the Toy Industry." *Playthings* (July 1982): 86.

Silverstone, Roger, and Eric Hirsch, eds. *Consuming Technologies: Media and Information in Domestic Spaces.* London: Routledge, 1992.

"Simulations: Building the Perfect Game." *Computer Gaming World* (January 1999): 185.

Sivanandan, A. "All That Melts into Air Is Solid: The Hokum of New Times." *Race and Class* 31, no. 3 (1989): 1–23.

Skelton, Tracey, and Gill Valentine, eds. *Cool Places: Geographies of Youth Culture.* London: Routledge, 1998.

Slater, Don, and Fran Tonkiss. *Market Society.* Cambridge, UK: Polity, 2001.

"Small Firms in Japan: Fabulous and Fabless." *The Economist* (29 March 1997). Available online at http://www.economist.com. Accessed 18 March 1999.

Smith, Adam. *An Inquiry into the Nature and Causes of the Wealth of Nations.* New York: Modern Library, 1937 [1776].

Smith, Tony. *Technology and Capital in the Age of Lean Production: A Marxian Critique of the "New Economy."* Albany, NY: State University of New York Press, 2000.

Smythe, Dallas W. *Dependency Road: Communications, Capitalism, Consciousness and Canada.* Norwood, NJ: Ablex, 1981.

Snider, Mike. "'Pocket Monsters' Pocket Sales." *USA Today,* 3 November 1999, D1. Available online at http://www.usatoday.com. Accessed 21 March 2002.

Software and Information Industry Association. Available online at http://www.siia.net/piracy/default.asp. Accessed 20 September 2002.

"Sony Plays the Video-Game Market." *Financial Post* (Toronto), 6–8 July 1996, C15.

Spigel, Lynn. Introduction to Raymond Williams. *Television: Technology and Cultural Form.* Hanover and London: University Press of New England and Wesleyan University Press, 1992.

Stafford, Brent. *Insert Coin: The Culture of Video Game Play.* Video documentary. Master's project, Simon Fraser University, Burnaby, Canada 1999.

Stephenson, William. "The Microserfs Are Revolting: Sid Meier's Civilization II." *Bad Subjects* 45 (October 1999). Available online at http://eserver.org/bs/45/stephenson.html.

Stapleton, Christopher. "Theme Parks: Laboratories for Digital Entertainment." In *Digital Illusion: Entertaining the Future with High Technology,* edited by Clark Dodsworth, Jr., 425–37.

Stern, Sydney, and Ted Schoenhaus. *Toyland: The High-Stakes Game of the Toy Industry.* Chicago: Contemporary Books, 1990.

Stevens, Elizabeth Lesley, and Ronald Grover. "The Entertainment Glut." *Business Week* (16 February 1998): 88–95.

Stiles, Chris. "Home Video Game Market Overview: July 1995." Asian Technology Information Program. Available online at http://www.cs.arizona.edu/japan/www/atip/public/atip.reports.95/atip95.43r.html. Accessed 14 March 2002.

Stoddard, Scott. "Sony's PlayStation 2 Could be Used to Launch Missiles, Japan says." *Globe and Mail* (Toronto), 17 April 2000, B1.

Stone, Allucquère Rosanne. *The War of Desire and Technology at the Close of the Mechanical Age.* Cambridge, MA: MIT Press, 1996.

Storey, John, ed. *What Is Cultural Studies? A Reader.* London: Arnold, 1996.

Storper, Michael. "The Transition to Flexible Specialization in the US Film Industry: External Economies, the Division of Labour, and the Crossing of Industrial Divides." In *Post-Fordism: A Reader,* edited by Ash Amin, 195–223.

Sutton, Neil. "Special Report: What You Didn't Know About Girl Gamers." Available online at http://www.myvideogames.com/features/feature20.asp. Accessed 24 March 2002.

Takahashi, Dean. "The Game of War." *Red Herring* (15 October 2001). Available online at http://www.redherring.com. Accessed 4 September 2002.

– "Let the Games Begin." *Red Herring* (15 April 2001): 80.

– "Games Get Serious." *Red Herring* (18 December 2000): 66.

– "Software Firm Crushed by Barbie." *Globe and Mail* (Toronto), 27 February 1999, C18.

Tapscott, Donald, David Ticoll, and Alex Lowy. *Digital Capital: Harnessing the Power of Business Webs.* Cambridge, MA: Harvard Business School Press, 2000.

Taylor, Paul. "Pentium II Aimed at Games Market." *Financial Post* (Toronto), 6–8 September 1997, 13.

Terranova, Tiziana. "Free Labor: Producing Culture for the Digital Economy." *Social Text* 18, no. 2 (2000): 33–57.

Tetzalf, David. "Yo-Ho-Ho and a Server of Warez: Internet Software Piracy and the New Global Information Economy." In *The World Wide Web and Contemporary Cultural Theory,* edited by Andrew Herman and Thomas Swiss, 77–99.

"The Movie: Press Release No. 2." PRNewsire. *Tomb Raider 2000 World Wide,* 2000. Available online at http://www.trww.com/movie_press2.shtml. Accessed 8 October 2000.

Thompson, Clive. "Why Your Fabulous Job Sucks." *Shift* (March 1999): 55–63.

– "Meet Lara Croft, Virtual Sensation." *Globe and Mail* (Toronto), 31 January 1998, C1, 14.

Toffler, Alvin. *The Third Wave.* New York: Bantam Books, 1981.

Total Control 5, March 1999.

Tremblay, Gaetan. "The Information Society: From Fordism to Gatesism." *Canadian Journal of Communication* 20 (1995): 461–82.

Turner, Chris. "Rockstar: The Cooling of Video Games." *Shift.Com* (1999). Available online at http://www.shift.com. Accessed 1 December 1999.

Uhlig, Robert. "Youth Grow Bored with Internet." *National Post* (Toronto), 1 December 2000, A1, 11.

United Nations. *Human Development Report.* New York: United Nations, 2000.

United States. Department of Commerce. "Falling through the Net IV: Towards Digital Inclusion." National Telecommunications and Information Administration. Washington, DC: US Department of Commerce (October 2000). Available online at http://www.ntia.doc.gov/ntiahome/digitaldivide. Accessed 24 March 2002.

United States. Department of Commerce. "Falling through the Net II: New Data on the Digital Divide." National Telecommunications and Information Administration. Washington, DC: US Department of Commerce, 1998. Available online at http://www.ntia.doc.gov/ntiahome/net2. Accessed 15 March 2002.

– Federal Trade Commission. "FTC Releases Report on the Marketing of Violent Entertainment to Children." Washington, DC: Federal Trade Commnission, 11 September 2000. Available online at http://www.ftc.gov/opa/2000/09/youthviol.htm. Accessed 26 March 2001.

Van de Ven, Andrew H. *Central Problems in the Management of Innovation.* Cambridge, MA: Marketing Science Institute, 1985.

"Video Game Company Moves." Financial pages, *Agence France Presse,* 1 June 1995. Available online at http://www.afp.com/english/home. Accessed 3 October 1999.

Walkerdine, Valerie, Angela Dudfield, and David Studdert. "Sex and Violence: Regulating Childhood at the Turn of the Millennium." Paper presented at "Research in Childhood: Sociology, Culture, and History," University of Southern Denmark, Odense, 28–31October 1999. Available online at http://www.hum.sdu.dk/center/kultur/bue/ric-papers/walk-regu.pdf.

Wark, McKenzie. "From Fordism to Sonyism: Perverse Readings of the New World Order." *New Formations* 15 (1991): 43–54.

– "The Information War." *21.c* 7 (Spring 1992). Available online at http://www.mcs.mq.edu.au/Staff/mwark/warchive/21*C/21c-cyberwar.html.

Warner Bros. "The Making of *Pokémon*: About the Phenomenon." 1999. Available online at http://pokemonthemovie.warnerbros.com/cmp/151fr.html. Accessed 23 March 2002.

Warshofsky, Fred. *The Patent Wars: The Battle to Own the World's Technology.* New York: John Wiley and Sons, 1994.

Wazir, Burhan. "Eating the Greens." *Observer* (London), 1 October 2000. Available online at http://www.commondreams.org/views/100100-106.htm. Accessed 23 March 2002.

Weber, Thomas E. "Maverick Programmers Prepare to Unleash Anarchy on the Web." *Wall Street Journal* (New York), 27 March 2000, B1.

Weinstein, Henry. "Ex-Workers Win Back Pay for Layoffs without Notice." *Los Angeles Times,* 4 June 1986, p. 2.

Weisman, Jordan. "The Stories We Played: Building BattleTech and Virtual World." In *Digital Illusion: Entertaining the Future with High Technology,* edited by Clark Dodsworth, Jr., 463–78.

Wen, Howard. "Why Emulators Make Video Game Makers Quake." *Salon.Com,* 4 June 1999. Available online at http://www.salon.com. Accessed 17 March 2002.

Wernick, Andrew. *Promotional Culture: Advertising, Ideology, and Symbolic Expression.* London: Sage, 1991.

– "Sign and Commodity: Aspects of the Cultural Dynamic of Advanced Capitalism." *Canadian Journal of Political and Social Theory* 15, nos. 1–3 (1991): 152–69.

Wheelwright, Geoff. "Game Buy: Deep Pocketed Monsters." *Financial Post* (Toronto), 19 July 1999, E1.

"Where There's MUD, There's Brass." *The Economist* (8 July 2000): 82.

Whitta, Gary. "Soldier of Fortune II: Double Helix." *Next Generation* 3, no. 11 (1 November 2000): 47.

Williams, Raymond. *Television: Technology and Cultural Form.* Hanover and London: University Press of New England and Wesleyan University Press, 1992.

– *The Long Revolution.* London: Chatto and Windus, 1961.

– *Communications.* Harmondsworth: Penguin, 1962.

– *Marxism and Literature.* Oxford: Oxford University Press, 1977.

– "Base and Superstructure in Marxist Cultural Theory." In Raymond Williams, *Problems in Materialism and Culture: Selected Essays,* 31–50.

– *Problems in Materialism and Cultue: Selected Essays.* London: Verso, 1980.

– *Towards 2000.* London: Chatto and Windus and The Hogarth Press, 1983.

Winston, Brian. *Media, Technology and Society: A History From the Telegraph to the Internet.* London: Routledge, 1998.

Woosely, Nate. "Conducent Entices Game Companies to Bid Goodbye to Retail." *Games Industry News.* Available online at http://www.conducent.com/press/gin/html. Accessed 12 June 2000.

Wylie, Margie, and Jeff Peline. "Back from Hard Knocks U." CNET News.com 20 September 1996. Available online at http://news.com.com. Accessed 14 March 2002.

Yankee Group. Press release. Boston, 11 November 1999. Available online at http://www.yankeegroup.com. Accessed 12 June 2000.

Ye, Sang. "Computer Insects." In *The China Reader: The Reform Era,* edited by Orville Schell and David Shambaugh, 291–6.

"Young at Heart." *The Economist* (20 June 1998): 70.

Yuen, Michael. "Games Aren't Just for Kids Anymore: An Insider's Look at Marketing Software." UC *Davis Innovator: Graduate School of Management.* Graduate School of Management, University of California, Davis, 1996.

Zhao, Yuezhi, and Dan Schiller. "A Dance with Wolves? China's Integration into Digital Capitalism." *Info* 3, no. 2 (April 2001): 137–51.

Zizek, Slavoj. "Welcome to the Desert of the Real." *Reconstructions* (15 September 2001). Available online at http://web.mit.edu/cms/reconstructions/interpretations/desertreal.html. Accessed 18 March 2002.

Zukin, Sharon. *Landscapes of Power: From Detroit to Disney World.* Berkeley: University of California Press, 1991.

– *The Cultures of Cities.* Oxford: Blackwell, 1995.

Index

academia: and military-industrial complex, 85

Activision: creation of, 97

advertising: Baudrillard on, 71; and cynicism, 220, 224; in interactive games, 235–6; interactive games on television, 223–5; and "mediatized marketplace," 40; in online games, 284; Sega's style of, 130–2; Sony's style of, 153–5. *See also* marketing, marketing (game)

After Dark, 264

Age of Empires: online fan community, 166–7; and "progress," 167–8

Aglietta, Michel, 64

Al Qaeda: computer simulation of, 289–90

anticorporate movement: in *State of Emergency*, 287; and globalization, 209, 286

antitrust: investigation into Nintendo, 123, 129–30

AOL/Time-Warner: online gaming interests of, 227, 272–3

Apple Computers, 96

Arakawa, Minoru, 110

arcade games: and Atari, 94; emergence of, 90–2; diminishing revenues of, 105. *See also* Playdium

AT&T: and online games, 137

Atari: acquired by Warner Communications, 97, 102; Jaguar console, 139–40; and 1984 industry meltdown, 104–6; origins of, and early days at, 94–6; and *Pong*, 91; sued by Nintendo, 112; treatment of designers at, 96–7. *See also* Bushnell

audience: "building" of, for games, 73; and "decoding," 43–4; fragmentation of, 64–5, 72; "liberated" by interactive media, 18; as market, 45, 66, 219

Baer, Ralph: and origins of console, 92–3

Barbie: *Barbie Fashion Designer* computer game, 259–60; doll, 112

Barbrook, Richard: on "Californian ideology," 85

BattleTech Centre: as location-based entertainment, 101

Battlezone: origins of, 96; as military-simulation tool, 100

Baudrillard, Jean: on marketing communication, 61, 70–1; on "simulacrum," 69

Bell, Daniel: on cultural intermediaries, 72; on "post-industrial" society, 5

Bertram, Ron: on Nintendo's marketing strategy, 224

branding: and gamer subculture, 57; and licensing, 227, 238; management of, in game industry, 131, 221–2; and Nintendo's colonization of children's culture, 125–6; and social movements, 286–7

Broderbund Software, 288

Bushnell, Nolan: and *Computer Space*, 90; leaving Atari, 102, 104; as "ur-entrepreneur," 94

Cairncross, Frances: as "information revolutionary," 9

Cameron, Andy: on "Californian ideology," 85

capitalism: alleged "friction-free" nature of, 9, 25; and communication systems, 38–40, 47; marketing communication in, 39–40, 71, 219–20; revolutionary nature of, 65; technological restructuring of, 64–5, 291; tension in, 76. *See also* "ideal commodity," post-Fordism

Carey, James: on Harold Innis, 291

Carmack, John: and id Software, 143

Cassell, Justine: on female gaming, 260

character-marketing: and licensing, 125–6; and "meta-genre," 238; and toy industry, 102. *See also* Mario

childhood: commodification of, 102, 243–5; concept of, 243; and Nintendo's commercialization of children's culture, 123–7

Children Now: their study of violence and sexism in games, 254, 256

China: piracy in, 214

Chomsky, Noam: on "media filtering," 40

circuit: concept of, 58–9, 296–7. *See also* three-circuits model

circuit of capital: process of, 39, 50. *See also* three-circuits model

Civilization, 44

class. *See* digital divide

Cockburn, Cynthia: on "circuit of technology," 52

Coleco: ColecoVision, 103; demise of, 105; Telstar, 102

Columbine shootings: and interactive games, 146–7, 247, 268

Combat Mission, 54–5

Command and Conquer: and September 11, 289

commodity: "commodity-sign," 71; "experiential commodities," 67–8; "fetishism" of, 197; interactive game as "experiential commodity," 183, 283–4; shortening lifecycle of, 66. *See also* "ideal commodity"

Commodore: and computer gaming, 93

communication: in advanced capitalism, 47; commercialization of, 32; consumption as, 71. *See also* media

computer game: see online gaming, video and computer games

computers: networking of, 100; and Pentagon, 85–6; and post-Fordism, 64–5; and "post-industrial society," 5; use of, for "fun," 88. *See also* personal computer

Computer Space: as first coin-operated arcade game, 90

Conducent: as adver-gaming company, 284

console (video game systems): Atari enters market of, 94–5; as gateway to Internet, 175, 187; household penetration of, 104, 150, 184–5; oligopolistic structure of business for, 171–6; origin of, 92–3; "perpetual innovation" of, 138–9, 222

consumer sovereignty: and "active audience," 44; and the interactive gamer, 14, 16–17, 57

consumer culture: and advertising, 71; and consumerism, 220; fluidification of consumption, 67–8; and mass society, 63; simulated in *The Sims*, 276–7

copyright. *See* intellectual property

cool hunters: and youth marketing, 233, 277

Counter-Strike: as consumer-led game modification, 252–3; and September 11, 289

Crawford, Chris: *The Art of Computer Game Design*, 98–9

Crash Bandicoot: and PlayStation branding, 154–5

creativity: and hegemony, 292; and innovation, 87

cultural intermediaries: as cadre of workers, 72–3; in game industry, 21, 45, 221; as market-managers, 219

cultural studies: perspectives in, 42–4; shortcomings of, 44–6

Cyan Studios: and *Myst*, 147, 149

Deer Hunter: and mainstreaming computer gaming, 159

demographics, of gaming audience: 12–13; and age, 222; and digital divide, 183–5; and gender, 248–9, 264–7; and PlayStation, 153, 156; and *Pokémon*, 243; and Sega, 130, 141; and xbox, 174. *See also* digital divide

Der Derian, James: and *Virtuous War*, 180–1

design, game: aesthetics of, 98–9; and "digital design," 21, 220–1, 295; goals of, 19–20; Nintendo's management of, 116–18; players as designers, 204–5, 252–3, 273–4; and marketing, 235–9; and "realism," 236, 250–1; and *Spacewar*, 87

development, game: conditions of work in, 199–201; and designers as "mini-gods," 96; "Easter eggs" in, 97–8, 120; and licensing, 113–14; and PC developers, 158; process and corporate structure of, 176–8; and standardization of game content, 179; unpaid workforce in, 201–5, 252–3

digital divide: and open architecture, 290; unequal access to game technology, 183–6, 190–1

Digital Equipment Corporation: and origins of interactive games, 85–6

Directx: 164–5

Donkey Kong: and copyright suit, 115; US launch of, 120

Doom: and Columbine shootings, 247; and commercialization of computer games, 143–6; level-editor in, 203–4; shareware marketing strategy of, 145–7

Dreamcast, Sega: launch and
    demise of, 173–4
DreamWorks: gaming interests of,
    138, 165–6, 230
Du Gay, Paul, 52, 221
Duke Nukem, 255
Dungeons and Dragons: and pro-
    gramming subculture, 89–90

Eidos: and Tomb Raider, 228
Electronic Arts: acquisitions, 161,
    271; affiliated labels, 271; losses,
    286; online gaming interests,
    188, 227, 272–3; origins of, 97;
    and PC game sales, 158; and
    September 11, 289; and The
    Sims, 271–3; and sports games,
    130, 226, 233
Ensemble Studios: and Age of
    Empires, 167
Entertainment Software Ratings
    Board, 135
Everquest: commodification of
    subculture of, 284–5; economics
    of, 188
E.T.: and synergy, 105

Fairchild Camera and Instrument,
    95
Famicom, Nintendo, 110
feminism: and critique of game cul-
    ture, 256. See also Purple Moon
    Company
film: synergies with games, 227–9,
    241–2
Final Fantasy: film spin-off, 228–9
first-person shooters: characteristics
    of, 143–4, 246, 251; developers
    of, 146; links to youth violence,
    146–7; as simulation device for
    military, 70
flexible specialization: in game
    manufacturing, 155–6

Ford, Henry, 62
Fordism: "ideal commodity" of, 74;
    as "regime of accumulation,"
    62–4
4Kids Entertainment: as Pokémon
    licensing agent, 241–2
Frankfurt School: and "the culture
    industry," 37–8
Friedman, Ted, 270

Game Boy, Nintendo: manufacture
    of, 207; and Pokémon, 239–43
GameCube, Nintendo, 174, 187
gameplay: "exploration" vs. "colo-
    nization," 124, 126–7; immer-
    sion in, 19–20; in information
    capitalism, 294; interactive
    nature of, 18, 44, 54; subject-
    positions offered in, 43; utopian
    aspects of, 285, 292–3; "virtual
    fantasy" in, 98–9
gamer: as consumer, 20, 57; inter-
    pellation of, 53; and rhetoric of
    "consumer sovereignty," 14;
    alleged liberation of, 16; as
    game-world creator, 146, 203–5,
    252–3, 274, 290; as market-
    research subject, 202; "played
    upon" by Sim Capital, 279;
    as subject of high-technology
    society, 55
GameWorks, 138
Garnham, Nicholas: on media
    in circuit of capital, 38–40; on
    video game industry, 13
Garriott, Richard: see Ultima
    Online
Gates, Bill: as ideologue of "low-
    friction" capitalism, 9; and
    "Internet Tidal Wave" memo,
    166; as monopolist, 164
gender: female exclusion from digi-
    tal play, 257–9; "girl games,"

258–61, 266–7; "grrl gamers," 262–3; historical factors of male-gamer bias, 91, 107, 248–9, 257; and international division of labour, 205; market(ing) logic of male-gamer bias, 249–51, 261, 267; and *Pokémon*, 243; representation of female game-character, 250, 256, 264; representation in game advertising, 131; and *The Sims*, 275. *See also* demographics, militarized masculinity

Genesis, Sega, 129–31, 137

genre: and computer game, 158; invention of, 96; "meta-genre," 236–9; narrowing diversity of, 254–5; and "platform games," 117; and violence, 253–5

globalization: and cultural hybridization in game culture, 190, 241–2; and division of labour in game industry, 205–9; of game industry operations, 13, 123, 136–7, 189–91; and game piracy, 213–15; of production and consumption, 64; and Sony, 152

"God Game": *The Sims* as, 270, 278

Golding, Peter, 45

Grand Theft Auto III, 41, 234

Grossman, David: on game violence, 247

Grove, Andy: on PC gamers, 157–8

Gunman Chronicles: as consumer-led modification, 253

hacking: and origin of interactive game, 86–8, 90; and *The Sims*, 281–3; assimilated into game development, 204–5, 252–3. *See also* piracy

Haddon, Leslie, 90

Half-Life: counterfeiting of, 214; as first-person shooter, 251; and consumer-led game modification, 252–3

Hall, Stuart, 42–3

Harry Potter: EA's licensing rights to, 227

Harvey, David, 60, 69, 190

Hawkins, Trip: as founder of Electronic Arts, 97; as founder of 3DO, 139–40

Hayes, Michael, and Stuart Dinsey, 11

Herman, Edward: on "media filtering," 40

Herz, J.C.: *Joystick Nation*, 76; on *Myst*, 149; on *The Sims*, 276; on *Ultima*, 161

Higinbotham, William A.: as inventor of interactive game, 92

Hingham Institute Study Group on Space Warfare: and origins of interactive game, 84

Hollywood: synergies with interactive games, 70, 104, 178–9, 227–8

Home Pong, 94–5

id Software, 143; as pioneer of shareware marketing, 145–7; as purveyor of virtual carnage, 250. *See also Doom, Quake*

"ideal-commodity": concept of, 24, 74; interactive game as, 75–7, 170–1

Idei, Nobuyuki: on Sony as digital empire, 175

"information age:" fusion of market ideology and technological determinism in, 11; militaristic roots of, 85; neo-Luddite critique of, 17; rhetoric of, 4 information capitalism: character of, 23, 51, 66–7; "free labour" in, 216; intensified commodification in, 50; "net slaves" in, 200–1; paradox of "play" in, 245; unresolved tensions in, 76–7. See also Sim Capital

"information revolution:" inadequate idea of, 7–10

informational principle: of post-Fordism, 64–5

Innis, Harold: on "bias" of communication, 31, 291; on "empire," 31–2; on "oligopoly of knowledge," 32

innovation: and creativity, 87; game industry outstripping military in, 181; process of, 55–6, 84–5; Sega's "revolution strategy" of, 129; subcultural context of, 88–90. See also perpetual-innovation

Intel: Pentium chip and PC gaming, 143, 157–8

intellectual property: contradictions of, 281–3, 290; Electronic Arts vs. Yahoo litigation, 281; as game industry concern, 114–15, 282; Nintendo vs. Atari litigation, 112; Nintendo vs. MCA Universal litigation, 115; and Nintendo's "lock-and-key" device, 111–12; in perpetual-innovation capitalism, 67. See also licensing, piracy

Interactive Digital Software Association (IDSA): formed by, 135; on gamer demographics, 184, 265–6; on piracy, 209–10

interactivity: characteristics of, 14; constraints on, 294–5; paradoxes of, 21–2; "three circuits" of, 53–8; utopian rhetoric of, 14–19, 54–5

IBM, 85

Internet: and early computer gaming, 142–3; origins of, 100; declining use of, 286; console as gateway to, 169, 175. See also online games

Internet economy: crash of, 286. See also "new economy"

Internet games. See online games

Internet Gaming Zone, Microsoft, 166

interpellation, 53

Jaguar system, 139–40

Jameson, Fredric, 70

Jane's Fleet Command, 288

Japan: "information society" and, 110; "Japan Panic," 122–3; youth in, 15; and Space Invaders, 96

Java Enterprise Computing, 7

Jenkins, Henry: on "girl games," 260–1; on Nintendo Power, 120; on Nintendo's impact on children's culture, 123–7

Jet Moto, 154

Katayama, Osamu, 118

Kee Games, 94

Kelly, Kevin: on customer-company relations, 251–2

Killer Instinct, 135

Kinder, Marsha: on synergy, 227; on toy marketing, 112; on violence, 133

Klein, Naomi: on branding, 227; on irony in marketing, 277; on September 11, 288–9

"knowledge workers": 5

Korea, South: popularity of games in, 190

Kroker, Arthur, 34–5

Kundnani, Arun, 66

Kutaragi, Ken: on PlayStation 1, 152; on PlayStation 2, 169, 175

labour: casualization of, in game development, 203; and conditions of work in game development, 199–201, 215–16; international division of, 205–6; size of game labour force, 198; unpaid work in game industry, 166–7, 201–5; unrest, 97–8, 207–9
Lanier, Jason, 290
Lara Croft: and representation of women in games, 263–4; and *Tomb Raider* spin-offs, 228
Lasn, Kalle, 287
Laurel, Brenda. *See* Purple Moon Company
Le Diberder, Alain and Frederic, 247
Lee, Martyn, 29. *See also* ideal commodity
legal battles: and *Ultima*, 162–3. *See also* intellectual property
Leiss, William: on "high-intensity marketplace," 40; *Under Technology's Thumb*, 6
Levy, Steven: on hackers, 86
Lewis, T.J.: *The Friction-Free Economy*, 13
licensing: and character-marketing, 227; of game engines, 146; impact on game development, 226–7, 237–8; and Mario character, 125–6; and playthings marketing, 102; and *Pokémon*, 239–41; and third-party game development, 113–14, 129–30
Lion King, 158
Lord of the Rings, The, 227
Lowenstein, Douglas: on piracy, 209–10, 212

Luddites, 17
Lunenfeld, Peter, 210–11

Magnavox: and Odyssey console, 92, 110
Main, Peter: on Nintendo marketing, 121, 126
MÄK Technologies, 182–3
Mario: as colonizer, 127; platform-game series, 117–18; and synergistic brand marketing, 125–6, 226
market: and "information revolution" ideology, 8–11; mediatization of, 40, 71; mediatized global marketplace, 50, 58–9
marketing communication: as bridge between production and consumption, 40, 219–20; and consumer culture, 71–4; and cultural analysis, 56; digitalization of, 291
marketing, game: and branding, 221–3; and console launch, 152–3; and "digital design," 21, 220–1; intensification of, 73–4; link to consolidation of ownership, 220, 236–8; and market research, 202, 252, 296; and narrowing game diversity, 237–8; Nintendo's strategy, 118–21, 125–6; Sega's strategy, 130–2; significance of, in game industry, 56–7; Sony's strategy, 153; spending, 174; synergistic nature of, 226–7; and violence, 249–51
Marx, Karl: and the circuit of capital, 38–9; and "fetishism of commodities," 197; and the "social character of labour," 211
massively multiplayer online games: characteristics of, 159–60; commercialization of, 188–9. *See also* *Everquest, Ultima Online*

mass media: commercialization of,
   38–40; as domestic technology,
   63; and "demassification," 5–6,
   14; link to game media, 18, 21,
   223; and marketplace, 40, 71–2
mass society: 5, 63
MIT: and origin of games, 84–6
Matrix, The, 70, 169
Mattel: enters game market, 102–3,
   105; and "girl games," 259, 261;
   "razor marketing theory" of, 112
maquiladoras: game manufacture
   in, 205, 207–9
Maxis, 271
media: and markets, 32, 40. *See
   also* Innis, McLuhan
media analysis: approach to, 22–4,
   49, 79–80
media theory: and Harold Innis and
   Marshall McLuhan, 31–7
Maxi-Switch, 207–8
MCA Universal: alliance with Sega,
   138; copyright battle with
   Nintendo, 115
McChesney, Robert, 40–1
McLuhan, Marshall: critiques of,
   36–7; on cultural disturbances of
   new media, 33–4; on "games,"
   35, 179; *The Mechanical Bride*,
   33; tension between optimism
   and pessimism, 34–5
McRobbie, Angela, 56, 79
Meier, Sid, 158
Menzies, Heather, 34, 290–1
Metal Gear Solid, 174
Microsoft: interests in PC game
   business, 163–7; and "new economy," 10; turns to PC game market, 157–8; and September 11,
   289. *See also* Gates and xbox
Midway Games, 94–6, 135
"militarized masculinity": concept
   of, 247–8, 254–6; as bias of

game experience, 265; and "war
   against terror," 289. *See also*
   gender, military
military: synergistic link to game
   industry, 179–82; and game
   violence, 248; and games as simulation device, 99–101, 180–3,
   289–90; and origin of interactive game, 85–6. *See also* "militarized masculinity"
Miller, Rand and Robyn, 147
Ms. Pac-Man, 96
Miyamoto, Sigeru: on *Mario*, 118
Monaco GP, 96
Morley, David, 122
Morris-Suzuki, Tessa: on
   "perpetual-innovation capitalism," 66–7
Mortal Kombat: violence controversy, 133–5
Mosco, Vincent, 45, 177
Murdock, Graham, 219
music industry: links to game
   industry, 97, 232–5
Myst: as alternative to violent
   games, 147–9; link to high-tech
   capitalism, 149

Namco, 96
Napster, 212
Negroponte, Nicholas: and military simulation-training, 100–1;
   as techno-utopian, 4, 14–15
"new economy": and dot.com fall-
   out, 191–2; inadequate idea of,
   7–10; interactive game exemplary of, 13; unpaid workforce
   in, 216. *See also* "information
   age," online games
new media: continuities with past,
   18, 21- 2
News Corporation, 273
Nichols, Bill, 292

Night Trap: violence controversy, 133–6

Nintendo: antitrust investigation into, 130; "branding" the game market, 125–7; "closing the loop" with target market, 120–1; high-intensity marketing apparatus of, 118–19; in-house development studio, 117; and international division of labour, 207–9; and "Japan Panic," 122–3; and "just-in-time" production, 206; and next-generation console war, 155, 187; origins and early days at, 109–11; profit-management strategy of, 111–14; quasi-monopolistic control of, 109; target market of, 249–50; as transnational corporation, 136, 190. *See also* Mario, *Pokémon*

Nintendo Entertainment System, 110–12

N64, 153–4

"Nintendo War," 288

NAFTA, 206

Odyssey (Magnavox), 92

Ohmae, Kenichi: on "Nintendo Kids," 15

online gaming: early versions of, 137, 160; fan communities of, 166–7, 252–3, 262, 274; growth and commercialization of, 159–61, 186–9; media empires, 166, 187–8, 272; as Net-economy laboratory, 147, 213, 283–5; and techno-utopianism, 16. *See also Counter-Strike, Doom, Ultima*

open architecture, 290

Origin Systems, 160–1, 163

Pac-Man, 96, 104

Paramount Pictures, 228

Pearce, Celia: on military-entertainment complex, 181; on male games, 250; on violent games, 251

peer-to-peer (P2P), 212

Periscope, 91–2

"perpetual-innovation economy": characteristics of, 66–7; dynamic in game industry, 73–4, 106–7, 222

Persian Gulf war, 180

personal computer: early use as game platform, 93–4; growing popularity as gaming platform, 142–3, 157–9, 172–3. *See also* video and computer games, online gaming

Ping-Pong, 94

piracy: and "black markets," 213–15; challenge posed to game industry by, 209–11, 281–3; contradictions of, 215–17; in early computer game culture, 142; and "warez" networks, 211–13

play: commodification of, 243–5, 283–4; and innovation, 87; McLuhan on, 35

Playdium, 229–32

PlayStation: and demographic segmentation, 153, 156; and "flexible specialization," 155; household penetration, 12; marketing strategy, 153–5; origins of, 152

PlayStation 2, 169, 187; in Sony's corporate strategy, 174–5

Pokémon: as synergistic media commodity, 239–43; and violence, 254

political economy, 37–41

Pong, 91

Poole, Steven: *Trigger-Happy*, 44

post-industrial society: thesis of, 5, 64
post-Fordism: dynamics of, 63–6,
72–3, 170–1; limitations of the
concept of, 76–7; and postmodernism, 68–9. *See also* ideal commodity
Postal, 255
postmodernism: dynamics of, 68–9;
link to marketing, 70–4
product placement, 235–6
Provenzo, Eugene: *Video Kids:
Making Sense of Nintendo*, 124,
250
publishing, game: concentration of
ownership in, 236–8; process and
corporate structure of, 178–9
Purple Moon Company, 260–1;
feminist critiques of, 261–2

Quake, 212, 250; and female
"*Quake* clans," 262

ratings, game: origins of, 135;
as marketing device, 250; and
violent games, 253–4
Raven Software, 246
REAL 3DO Multiplayer: origins and
demise of, 139–40
Regulation School, 62. *See also*
post-Fordism
Rifkin, Jeremy: on commodifica-
tion of play, 244, 283–4
Riven, 149
Robins, Kevin, 6, 22, 122
Rockett's New School, 261
Rockstar Games: marketing strat-
egy of, 234–5; and *State of
Emergency*, 287
role-playing games: and program-
ming subculture, 89–90; and *The
Sims*, 275
Romero, John, 143; on *Deer
Hunter*, 159; on female exclu-
sion, 257

Rosen, David, 91, 128–9
Rushkoff, Douglas: as techno-
utopian, 15–17
Russell, Steve: and origins of inter-
active game, 80, 84, 88
Russia: piracy in, 215

Sanders Associates, 92
Saturn, Sega, 132, 141–2
Schiller, Herbert, 38
Schumpeter, Joseph, 149
science-fiction: and programming
subculture, 88–90
Sega: European operations of, 136–
7; loser in console-war, 140–2;
marketing pirated by Sony,
153; marketing strategy, 130–2;
origins of, 128–30; and
"perpetual-innovation," 129;
spin-off ventures, 137–8, 230;
target audience, 141
Sega 32-X, 132, 141
Seiter, Ellen: on *Pokémon*, 239
semiotics, 42–3, 45, 275
September 11: links to interactive
game, 288–90
Shapiro, Carl, 129
shareware: as game marketing strategy, 145–7
Sheff, David, 120–1
Sierra Games, 252–3
Sim Capital: concept of, 278–9; and
dissent, 286–8; dynamics of,
291–3
SimCity, 44, 89
SimCopter, 97–8
Sims, The: harnessing players' creativity, 274; marketing strategy,
271–2; and piracy, 281–2; popularity of, 269–70; as simulated
consumer society, 276–7; as
socialization for digital capital,
278–9; utopian aspects of, 285,
292

SIMNET: 100–1
Smith, Adam, 7–8
SoftImage, 165
Soldier of Fortune: violence controversy, 246
Sonic the Hedgehog, 130
Sony: and anti-corporate movement, 287; and "flexible specialization," 155–6; enters the game industry, 151–2; marketing strategy, 153–5; place of PlayStation 2 in empire of, 157, 175; target audience, 153. *See also* Idei, PlayStation
Sony Net Yaroze, 204–5
Space Invaders, 95–6
Spacewar: and origins of, 80, 84, 87–8, 248
Spider-Man 2, 289
Star Wars, 102, 158
State of Emergency, 287
Stone, Allucquère Rosanne: on social impact of games, 13; and programming subculture, 87
Street Fighter II, 131
subject-positions: offered by games, 53, 275, 295
Super Mario Bros., 117–18, 126
Square, 229

Taiwan: piracy in, 213
Tajiri, Satoshi: and *Pokémon*, 240, 245
Tank, 94
technology: blind faith in, 6; "disappearance" of, in game design, 19; and interactivity, 14. *See also* "information age"
technological determinism: critique of, 6, 17–18, 46–7; and market ideology, 11
1080° Snowboarding, 236
television: commercialization of, 47–8, 80; fragmentation of audience, 64–5; linked to gaming, 18; and origin of interactive game, 92–3; and targeting game consumers, 223–5
Tengen, 112
Tennis for Two, 92
Terranova, Tiziana, 216
testing, game, 203
three-circuits model: and alternatives, 292, 296–7; as analytical framework, 23, 50–3, 58–9, 270–1; applied to the interactive game, 53–8; contradictions in, 280–6; subjectivities in, 295
3DO console, 139
Titanic: Adventure in Time, 11
Toffler, Alvin: on "demassification," 5–6, 239; on the video game, 13–14
Tomb Raider: cross-platform development, 158; spin-offs, 228–9. *See also* Lara Croft
Tommy Hilfiger, 235–6
toy industry: interest in interactive game industry, 102–3, 105

UbiSoft, 289
Ultima Online: 160–2; class-action suit, 162–3; marketization of subculture, 284–5
Unreal Tournament, 70
utopia: and gameplay, 285, 292–3

Valve, 252–3
Varian, Hall, 129
video and computer games: commercialization of, 90–4; continuities with mass-mediated culture, 18, 21; explosive growth of, 11–13; fuelling techno-utopianism, 15–17; global market of, 13, 189–91; as "ideal commodity" of information capitalism, 75–7, 291–2; industry marketing

apparatus, 56–7; industry structure, 170–9; integration into e-capitalism, 176, 291; narrowing diversity of, 179, 236–8; origins of, 85–90; and ownership consolidation, 177–9; as perpetual-innovation industry, 73–4, 149–50; revenue model in console business, 112–14; and the "three circuits," 53–8

violence, game: alternatives to, 148–9, 255–6, 280–1; amplified in gamer subculture, 252–3; campaigns against, 267–8; controversies and investigations, 134–6, 246, 254; historical factors of, 248–9; market and technological logic of, 249–51; marketing of, 132–3; and "media effects," 247, 268; and ratings, 135. *See also* first-person shooters, militarized masculinity

war: as early game theme, 96. *See also* military, militarized masculinity

Warner Communications: acquires Atari, 97, 102, 104; sells Atari, 105–6. *See also* AOL/Time-Warner

Wernick, Andrew, 71–2, 219

Williams, Raymond: approach to media analysis, 28, 49; on "flow," 48; on "mass" as inadequate term, 49; on "media effects," 247; on "politics of hope," 298; on "technological determinism," 46–7; updating for information capitalism, 60–1

Wipeout XL, 234

Wired, 10–11

World Trade Organization, 287

Wolfenstein, 143–4

work. *See* information capitalism, labour

Wright, Will, 270–1

xbox, Microsoft, 174–6, 187

Yahoo: lawsuit against, 281

Yamauchi, Hiroshi, 110; on contracting out, 206; on games as "no borders" business, 190; and "Japan Panic," 123; on Nintendo's "in-house" software, 117, 199

youth: advertising savvy, 220, 224; as consumer market, 221; and cooptation, 232–3; dissent, 286–7; and irony, 277; "liberated" by new media, 14–17; violence, 146–7, 268

Zizek, Slavoj, 288